AF806142

Clean Energy Production Technologies

Series Editors

Neha Srivastava, Department of Chemical Engineering and Technology, IIT (BHU) Varanasi, Varanasi, Uttar Pradesh, India

P. K. Mishra, Department of Chemical Engineering and Technology, IIT (BHU) Varanasi, Varanasi, Uttar Pradesh, India

The consumption of fossil fuels has been continuously increasing around the globe and simultaneously becoming the primary cause of global warming as well as environmental pollution. Due to limited life span of fossil fuels and limited alternate energy options, energy crises is important concern faced by the world. Amidst these complex environmental and economic scenarios, renewable energy alternates such as biodiesel, hydrogen, wind, solar and bioenergy sources, which can produce energy with zero carbon residue are emerging as excellent clean energy source. For maximizing the efficiency and productivity of clean fuels via green & renewable methods, it's crucial to understand the configuration, sustainability and techno-economic feasibility of these promising energy alternates. The book series presents a comprehensive coverage combining the domains of exploring clean sources of energy and ensuring its production in an economical as well as ecologically feasible fashion. Series involves renowned experts and academicians as volume-editors and authors, from all the regions of the world. Series brings forth latest research, approaches and perspectives on clean energy production from both developed and developing parts of world under one umbrella. It is curated and developed by authoritative institutions and experts to serves global readership on this theme.

Manish Srivastava • P. K. Mishra
Editors

NanoBioenergy: Application and Sustainability Assessment

Springer

Editors
Manish Srivastava
Department of Chemical Engineering and Technology
Indian Institute of Technology (BHU)
Varanasi, Uttar Pradesh, India

P. K. Mishra
Department of Chemical Engineering and Technology
Indian Institute of Technology (BHU)
Varanasi, Uttar Pradesh, India

ISSN 2662-6861 ISSN 2662-687X (electronic)
Clean Energy Production Technologies
ISBN 978-981-19-6233-2 ISBN 978-981-19-6234-9 (eBook)
https://doi.org/10.1007/978-981-19-6234-9

© The Editor(s) (if applicable) and The Author(s), under exclusive license to Springer Nature Singapore Pte Ltd. 2023
This work is subject to copyright. All rights are solely and exclusively licensed by the Publisher, whether the whole or part of the material is concerned, specifically the rights of translation, reprinting, reuse of illustrations, recitation, broadcasting, reproduction on microfilms or in any other physical way, and transmission or information storage and retrieval, electronic adaptation, computer software, or by similar or dissimilar methodology now known or hereafter developed.
The use of general descriptive names, registered names, trademarks, service marks, etc. in this publication does not imply, even in the absence of a specific statement, that such names are exempt from the relevant protective laws and regulations and therefore free for general use.
The publisher, the authors, and the editors are safe to assume that the advice and information in this book are believed to be true and accurate at the date of publication. Neither the publisher nor the authors or the editors give a warranty, expressed or implied, with respect to the material contained herein or for any errors or omissions that may have been made. The publisher remains neutral with regard to jurisdictional claims in published maps and institutional affiliations.

This Springer imprint is published by the registered company Springer Nature Singapore Pte Ltd.
The registered company address is: 152 Beach Road, #21-01/04 Gateway East, Singapore 189721, Singapore

Preface

This book entitled, *NanoBioenergy: Application and Sustainability Assessment*, provides ecofriendly nanomaterials as green catalysts to enhance biofuels. Nanomaterials work as catalysts in various steps of different biofuels production options in order to improve the productivity of the bioenergy. Nevertheless, synthesis cost of the nanomaterial may be the hindrance to limit the biofuels production process. Therefore, green synthesis of nanomaterial may play a potential role to improve sustainability of overall biofuels process. Based on this concept, this book has been divided into nine focused chapters in which Chaps. 1 and 2 explore feasibility and applications of nanomaterial to improve biofuels production. Whereas Chaps. 3 and 4 deal with the possibility of nanomaterial application in gaseous biofuels and biosynthesis of nanomaterial through microbial process. Further, Chaps. 5 and 6 are focused on green synthesis of iron nanomaterial for biohydrogen and other biofuels applications while Chaps. 7, 8, and 9 discuss green synthesis of nanomaterials using waste biomass and microbial cultures and contributions of nanomaterials in microbial fuel cells application. The book certainly sets a potential milestone that overcomes the high production cost and low productivity issues related to waste to bioenergy production.

Varanasi, Uttar Pradesh, India

Manish Srivastava
P. K. Mishra

Acknowledgments

The editors are thankful to all the academicians and scientists whose contributions have enriched this volume. We also express our deep sense of gratitude to our parents whose blessings have always prompted us to pursue academic activities deeply. It is quite possible that in a work of this nature, some mistakes might have crept in text inadvertently and for these we owe undiluted responsibility. We are grateful to all authors for their contribution to this book. We are also thankful to Springer Nature for giving this opportunity to editors and Department of Chemical Engineering and Technology, IIT (BHU) Varanasi, Uttar Pradesh, India, for all technical support. We thank them from the core of our heart.

Contents

Editors and Contributors

About the Editors

Manish Srivastava is currently working as SERB-Research Scientist in the Department of Chemical Engineering and Technology IIT (BHU), Varanasi, India. He has worked as DST INSPIRE faculty in the Department of Physics and Astrophysics, University of Delhi, India during June 2014 to June 2019. He has published 46 research articles in peer-reviewed journals, edited 8 books for publishers of international renown, authored several book chapters, and filed one patent. He worked as a postdoctoral fellow in the Department of BIN Fusion Technology, Chonbuk National University, South Korea from August 2012 to August 2013. He was an assistant professor in the Department of Physics, DIT School of Engineering, Greater Noida, from July 2011 to July 2012. He received his PhD in Physics from the Motilal Nehru National Institute of Technology, Allahabad, India, in 2011. Presently, he is working on the synthesis of graphene-based metal oxide hybrids and their applications as catalysts. His areas of interest are synthesis of nanostructured materials and their applications as catalyst for the development of electrode materials in energy storage, biosensors, and biofuels production.

P. K. Mishra is currently Professor and Head in the Department of Chemical Engineering and Technology, Indian Institute of Technology (BHU), Varanasi, India. He has authored/co-authored over 60 technical papers published in reputed national/international journals and supervised more than 20 doctoral students. He has received several awards and honours and has 5 patents with 1 Technology Transfer. He is Fellow of Institution of Engineers India. He has received several awards and honours at National/International levels.

Contributors

Ghufran Ahmed ICMR Rajendra Memorial Research Institute, Patna, Bihar, India

Shanthipriya Ajmera Department of Microbiology, Palamuru University, Mahabubnagar, Telangana, India

Zeenat Arif Department of Chemical Engineering and Technology, Indian Institute of Technology (BHU), Varanasi, India

Debashree Borthakur Assam Don Bosco University, Guwahati, Assam, India

Ramesh Chandra Department of Zoology, S S College, Shahjahanpur, Uttar Pradesh, India

Sai Prashanthi Govumoni Department of Microbiology, St. Pious X Degree and PG College for Women, Hyderabad, Telangana, India

P. Kumar Department of Chemical Engineering and Technology, Indian Institute of Technology (BHU), Varanasi, India

Sachin Kumar Punjab Agricultural University, Ludhiana, Punjab, India

Kundan Kunal ICMR Rajendra Memorial Research Institute, Patna, Bihar, India

Jyoti Mishra Department of Chemistry (Environmental Science and Technology Program), ITER, Siksha'O'Anusandhan (Deemed to be) University, Bhubaneswar, Odisha, India

P. K. Mishra Department of Chemical Engineering and Technology, Indian Institute of Technology (BHU), Varanasi, India

B. S. Naveen Prasad Chemical Engineering Section, Engineering Department, University of Technology and Applied Sciences (Salalah College of Technology), Salalah, Oman

M. Noorjahan Department of Chemistry, Palamuru University, Mahabubnagar, Telangana, India

Dharm Pal Department of Chemical Engineering, National Institute of Technology Raipur, Raipur, Chhattisgarh, India

Gaurav Kumar Pandit Department of Botany, Patna University, Patna, Bihar, India

Dhruti Sundar Pattanayak Department of Chemical Engineering, National Institute of Technology Raipur, Raipur, Chhattisgarh, India

Sindhura Podduturi Telangana Social Welfare Residential Degree College for Women, Warangal, Telangana, India

Souvik Roy Post-Graduate Department of Biotechnology, St. Xavier's College (Autonomous), Kolkata, West Bengal, India

Tanvi Sahni Punjab Agricultural University, Ludhiana, Punjab, India

Naresh Kumar Sahoo Department of Chemistry (Environmental Science and Technology Program), ITER, Siksha'O'Anusandhan (Deemed to be) University, Bhubaneswar, Odisha, India

B. Saikat Chemical Engineering Section, Engineering Department, University of Technology and Applied Sciences (Salalah College of Technology), Salalah, Oman

Loveleen Kaur Sarao Punjab Agricultural University, Ludhiana, Punjab, India

Sreedevi Sarsan Department of Microbiology, St. Pious X Degree and PG College for Women, Hyderabad, Telangana, India

Naresh K. Sethy Department of Chemical Engineering and Technology, Indian Institute of Technology (BHU), Varanasi, India

Shanvi Department of Botany, Patna University, Patna, Bihar, India

Veer Singh School of Biochemical Engineering, IIT (BHU), Varanasi, Uttar Pradesh, India

S. Sivamani Chemical Engineering Section, Engineering Department, University of Technology and Applied Sciences (Salalah College of Technology), Salalah, Oman

J. Immanuel Suresh Department of Microbiology, The American College, Madurai, Tamil Nadu, India

Ritesh Kumar Tiwari Department of Botany, Patna University, Patna, Bihar, India

Diksha Verma Punjab Agricultural University, Ludhiana, Punjab, India

Manisha Verma School of Biochemical Engineering, IIT (BHU), Varanasi, Uttar Pradesh, India

Chapter 1
Application of Nanomaterials for Renewable Energy Production

Gaurav Kumar Pandit, Ritesh Kumar Tiwari, Shanvi, Manisha Verma, Veer Singh, Kundan Kunal, Ghufran Ahmed, and Ramesh Chandra

Abstract The ever-increasing demand for energy due to the rapidly increasing industrialization and urbanization compels the research community to devise ways to transition from non-renewable sources of energy to renewable energy sources. The burning of fossil fuels is the primary source of energy that we are utilizing today. Apart from the fact that it is not sustainable and is likely to diminish by 2050 (if we continue using it at the same rate), it also poses severe adverse threats to the environment due to harmful greenhouse gases (GHGs). Thus, we must look for ways to utilize renewable energy in ways so that it can fulfill our energy demands without causing harm to the environment. The efficiency of production and storage of renewable energy needs to be enhanced. Nanotechnology is one such field that is being explored and studied extensively lately due to its practical applications in renewable energy. This chapter discusses the primary classification of nanomaterials and their applications in various renewable energy generation and storage, such as solar energy, hydrogen energy.

Keywords Non-renewable energy · Renewable energy · Fossil fuels · Nanotechnology · Solar energy · Wind energy · Hydrogen energy · Hydrogen economy

G. K. Pandit (✉) · R. K. Tiwari · Shanvi
Department of Botany, Patna University, Patna, Bihar, India

M. Verma · V. Singh
School of Biochemical Engineering, IIT (BHU), Varanasi, Uttar Pradesh, India

K. Kunal · G. Ahmed
ICMR Rajendra Memorial Research Institute, Patna, Bihar, India

R. Chandra
Department of Zoology, S S College, Shahjahanpur, Uttar Pradesh, India

© The Author(s), under exclusive license to Springer Nature Singapore Pte Ltd. 2023

M. Srivastava, P. K. Mishra (eds.), *NanoBioenergy: Application and Sustainability Assessment*, Clean Energy Production Technologies,
https://doi.org/10.1007/978-981-19-6234-9_1

1.1 Introduction

Energy is very crucial for the holistic progress of any country (Stern 2011; Barnes et al. 2011). The energy demand increases exponentially due to the rapid increase in industrialization and urbanization (Graham 2009; Mazur 1994). The significant portion of energy that we are utilizing today comes from the burning of fossil fuels. But the burning of fossil fuels has many deleterious effects on the environment attributed to the release of harmful greenhouse gases (GHGs). It also adversely impacts human health (Kataki et al. 2017). It is cited as the primary contributor to the increasing global warming. As stated by the US Energy Information Administration, the all-consuming of fossil fuels in 2016 caused a substantial 76% of the US greenhouse gas emissions. Also, fossil fuels are representatives of non-renewable sources of energy; that is, they cannot be utilized sustainably. If we continue using fossil fuels at the same pace for our energy needs, it is reported that we will fall short of this energy source by 2050 (Satyanarayana et al. 2011; Demirbas 2009). Thus, it is very imperative that we look for other sources of energy having some of the essential attributes of sustainability and minimal harmful effect on the environment. Renewable sources of energy are the potential alternatives having the characteristics above. Renewable energy is any energy source that can fulfill the existing energy demands without compromising the energy needs of the future as well. In other words, they are a form of a sustainable source of energy. Renewable energy sources such as solar energy, hydrogen energy, wind energy, geothermal energy, and biomass energy have the potential to generate electricity, heat, and light, which can be utilized for various purposes without having deleterious impacts on the environment. Nowadays, much interest and research have been attentive on the practice of Nanotechnology in the renewable energy field. Nanotechnology is basically the science of materials in the nanoscale (diameter having less than 100 nm mostly). Nanomaterials, owing to their very small size, confer many technological and engineering advantages for the parts or equipment associated with renewable energy generation. Various beneficial aspects of nanomaterials have been explored in renewable energy generation, such as wind energy, solar energy, hydrogen energy, etc., and efforts are being made to transition the use of nanotechnology for renewable energy production to a commercial scale. Let us know more about the use of nanotechnology in renewable energy generation.

1.2 Classification of Nanomaterials

Nanomaterials can be classified according to various parameters. Broadly, they can be classified into (Mageswari et al. 2016): (1) Nanoparticles, (2) Nanoclays, and (3) Nanoemulsions.

1.2.1 Nanoparticles

Nanostructures and composites are the two forms in which nanoparticles can exist. Their size range from 1 to 100 nm (Hasan 2015). They can be of various shapes, sizes and can be composed of different types of materials. Based on the materials they are composed of, nanoparticles can be (Mageswari et al. 2016; Jeevanandam et al. 2018; Ealia and Saravanakumar 2017)

1.2.1.1 Organicnanoparticles

These nanoparticles are favorable choices for drug delivery. Some of the characteristics which make them a favorable candidate for this purpose are: (1) They are biodegradable; (2) They are not toxic by nature; (3) Some of them form a hollow core and are called nanocapsules. They are believed to be sensitive towards light and heat (Tiwari et al. 2008), such as liposomes and micelles; (4) Suitable for targeted drug delivery. Besides their general characteristics such as morphology, size, etc., their field of application is also determined by their drug-carrying capacity, drug delivery system (whether encapsulated or adsorbed) as well as stability. Some examples are liposomes, ferritin, micelles, and dendrimers.

1.2.1.2 Inorganic Nanoparticles

They do not contain carbon. They comprise metal and its oxides-based nanoparticles.

1.2.1.3 Carbonnanoparticles

It consists of carbon entirely. Examples: Fullerenes, carbon nanofibers, carbon nanotubes (CNT), etc.

1.2.2 Dimension-Based Nanomaterials Classification

The nanomaterials classification based on dimension is achieved by considering the pattern of the electron trail alongside the various dimensions in the nanomaterials. Pokropivny and Skorokhod proposed this system of classification in 2007 (Pokropivny and Skorokhod 2007). Based on the dimension, nanoparticles can be of different types (Jeevanandam et al. 2018; Mageswari et al. 2016):

1.2.2.1 0D

In these types of nanomaterials, the electrons movement is enmeshed in all three dimensions, or they are confined within the dimensionless space. Examples are fullerenes, molecules, metal carbides, etc.

1.2.2.2 1D

The electrons movement in this type is restricted in one dimension, in the X-direction. Examples include nanotubes, filaments, fibers, etc.

1.2.2.3 2D

The movement of electrons is confined in the X-Y plane. Examples- Layers.

1.2.2.4 3D

The movement of electrons can occur along the X, Y, and Z directions (Siegel 1993).

1.2.3 Classification Based on the Origin of Nanomaterials

1.2.3.1 Natural

They are naturally present in the Earth's sphere, i.e., atmosphere comprising of hydrosphere, lithosphere, troposphere, and even the biosphere. It is noteworthy to mention here that the biosphere includes all the microorganisms and the higher organisms, which include humans (Sharma et al. 2015; Hochella et al. 2015). Either natural processes or some sort of anthropogenic activity serves to produce these nanomaterials.

1.2.3.2 Synthetic

They are fabricated or engineered and are generated by processes that may be physical, biological, chemical, or hybrid methods such as mechanical grinding, etc. One of the significant challenges with these nanomaterials is difficulty in assessing whether the current knowledge is sufficient in forecasting their behavior or if they exhibit any environment-related activity that is distinct from the nanomaterials occurring naturally (Wagner et al. 2014).

1.2.4 Nanoclays

They are the other types of nanomaterials, the preparation of which exploits the hydrophilic or charged characteristics of clay molecules. The charged groups can be ammonium, aryl/alkyl, phosphonium, or imidazolium, either in the aqueous or solid-state. X-ray diffraction, gravimetric analysis, Fourier transform infrared (FTIR) spectroscopy, and inductively coupled plasma can be used for chemical characterization.

1.2.5 Nano-Emulsion

These types of nanomaterials are in the form of soft materials and are generated by dispersing the solid materials, droplets, and polymers in the viscous liquid. It is generally synthesized by using either of the two methods:

1. High-energy emulsification includes microfluidizer, ultrasonication, and high-pressure homogenization.
2. Low energy emulsification includes techniques like phase inversion temperature, solvent displacement, and phase inversion composition.

The given figure (Fig. 1.1) shows the general classification of nanomaterials.

1.3 Synthesis of Nanomaterials

The route for the synthesis of nanomaterials can be physical, chemical, or biological (Mageswari et al. 2016).

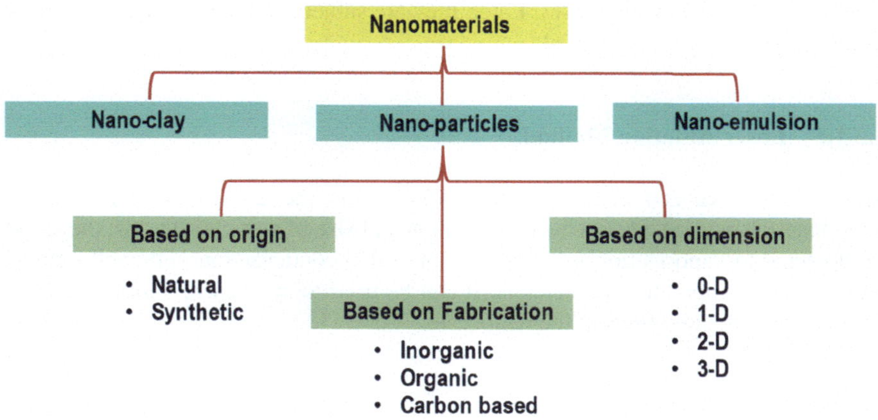

Fig. 1.1 Types of nanomaterials

1.3.1 Physical Methods

These include physical techniques to synthesize nanomaterials. One of the main advantages of physical methods over chemical methods is that nanomaterials have a uniform distribution as they are devoid of solvents. Some of the techniques include:

1.3.1.1 Laser

It uses Transmission Electron Microscopy to emit high-energy electron beams and irradiate surfaces to synthesize nanomaterials of various forms. Examples are carbon nanocapsules, nanotubes, etc.

1.3.1.2 Arc-Discharge

Two methods, including higher frequency plasma and direct current arc plasma, can be applied for this technique. This is mainly used to synthesize fullerenes.

1.3.1.3 Combustion

In this technique, the activation energy barrier is overcome by the heat generated during the exothermic reaction. It is fast and has the potential to form a wide range of ceramic oxides.

1.3.1.4 Evaporation–Condensation

In this technique, first, the metals, ceramics, or alloys are allowed to evaporate and react with each other using gases. Later, they are subjected to condensation, leading to the formation of nanomaterials.

1.3.1.5 Laser Ablation Method

This is an advanced technique that allows for the controlled synthesis of nanomaterials with respect to the composition and size of particles. In this, the target is subjected to vaporization followed by controlled condensation with well-defined pressure and temperature parameters. It can be used to synthesize various nitrides, carbides, and metal oxides at the nanoscale.

1.3.2 Chemical Methods

These methods allow for the extensive quantity synthesis of nanomaterials in a controlled manner (Hyeon 2003). There are many methods for synthesizing nanomaterials by chemical means (Chaki et al. 2015; Umer et al. 2012; Anbarasu et al. 2015; Yu et al. 2009).

1.3.2.1 Chemical Reduction

It is the most applied technique for the synthesis of nanomaterials in the form of colloids. It involves the chemical reduction of inorganic as well as organic reducing agents.

1.3.2.2 Oxidation

The process of oxidation can be used to form nanomaterials of alloys, metals, or oxides, either in water or some other organic solutions.

1.3.2.3 Microemulsion

It involves the separation between two immiscible phases in space, that is, between reducing agent and metal in two-phase aqueous organic systems. Quaternary ammonium salts are used to mediate the interface between two phases. Stabilization of the clusters of the metal is achieved as the stabilizer molecules are capped in the non-polar aqueous medium, which is then transferred to the organic phase.

1.3.2.4 Sol-Gel Process

In this method, the first formation of sol is achieved by dispensing precursors in suitable solvents, which upon drying, are put in solid (gels) to form a polymeric network. When the gel is subsequently dried to be subjected to operations such as sintering and calcination, ceramic products are formed. Other chemical synthesis methods include polymerization, microwave-assisted synthesis, UV-initiated photoreduction, irradiation, etc.

Some drawbacks of physical and chemical methods of nanomaterial synthesis (Mageswari et al. 2016; Hebbalalu et al. 2013):

1. Complex protocols.
2. High operational costs.
3. The presence of minor toxic components makes their biological use questionable.

Table 1.1 Synthesis of nanomaterials using various biological agents

Biological agent	Examples	Types of nanoparticles
Plants	*Azadirachta indica,* Lemon grass, *Garcinia mangostana,* etc.	Gold, Silver, Copper, Aluminium oxide, etc.
Bacteria	*E.coli,* sulfate-reducing bacteria, *Bacillus subtilis,*etc.	Cadmium sulfide, Magnetite, Titanium oxide, etc.
Fungi	*Aspergillus flavus, F.oxysporum, Agaricus bisporus,*etc.	Silver, Lead sulfide, Zinc sulfide, Silver, etc.
Yeast	*Saccharomyces cerevisiae,* Extremophilic yeast, *Yarrowia lipolytica,* etc.	Titanium oxide, Manganese oxide, Gold, Silver, etc.
Actinomycetes	*Rhodococcus* sp., *Nocardiopsis* sp., *S. albidoflavus,* etc.	Gold, Silver, etc.

1.3.3 Biological Synthesis

Biological production of nanomaterials is more environment-friendly and thus preferred more over physical and chemical modes (Hebbalalu et al. 2013; Sastry et al. 2003; Kruis et al. 2000; Ahmad et al. 2003).

The given table (Table 1.1) enlists some of the biological agents with examples for the synthesis of nanoparticles.

1.4 Nanomaterials: Applications in Renewable Energy

The advancements in nanotechnology, together with the fact that nanomaterials hold several benefits to be applied for renewable energy, have in all opened a new domain of research. Some of the attributes of nanomaterials that make them a preferred choice for various renewable energy applications are (Hussein 2015):

1. They provide greater capacity for energy storage and efficiency for lighting and heating.
2. The energy so generated with the use of nanotechnology can help curtail pollution.

The nanomaterials, having their one or more dimensions at the nanoscale, tend to disobey conventional rules of physics and thus express remarkable properties compared to their larger entities. Some of the advantageous features and potential benefits they show at the nanoscale are their strength, electrical conduction capacity, and reactivity increase.

1.4.1 Solar Energy

Solar energy is one of the primary sources which can be harnessed to produce renewable energy. The sunlight that is reflected on the Earth's surface annually surpasses the total resources that we use. It is noteworthy to mention here that the sunlight of 1 h equates to more than the total energy consumed annually (Vayssieres 2009). Thus, it is of utmost relevance to utilize this most excellent energy source to meet the energy requirements efficiently and inexpensively. There are two ways by which we can use solar energy (Ghasemzadeh and Shayan 2020; Esmaeili Shayan et al. 2020):

1. To directly produce electricity utilizing sunlight.
2. Solar thermal energy can be used in high-temperature power plants to generate electricity. Or for ventilation in houses and processing of hot water when used in low-temperature power plants.

Nanotechnology can be used to improve heat and electricity generation. Nanotechnology can be exploited to increase the efficiency of solar cells in many ways (Sarbu et al. 2017):

1. Nanotechnology can be used to enhance the storage of solar power.
2. The efficiency of solar cells can be boosted using nanowires.
3. Sunlight absorption and retention can be enhanced.

1.4.1.1 Improved Absorption and Capture of Solar Energy

Nanomaterials having the capacity to emit and capture light such as Silver, gold, or quantum dots, and fluorescent nanofibers may be employed to advance the functioning of solar cells (Esmaeili Shayan and Najafi 2019). Based on their dimensions, these nanoparticles can absorb different wavelengths and become excited. After this these nanoparticles emit the absorbed energy in the form of radiation with a different wavelength or from the earliest one. Owing to their photoelectric properties, which help in the conversion of solar light to electricity, however, they tend to absorb wavelengths that are beyond the visible spectrum. A fascinating improvement of 64% in the performance of solar energy is observed when projected simulating the quantum-based cell theory. Quantum dots of silver sulfide and silver selenide tend to immensely enhance the response of the solar cell to light. It has been observed that almost a 4% increase in the production of electrical energy is seen with these quantum dots when compared to routine pigment-sensitive cells. Hence, such quantum dots are very relevant to increase the efficiency of solar cells.

1.4.1.2 Nanofluids

Nanofluids, owing to their high heat transfer coefficient in heat exchangers or engines, serve to rise the economy and performance. Many businesses and academic institutes are adopting solar batteries and heaters.

1.4.1.3 Photocatalysts

These are stable semiconductors generating an electron-hole pair as they collect photons, thus interfering with the molecules at the surface level. Some of the advantages of nano-photocatalysts with respect to solar cells are:

1. The performance of the solar cells can be enhanced dramatically as the absorption of light can be improved attributed to their property of absorbing specific light spectra.
2. The absorption of sunlight and the performance of the solar cells also get improved as they render a clean atmosphere free from air pollutants and obstacles to light. This is due to their anti-dust, self-cleaning, and anti-steam properties.
3. They also help in improving the energy-transfer capacities.

1.4.1.4 Nanotechnology in the Storage of Power

Nanotechnology is also being increasingly used for the storage of power. Many factors such as environmental conditions, including temperature, hours, photoperiod, and atmospheric patterns, tend to impact solar power generation systems production, rendering the consistent supply of output not feasible in such cases (Achkari and El Fadar 2020). Owing to the fact that the ordinary or conventional batteries have certain loopholes such as inefficient capacity, heavyweight and poor performance and therefore, Lithium batteries are in trend nowadays (Walker 2013). Lithium-ion batteries utilize organic solvents for the purpose of electrolytes instead of gas used in conventional batteries. But liquid electrolytes have strong electrical resistance, and nanomaterials are being utilized to increase the electrolyte's efficiency. Nanotechnology aims to boost the conductivity by sixfolds by adding powers in the form of nanoparticles (silicon oxide, zirconium oxide, etc.) to non-aqueous electrolytes.

The given figure (Fig. 1.2) displays some of the general approaches in which nanotechnology can be applied in solar energy sector.

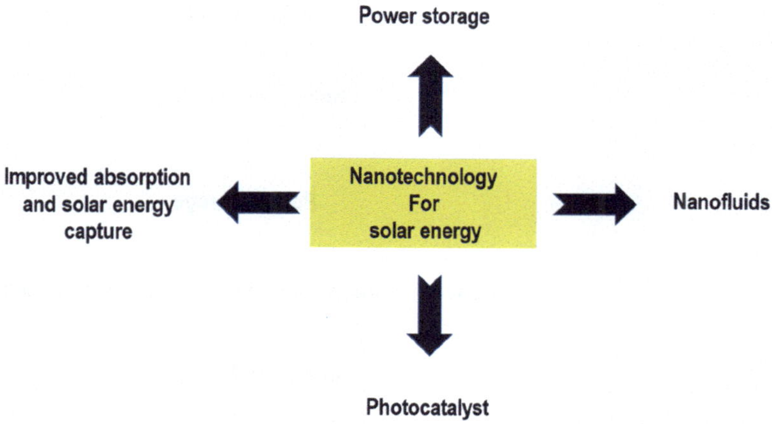

Fig. 1.2 Approaches for nanotechnology applications in solar energy

1.4.2 Renewable Hydrogen Energy and Use of Nanotechnology

Hydrogen energy is also being explored as a clean source of renewable energy for it only produces water upon consumption in fuel cells. Solar water splitting has been considered as one of the most efficient ways of producing hydrogen energy. The hydrogen so produced can then be utilized by fuel cells for the generation of electricity, with water being the only emission. This also paves the way for the effective utilization of solar energy and its conversion (Mao et al. 2012). For the purpose of ensuring efficient and optimum utilization of solar energy for hydrogen energy production, the efficacy of solar water splitting systems must be increased. Also, the storage capacity and performance of fuel cells should be enhanced to make hydrogen energy the primary contributor to the prospective renewable energy-based economy.

The given figure (Fig. 1.3) shows the general framework for production of hydrogen energy using solar water splitting-

1.4.2.1 Nanomaterials-Based Electrodes for Photo-Electrochemical Water Splitting

Water splitting by photo-electrochemical (PEC) means been the fascinating way of hydrogen energy production amongst other techniques. Fujishima and Honda first pioneered Solar energy-induced splitting of water in PECs, wherein they used TiO_2 anode and Pt cathode to generate oxygen and hydrogen, respectively (Fujishima and Honda 1972). In this system, once the anodes are irradiated by sunlight, and if the irradiated energy has energy larger than its bandgap, then electrons are generated in

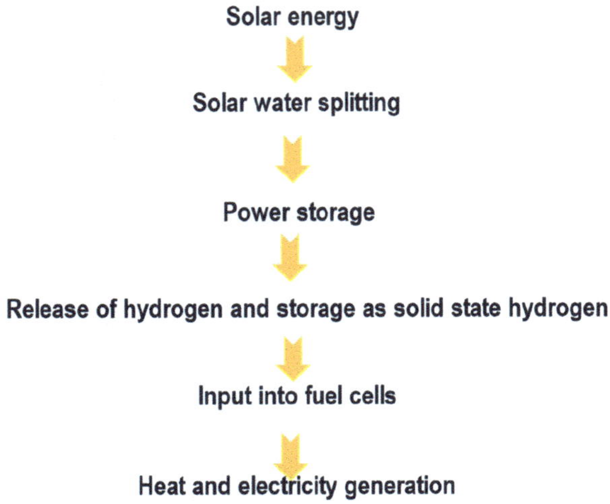

Fig. 1.3 A general framework for hydrogen energy production using solar water splitting

the conduction band, and holes are generated in the valence band respectively (Chen et al. 2010). As a result of this, water gets oxidized at the TiO_2 anode liberating oxygen, and the photogenerated electrons are transferred to the Pt cathode to produce hydrogen. However, oxygen production at the photoanode is kinetically limited for splitting of water in the PEC process, and so, nanostructured photoanodes were designed for PEC water splitting leading to the production of oxygen.TiO_2, which is the most widely used semiconductor for PEC-based water splitting, has a very large bandgap of almost 3.2 eV (Chen and Mao 2007; Ni et al. 2007). This renders the TiO_2 anode inefficient and incapable of absorption of light in the visible and infrared range. Thus, photoanodes were designed to introduce either a donor or acceptor level in the forbidden gap to narrow the bandgap of TiO_2. This makes the photoanode TiO_2 sensible to visible light (Chen et al. 2010; Chen and Mao 2007). For instance, a TiO_2 doped nanocrystalline film showed a more excellent efficient water splitting phenomenon with 11% accounting for total energy conversion and 8.35% of photoconversion efficiency, attributed to the enhanced capability to absorb visible light (Khan et al. 2002). The morphology of TiO_2 is also of considerable importance since the morphology impacts the capability of transfer of charge. TiO_2 nanotube arrays were designed and examined by Grimes and co-workers for water splitting by PEC, and it was found that they render more efficient charge separation owing to their architecture (Rani et al. 2010; Mor et al. 2005, 2007). The photoelectric conversion efficiency of almost 16.5% could be obtained under UV illumination when nanotubes of 24 μm in length were fabricated electrochemically in an ethylene glycol-based electrolyte (Mao et al. 2012).

1.4.2.2 Nano-Photocatalysts and Hydrogen Production

Bard in 1979 designed a photocatalytic water splitting system based on the concept of photoelectrical splitting of water. He utilized particles or powders of semiconductors as photocatalysts (Bard 1979). Electrons and holes photogenerated at the conduction and valence bands, respectively, are transferred to photocatalysts, where they participate in a redox reaction leading to oxygen and hydrogen production.

Some of the fundamental characteristics to be considered for photocatalysts are (Mao et al. 2012):

1. Relevant band gaps and structures to ensure optimum absorption of sunlight are needed to drive oxygen and hydrogen production.
2. Efficient transferability of holes and electrons.
3. High catalytic reactivity of surface for half-reactions.

Extensive research and efforts have been put in the past decades to meet the specific requirements and ensure efficient generation of hydrogen from water (Chen et al. 2010; Shen et al. 2011; Osterloh 2008; Shen and Mao 2012; Maeda and Domen 2010).

1. **Surface Layer Disorders**

   As mentioned earlier, TiO_2 is the most widely studied photocatalyst, but it has a wide band gap, limiting its absorption efficiency. To narrow its bandgap, doping with ions is done, but this also has a drawback. The energy levels created by doping can then act as recombination centers for photoinduced charges and deleteriously impact the photocatalytic activity of doped TiO_2 (Chen and Mao 2007). To circumvent this issue, a new approach of surface layer disorder was envisioned for the enhanced absorption of solar energy by TiO_2 nanocrystals (Chen et al. 2011). Surface disordered black TiO_2 nanocrystals obtained by hydrogenation of anatase TiO_2 nanocrystals at 200 °C for a duration of 5 days in 20.0- bar H_2 atmosphere, produced hydrogen at a constant rate. The hydrogen production rate so observed (10 mmol h^{-1} g^{-1} of photocatalysts) and efficiency as close to 24% of solar energy conversion is almost two times greater than most semiconductor photocatalysts' yield (Chen et al. 2010). This observation is attributed to the efficient energy harvesting from UV to near infrared by the surface disordered black TiO_2 and the retardation in recombination of charge (Mao et al. 2012).
2. **Cocatalysts**

   Loading of cocatalysts onto photocatalysts is considered as an efficient strategy for optimal water splitting due to increased photocatalytic activity for the formation of hydrogen or oxygen. Different materials have been developed as proficient cocatalysts in the past few decades, such as sulfides and oxides of metal, transition metals, etc.

1.4.2.3 Solid-State Nanomaterials (Hydrogen Storage)

Storage of hydrogen is very crucial to realize the full potential of hydrogen energy economy or renewable energy economy. It can be stored in various ways—as a cryogenic liquid, pressurized gas, or inappropriate solid-state materials such as carbon materials, metal organics, or metal hydrides. Of all these ways, solid-state hydrogen storage is the most efficient and technically feasible approach. Material specified surface interactions determine the capacity of storage and kinetics in case of solid-state storage, either through strong chemisorption or weaker dispersed physisorption (Liu et al. 2012; Abbasi and Abbasi 2011; Froudakis 2011; Dagdougui 2012; Ding and Yakobson 2011). Metal hydrides (reversible chemisorption), particularly MgH_2, are essential representatives of the candidates for solid-state hydrogen storage attributed to their high capacity of hydrogen storage and comparatively high weight percentage (7.6%) of hydrogen in MgH_2 besides the fact that magnesium is also available abundantly (Mao et al. 2012; Sakintuna et al. 2007).

However, there are certain limitations with these metal hydrides:

1. Poor reaction kinetics
2. High thermodynamic stability
3. High enthalpy of formation of bulk MgH_2 rendering it incapable of producing hydrogen at temperatures below 300 °C

And so, it is needed to reduce the enthalpy to make MgH_2 a potent candidate for storage systems. Alloying Mg with metals such as Al or Ni can help in circumventing this enthalpy issue. However, they tend to substantially decrease hydrogen weight percentage (Bouaricha et al. 2000; Hirata et al. 1983). A better option is the sandwiched Mg nanoparticle layer as Pd/Mg/Pd thin films deposited by the PLD approach. This nanoparticle film proves beneficial in the reduction of enthalpy of formation, and so, the thermodynamic barrier to the formation of hydride can be catered (Barcelo et al. 2010).

1.4.3 PEMFCs and Nano-Electrocatalyst

Polymer Electrolyte Membrane (or Proton Exchange Membrane) Fuel Cells are very potent in the conversion of clean energy for they can generate electricity without pollution and combustion at an astounding conversion efficiency of 70% by harnessing hydrogen's chemical energy. This renders them as a highly eligible candidate to replace combustion engines both for stationary and mobile purposes (Yuan et al. 2012). In this system, ionization of hydrogen occurs at the anode liberating protons and electrons, which subsequently recombine at the cathode and reduce oxygen to water. Electrocatalysts are needed to improve the kinetics of reduction of oxygen at cathode, which otherwise, because of slow kinetics, can significantly cause voltage loss in PEMFCs. Pt is regarded as a good electrocatalyst for this purpose; however, owing to its scarcity and very high cost, efforts have been made in the past few decades to devise low-Pt alloys for the reduction in the use of Pt in fuel cells (Chrzanowski and Wieckowski 1998). Some of the non-Pt electrocatalysts include metal carbides, metal chalcogenides, metal oxides, metal nitrides, and macrocycles. Attributed to the high resistance against corrosion and wear, non-Pt chromium nitrides electrocatalysts such as CrN (highly crystalline CrN nanoparticles) and Cr_2N are the prospective electrocatalysts (Volz et al. 1998; Lackner et al. 2006).

1.4.4 Nanotechnology and Wind Energy

Wind energy is also a type of renewable energy in which wind turbines are used to convert the wind's kinetic energy into another form such as electrical or mechanical energy, which can then be utilized for various functional applications. Wind energy is believed to be less detrimental to the environment when compared to other energy sources, for wind turbines do not burn the fuels for the generation of electricity. For the optimum utilization of wind power plants, wind speed should be a minimum of 13–15 m/s and 10 m/s for the smaller ones. However, the wind speed is always not uniform and optimum, which limits the efficiency of these plants up to 30–60% (Hussein 2015). It is noteworthy to mention here that the square of the length of the wind turbine and the energy production are proportionally related (Dalili et al. 2009; Sherif et al. 2005). This is where nanotechnology steps in. Nano-composites enable the fabrication of a longer blade with greatly enhanced strength. This can be attributed to their superb and efficient stiffness and strength-to-weight ratios. Nanotechnology via low-friction coatings and nano-lubricants also serves to reduce the loss of energy due to various tribological problems such as scuffing, wear, etc. (Hussein 2015).

1.4.5 Nanotechnology in Other Renewable Energy Sources

1.4.5.1 Biofuels

- It may successfully lead to the generation of biofuels from solid wastes (Mahmood and Hussain 2009). Nanotechnology also serves to increase the yield of biodiesel production. It has been observed that KF/CaO nanocatalyst led to biodiesel production with a yield of 96.8%. It has also been stated that it could also be used for the conversion of oil with high acid value into biodiesel efficaciously (Wen et al. 2010). It has also been stated that a more "greener" production of biodiesel could be obtained with the use of nanocatalyst (Konwar et al. 2014). Sajith et al. (2010) concluded that the emission of nitrogen oxides compounds and hydrocarbons could substantially be reduced with the addition of particles of cerium oxide in the nano range on biodiesel (Sajith et al. 2010). Hipólito et al. (2014) stated that they could obtain a biodiesel yield of 97–100% using STNT following a chemical reaction of 8 h duration (Hipólito et al. 2014).

1.4.5.2 Geothermal Energy

Nanofluids can prove to be crucial in cooling the pipes exposed to very high temperatures during the extraction of geothermal energy from the crust of the Earth. They can also be utilized for cooling purposes in components such as electronics and sensors in drilling machines (Hussein 2015).

1.5 Conclusion and Future Perspectives

It can evidently be stated that nanotechnology is playing a significant part in producing various renewable energies. Nanomaterials greatly influence the absorption capacities of various devices involved in the conversion and utilization of renewable energy. Various aspects of nanotechnology are being explored and exploited lately. As the need for clean energy generation increases due to multifarious problems associated with conventional energy sources, i.e., burning fossil fuels, the demand for renewable energy production is also increasing exponentially. Nanotechnology finds its wide applications in almost all renewable energy sources—solar energy, hydrogen energy, biofuels, geothermal energy, wind energy, etc. Nanomaterials can be efficiently used for the storage of renewable energies, such as solar energy and hydrogen energy. They also boost the efficiency of renewable energy-producing devices; for example, nano-composites can be used to fabricate longer blades with enhanced strength, which will ultimately increase wind energy generation efficiency.

1. Nanotechnology is one of the critical factors for the complete realization of the "Hydrogen Economy."
2. It is mandatory to study, research, and explore more about nanotechnology so that its shortcomings can be managed and its transition to commercial status can be accomplished entirely soon.

Acknowledgments Authors are grateful to the Patna University, IIT (BHU), Rajendra Memorial Research Institute, Patna, and S S College Shahjahanpur for providing necessary facilities for this work.

References

Abbasi T, Abbasi SA (2011) 'Renewable' hydrogen: prospects and challenges. Renew Sustain Energy Rev 15(6):3034–3040

Achkari O, El Fadar A (2020) Latest developments on TES and CES technologies—energy and environmental issues, applications and research trends. Appl Therm Eng 167:114806. https://doi.org/10.1016/j.applthermaleng.2019.114806

Ahmad A, Senapati S, Khan MI et al (2003) Extracellular biosynthesis of monodisperse gold nanoparticles by a novel extremophilic actinomycete, Thermomonospora sp. Langmuir 19: 3550–3553. https://doi.org/10.1021/la026772l

Anbarasu M, Anandan M, Chinnasamy E et al (2015) Synthesis and characterization of polyethylene glycol (PEG) coated Fe3O4 nanoparticles by chemical co-precipitation method for biomedical applications. Spectrochim Acta A Mol Biomol Spectrosc 135:536–539. https://doi.org/10.1016/j.saa.2014.07.059

Barcelo S, Rogers M, Grigoropoulos CP, Mao SS (2010) Hydrogen storage property of sandwiched magnesium hydride nanoparticle thin film. Int J Hydrog Energy 35(13):7232–7235. https://doi.org/10.1016/j.ijhydene.2010.01.130

Bard AJ (1979) Photoelectrochemistry and heterogenous photocatalysis at semiconductors. J Photochem 10(1):59–75. https://doi.org/10.1016/0047-2670(79)80037-4

Barnes DF, Khandker SR, Samad HA (2011) Energy poverty in rural Bangladesh. Energy Policy 39(2):894–904. https://doi.org/10.1016/j.enpol.2010.11.014

Bouaricha S, Dodelet JP, Guay D, Huot J, Boily S, Schulz R (2000) Hydriding behavior of Mg-Al and leached Mg-Al compounds prepared by high-energy ball-milling. J Alloys Compd 297(1–2):282–293. https://doi.org/10.1016/S0925-8388(99)00612-X

Chaki SH, Malek TJ, Chaudhary MD et al (2015) Magnetite Fe3O4 nanoparticles synthesis by wet chemical reduction and their characterization. Adv Nat Sci Nanosci Nanotechnol 6:035009. https://doi.org/10.1088/2043-6262/6/3/035009

Chen X, Mao SS (2007) Titanium dioxide nanomaterials: synthesis, properties, modifications, and applications. Chem Rev 107(7):2891–2959. https://doi.org/10.1021/cr0500535

Chen X, Shen S, Guo L, Mao SS (2010) Semiconductor- based photocatalytic hydrogen generation. Chem Rev 110(11):6503–6570. https://doi.org/10.1021/cr1001645

Chen X, Liu L, Yu PY, Mao SS (2011) Increasing solar absorption of photocatalysis with black hydrogenated titanium dioxide nanocrystals. Science 331(6018):746–750. https://doi.org/10.1126/science.1200448

Chrzanowski W, Wieckowski A (1998) Surface structure effects in platinum/ruthenium methanol oxidation electrocatalysis. Langmuir 14(8):1967–1970. https://doi.org/10.1021/la980184j

Dagdougui H (2012) Models, methods, and approaches for the planning and design of the future hydrogen supply chain. Int J Hydrog Energy 37(6):5318–5327. https://doi.org/10.1016/j.ijhydene.2011.08.041

Dalili N, Edrisy A, Carriveau R (2009) A review of surface engineering issues critical to wind turbine performance. Renew Sustain Energy Rev 13(2):428–438. https://doi.org/10.1016/j.rser.2007.11.009

Demirbas A (2009) Global renewable energy projections. Energy Sources Part B Econ Plan Policy 4(2):212–224. https://doi.org/10.1080/15567240701620499

Ding F, Yakobson BI (2011) Challenges in hydrogen adsorptions: from physisorption to chemisorption. Front Phys 6:142–150. https://doi.org/10.1007/s11467-011-0171-6

Ealia SAM, Saravanakumar MP (2017) A review on the classification, characterization, synthesis of nanoparticles and their application. IOP Conf Ser Mater Sci Eng 263(3):032019. IOP Publishing

Esmaeili Shayan M, Najafi G (2019) Energy-economic optimization of thin layer photovoltaic on domes and cylindrical towers. Int J Smart Grid 3:84–91

Esmaeili Shayan M, Najafi G, Gorjian S (2020) Design principles and applications of solar power systems (in Persian). 1st edn. ACECR Publication-Amirkabir University of Technology Branch, Tehran

Froudakis GE (2011) Hydrogen storage in nanotubes and nanostructures. Mater Today 14(7–8): 324–328. https://doi.org/10.1016/S1369-7021(11)70162-6

Fujishima A, Honda K (1972) Electrochemical photolysis of water at a semiconductor electrode. Nature 238:37–38. https://doi.org/10.1038/238037a0

Ghasemzadeh F, Shayan ME (2020) Nanotechnology in the service of solar energy systems. In: Nanotechnology and the environment. IntechOpen. https://doi.org/10.5772/intechopen.93014

Graham Z (2009) Peak people: the interrelationship between population growth and energy resources. Energy Bulletin, Leederville

Hasan S (2015) A review on nanoparticles: their synthesis and types. Res J Recent Sci 4:9–11. issn:2277-2502

Hebbalalu D, Lalley J, Nadagouda MN, Varma RS (2013) Greener techniques for the synthesis of silver nanoparticles using plant extracts, enzymes, bacteria, biodegradable polymers, and microwaves. ACS Sustain Chem Eng 1:703–712. https://doi.org/10.1021/sc4000362

Hipólito PH, Castillejos MG, Klimova EM, Flores NJ (2014) Biodiesel production with nanotubular sodium titanate as a catalyst. Catal Today 6(4–11):220–222. https://doi.org/10.1016/j.cattod.2013.09.003

Hirata T, Matsumoto T, Amano M, Sasaki Y (1983) Dehydriding reaction kinetics in the improved intermetallic Mg_2Ni-H system. J Less Common Metals 89(1):85–91. https://doi.org/10.1016/0022-5088(83)90252-7

Hochella MF, Spencer MG Jr, Jones KL (2015) Nanotechnology: nature's gift or scientists' brainchild? Environ Sci Nano 2:114–119. https://doi.org/10.1039/C4EN00145A

Hussein AK (2015) Applications of nanotechnology in renewable energies—a comprehensive overview and understanding. Renew Sustain Energy Rev 42:460–476. https://doi.org/10.1016/j.rser.2014.10.027

Hyeon T (2003) Chemical synthesis of magnetic nanoparticles. Chem Commun:927–934. https://doi.org/10.1039/B207789B

Jeevanandam J, Barhoum A, Chan YS, Dufresne A, Danquah MK (2018) Review on nanoparticles and nanostructured materials: history, sources, toxicity and regulations. Beilstein J Nanotechnol 9:1050–1074. https://doi.org/10.3762/bjnano.9.98. PMID: 29719757; PMCID: PMC5905289

Kataki R et al (2017) Biohydrogen production scenario for Asian countries. In: Singh A, Rathore D (eds) Biohydrogen production: sustainability of current technology and future perspective. Springer, New Delhi, pp 207–235. https://doi.org/10.1007/978-81-322-3577-4_10

Khan SUM, Al-Shahry M, Ingler WB Jr (2002) Efficient photochemical water splitting by a chemical modified n-TiO_2. Science 297(5590):2243–2245. https://doi.org/10.1126/science.1075035

Konwar LJ, Boro J, Deka K (2014) Review on latest developments in biodiesel production using carbon-based catalysts. Renew Sustain Energy Rev 29:546–564. https://doi.org/10.1016/j.rser.2013.09.003

Kruis FE, Maisels A, Fissan H (2000) Direct simulation Monte Carlo method for particle coagulation and aggregation. AICHE J 46:1735–1742. https://doi.org/10.1002/aic.690460905

Lackner JM, Waldhauser W, Major B, Morgiel J, Major L, Takahashi H, Shibayama T (2006) Growth structure and growth defects in pulsed laser deposited Cr-CrN_X-CrC_XN_{1-X} multilayer coatings. Surf Coat Technol 200(11):3644–3649. https://doi.org/10.1016/j.surfcoat.2005.03.032

Liu PF, Chu JK, Hou SJ, Xu P, Zheng JY (2012) Numerical simulation and optimal design for composite high-pressure hydrogen storage vessel. Renew Sustain Energy Rev 16(4):1817–1827. https://doi.org/10.1016/j.rser.2012.01.006

Maeda K, Domen K (2010) Photocatalytic water splitting: recent progress and future challenges. J Phys Chem Lett 1(18):2655–2661. https://doi.org/10.1021/jz1007966

Mageswari A, Srinivasan R, Ramesh N, Gothandam KM (2016) Nanomaterials: classification, biological synthesis and characterization. Nanoscience in food and agriculture, vol. 3. Sustainable agriculture reviews 23. https://doi.org/10.1007/978-3-319-48009-1_2

Mahmood T, Hussain S (2009) Nanobiotechnology for the production of biofuels from spent tea. Afr J Biotechnol 9(6):858–868. https://doi.org/10.5897/AJB091555

Mao SS, Shen S, Guo L (2012) Nanomaterials for renewable hydrogen production, storage and utilization. Progress Natural Sci Mater Int 22(6):522–534. https://doi.org/10.1016/j.pnsc.2021.12.003

Mazur A (1994) How does population growth contribute to rising energy consumption in America? Popul Environ 15:371–378. https://doi.org/10.1007/BF02208318

Mor GK, Shankar K, Paulose M, Varghese OK, Grimes CA (2005) Enhanced photocleavage of water using titania nanotube arrays. Nano Lett 5(1):191–195. https://doi.org/10.1021/nl048301k

Mor GK, Prakasam HE, Varghese OK, Shankar K, Grimes CA (2007) Vertically oriented Ti-Fe-O nanotube array films: toward a useful material architecture for solar spectrum water photoelectrolysis. Nano Lett 7(8):2356–2364. https://doi.org/10.1021/nl0710046

Ni M, Leung MKH, Leung DYC, Sumathy K (2007) A review and recent developments in photocatalytic water- splitting using TiO_2 for hydrogen production. Renew Sustain Energy Rev 11(3):401–425

Osterloh FE (2008) Inorganic materials as catalysts for photochemical splitting of water. Chem Mater 20(1):35–54. https://doi.org/10.1021/cm7024203

Pokropivny VV, Skorokhod VV (2007) Mater Sci Eng C 27:990–993. https://doi.org/10.1016/j.msec.2006.09.023

Rani S, Roy SC, Paulose M, Varghese OK, Mor GK, Kim S, Yoriya S, LaTempa TJ, Grimes CA (2010) Synthesis and applications of electrochemically self- assembled titaniananotube arrays. Phys Chem Chem Phys 12:2780–2800. https://doi.org/10.1039/B924125F

Sajith V, Sobhan C, Peterson G (2010) Experimental investigations on the effects of cerium oxide nanoparticle fuel additives on biodiesel. Adv Mech Eng: 1–6. https://doi.org/10.1155/2010/581407

Sakintuna B, Darkrim FL, Hirscher M (2007) Metal hydride materials for solid hydrogen storage: a review. Int J Hydrog Energy 32(9):1121–1140. https://doi.org/10.1016/j.ijhydene.2006.11.022

Sarbu I, Sebarchievici C, Sarbu I et al (2017) Solar collectors. Sol heating & cooling system: 29–97

Sastry M, Ahmad A, Islam Khan M, Kumar R (2003) Biosynthesis of metal nanoparticles using fungi and actinomycete. Curr Sci 85:162–170

Satyanarayana KG, Mariano AB, Vargas JVC (2011) A review on microalgae, a versatile source for sustainable energy and materials. Int J Energy Res 35(4):291–311. https://doi.org/10.1002/er.1695

Sharma VK, Filip J, Zboril R, Varma RS (2015) Natural inorganic nanoparticles—formation, fate, and toxicity in the environment. Chem Soc Rev 44(23):8410–8423. https://doi.org/10.1039/c5cs00236b. Epub 2015 Oct 5. PMID: 26435358.

Shen S, Mao SS (2012) Nanostructure designs for effective solar- to- hydrogen conversion. Nanophotonics 1(1):31–50. https://doi.org/10.1515/nanoph-2012-0010

Shen S, Shi J, Guo P, Guo L (2011) Visible- light-driven photocatalytic water splitting on nanostructured semiconducting materials. Int J Nanotechnol 8(6–7):523. https://doi.org/10.1504/IJNT.2011.040192

Sherif SA, Barbir F, Veziroglu TN (2005) Wind energy and the hydrogen economy—review of the technology. Sol Energy 78(5):647–660. https://doi.org/10.1016/j.solener.2005.01.002

Siegel RW (1993) Synthesis and properties of nanophase materials. Mater Sci Eng A 168:189–197. https://doi.org/10.1016/0921-5093(93)90726-U

Stern DI (2011) The role of energy in economic growth. Ann N Y Acad Sci 1219:26–51. https://doi.org/10.1111/j.1749-6632.2010.05921.x

Tiwari DK, Behari J, Sen P (2008) Application of nanoparticles in waste water treatment. World Appl Sci J 3(3):417–433. issn:1818-4952

Umer A, Naveed S, Ramzan N (2012) Selection of a suitable method for the synthesis of copper nanoparticles. Nano Brief Rep Rev 7:1230005–1230020

Vayssieres L (2009) On solar hydrogen and nanotechnology. Wiley. Singapore

Volz K, Kiuchi M, Ensinger W (1998) Structural investigations of chromium nitride films formed by ion beam-assisted deposition. Surf Coat Technol 108–109:303–307. https://doi.org/10.1016/S0257-8972(98)00565-9

Wagner S, Gondikas A, Neubauer E, Hofmann T, von der Kammer F (2014) Spot the difference: engineered and natural nanoparticles in the environment—release, behavior, and fate. Angew Chem Int Ed Engl 53(46):12398–12419. https://doi.org/10.1002/anie.201405050. Epub 2014 Oct 27. PMID: 25348500

Walker A (2013) Solar energy: technologies and project delivery for buildings. Wiley, Hoboken. isbn:978-1-118-13924-0

Wen L, Wang Y, Lu D, Hu S, Han H (2010) Preparation of KF/CaO nanocatalyst and its application in biodiesel production from Chinese tallow seed oil. Fuel 89(9):2267–2271. https://doi.org/10.1016/j.fuel.2010.01.028

Yu CH, Tam K, Tsang ESC (2009) Chemical methods for preparation of nanoparticles in solution. Handb Met Phys 5:113–141

Yuan W, Tang Y, Yang X, Wan Z (2012) Porous metal materials for polymer electrolyte membrane fuel cells—a review. Appl Energy 94:309–329. https://doi.org/10.1016/j.apenergy.2012.01.073

Chapter 2
Applications of Nanomaterials in Liquid Biofuels Production

B. S. Naveen Prasad, S. Sivamani, and B. Saikat

Abstract Liquid biofuels are mainly used as fuels in power generation and transportation sectors. The common liquid biofuels include biooil, bioethanol, and biodiesel. Nanomaterials are gaining more significance today mainly because of their large surface area and reusability. The present review focuses on the application of nanomaterials in the production of liquid biofuels. Woody biomass, agricultural and agro-industrial residues, aquatic and marine biomass, animal residues, and waste generated from large-scale commercial activities are the feedstocks to produce liquid biofuels. Biooil is produced by either pyrolysis or liquefaction of biomass, whereas bioethanol and biodiesel are produced by fermentation and transesterification of biomass, respectively.

Keywords Nanomaterials · Biooil · Bioethanol · Biodiesel

2.1 Introduction

Liquid biofuels are alternative fuels to conventional petroleum-based fuels that are used mainly for power generation and transportation (Salvi et al. 2013). Liquid biofuels can be produced by any of the following processes: Physicochemical, thermochemical, and biochemical processes (Fig. 2.1). Physicochemical processes involve extraction of oil from vegetable seeds or animal fats and transesterification of oils to biodiesel (Akbar et al. 2009). Thermochemical conversion converts biomass to liquid biofuels through liquefaction and pyrolysis (Zhang et al. 2010). Biochemical processes convert sugary feedstocks to bioethanol through fermentation (das Neves et al. 2007). Liquid biofuels include vegetable and animal oils, biodiesel, biooil, and bioethanol (Table 2.1). Biooil is produced by either pyrolysis or liquefaction of biomass, whereas bioethanol and biodiesel are produced by fermentation and transesterification of biomass, respectively (Demirbas 2011).

B. S. Naveen Prasad (✉) · S. Sivamani · B. Saikat
Chemical Engineering Section, Engineering Department, University of Technology and Applied Sciences (Salalah College of Technology), Salalah, Oman

© The Author(s), under exclusive license to Springer Nature Singapore Pte Ltd. 2023
M. Srivastava, P. K. Mishra (eds.), *NanoBioenergy: Application and Sustainability Assessment*, Clean Energy Production Technologies,
https://doi.org/10.1007/978-981-19-6234-9_2

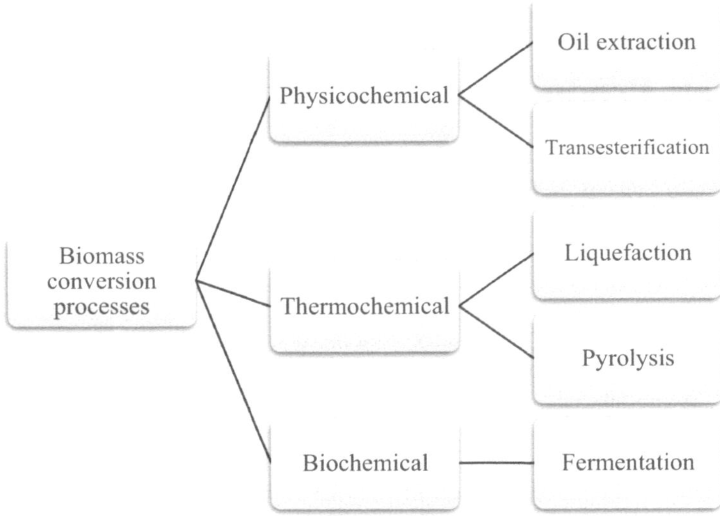

Fig. 2.1 Biomass conversion processes for liquid biofuels production

Table 2.1 List of liquid biofuels and their production processes

Liquid biofuel	Production process
Vegetable oil	Solvent extraction
Animal oil	Melting
Biodiesel	Transesterification
Biooil	Liquefaction and pyrolysis
Bioethanol	Simultaneous hydrolysis and fermentation

Vegetable oils are produced by solvent extraction, whereas animal oils are produced by melting of fats (Srinivasan and Jambulingam 2018).

2.2 Biomass Conversion Processes for Liquid Biofuels Production

Extraction is the term used to describe the process of extracting triglycerides from oilseeds (Yusuf 2018). This is accomplished by utilizing a variety of chemical, biochemical, and mechanical techniques designed to maximize yields while minimizing product quality degradation. It is the most critical step in the oil processing process and is determined by the part of the kernel, seed, or pulp that contains the oil (Subroto et al. 2015). In most industries, such as food and pharmaceuticals, the separation of bioactive compounds of interest requires the use of a large amount of solvent (water, n-hexane, ethanol, chloroform, methanol, petroleum ether, etc.). Researchers have spent decades focusing their efforts on developing more efficient

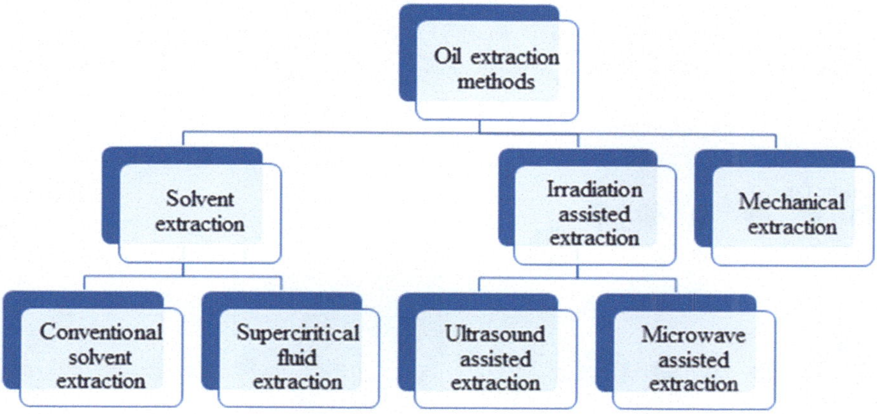

Fig. 2.2 Oil extraction methods

and environmentally friendly methods of extraction (Sheikh and Kazi 2016). Extractions have progressed from conventional solvent and mechanical extractions to more advanced non-traditional techniques such as Supercritical Fluid Extraction (SFE), ultrasound-assisted extraction (UAE), and microwave-assisted extraction (MAE) (Fig. 2.2). Grinding is a critical operation in the preparation of materials for oil extraction, regardless of the method used, because it can result in high oil yields and have a beneficial effect on downstream operations when performed properly (Alonge and Jackson 2018; Nde and Foncha 2020).

2.2.1 Processing of Animal Fat

The process begins with a series of purification steps, followed by modification to create more usable products and finally packaging (Sharma et al. 2013). Settling and degumming are used to remove proteins from animals or plants, carbohydrates residues, phosphatides, and water. Settling entails the passive storage of heated fats in conical-bottomed tanks (Chakrabarty 2003). Degumming is a process that removes phospholipids by adding water at a concentration of 1–3% at temperatures between 60 and 80° for 30–60 min. A small amount of acid is frequently added to water to raise the hydrogen content of the phospholipids. The term "neutralization" refers to the process of washing oils with strong alkaline water solutions to remove nonglyceride fatty materials (sodium hydroxide). Bleaching is a critical step in the production of fatty acids. Frequently, crude oils contain pigments that give off undesirable colors (carotenoids, gossypol, etc.) or promote lipid oxidation (chlorophyll). Bleaching darkens fats and oils by adsorbing colorants on bleaching earth and/or charcoal or by chemical reactions involving their oxidation or reduction. Deodorization is a critical step in the oil refining process due to consumer demand

Fig. 2.3 Processing of animal fat

Fig. 2.4 Transesterification of vegetable oil or animal fat

for extremely bland or flavorless fats and shortenings (Shahidi 2005). Fractionation is the process by which a fat or oil partially crystallizes at a given temperature (Tirtiaux 1983). For a period of time, the fat is kept at the crystallization temperature to allow for the formation of equilibrium or near equilibrium between crystallizing and non-crystallizing triacylglycerols. Hydrogenation, the direct addition of hydrogen to fatty acid double bonds, is used to modify large amounts of fats and oils. This process modifies the molecular structure and the geometry, number, and location of double bonds (Figs. 2.2 and 2.3).

2.2.2 *Transesterification*

The conventional biodiesel production process involves the transesterification of vegetable oils, animal fats, or waste cooking oils. Transesterification is a chemical reaction in which a glyceride reacts with an alcohol (typically methanol or ethanol) in the presence of a catalyst to produce fatty acid alkyl esters and an alcohol (Fig. 2.4). Transesterification can be carried out on any fatty acids derived from plants or animals, as well as used cooking oils (UCO). Rapeseed, sunflower, soy, and oil palm are the most frequently used vegetable oils. Transesterification is a reversible reaction that occurs when the reactants—fatty acids, alcohol, and catalyst—are combined. Catalysts can be used with either a strong base or a strong acid. Sodium or potassium methanolate is primarily used on an industrial scale. Raw biodiesel and raw glycerol are the end products of the transesterification process. These raw materials are then cleaned in a subsequent procedure. FAME (fatty acid methyl ester) biodiesel is produced when methanol is used as the alcohol (Sivamani et al. 2021a, b). Purified glycerol is suitable for use in the food, cosmetic, and oleochemical industries. Additionally, glycerol can be utilized as an anaerobic digestion substrate.

2.2.3 *Liquefaction*

The indirect liquefaction process is a promising technology comprised of two stages. A thermochemical gasification process is used as the first stage. After the raw material reacts with air or steam in this process, syngas is produced. CO, CO_2, H_2, and H_2O are the principal components of syngas. The second stage employs the time-honored Fischer–Tropsch (F–T) reaction. The mixture would be used in the F–T process to generate a variety of chemicals, including methyl alcohol, dimethyl ether, and ethyl alcohol, but little research has been conducted on the higher alcohols derived from biomass syngas. The most significant challenges are developing a novel catalytic reactor for the typically smaller scale of biomass conversion processes and developing catalysts for specific chemicals based on the molar ratio of H_2 to CO (Chornet and Overend 1985). To illustrate the indirect liquefaction process, we will use the synthesis of ethyl alcohol as an example.

Direct liquefaction (particularly hydrothermal processing) occurs in a non-oxidative atmosphere, where biomass is fed into a unit at lower temperatures as an aqueous slurry, with the product being biocrude in the liquid state. These processes are primarily concerned with the production of a liquid product that is a hydrocarbon with an atomic H:C ratio of 2 and a boiling point of 170–280 °C. Numerous processes have been developed in the coal-to-liquids conversion process. The primary goals of converting coal and biomass to liquids are to generate liquids, remove undesirable components (e.g., sulphur, oxygen, nitrogen, minerals), and create a material with a higher energy density that will flow (Behrendt et al. 2008). The combination of thermal decomposition and hydrogenation under pressure is one of the primary processes for directly converting coal to liquids.

2.2.4 *Pyrolysis*

Pyrolysis is the process of heating an organic material, such as biomass, without oxygen. Due to the absence of oxygen, the material does not burn, but the chemical compounds that comprise it (cellulose, hemicellulose, and lignin) thermally decompose into combustible gases and charcoal. The majority of these combustible gases can be condensed into a combustible liquid called pyrolysis oil (biooil), though certain gases (CO_2, CO, H_2, light hydrocarbons) are permanent. Thus, pyrolysis of biomass produces three products: a liquid called biooil, a solid called biochar, and a

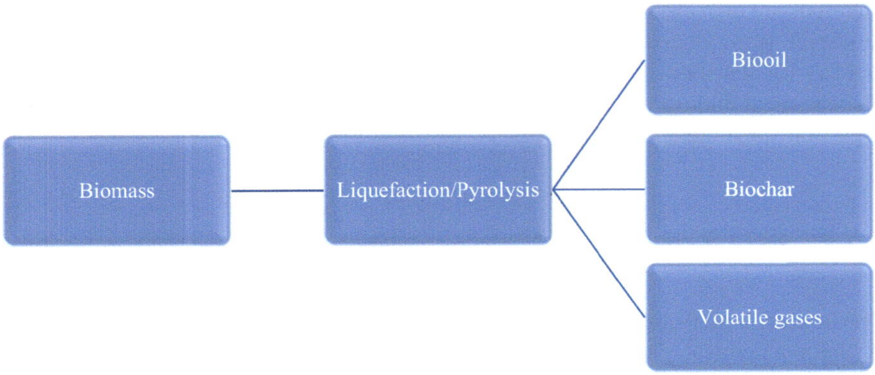

Fig. 2.5 Liquefaction/pyrolysis for liquid biofuels production

gaseous called biochar (syngas). The percentage of these products is determined by a number of variables, including the composition of the feedstock and process parameters (Fig. 2.5). However, all other factors being equal, the yield of biooil is optimized when the pyrolysis temperature is around 500 °C and the heating rate is high (i.e., 1000 °C/s), i.e., when fast pyrolysis conditions are used (Aswin et al. 2020). Under these conditions, biooil yields 60–70 wt. % can be obtained from a typical biomass feedstock, while biochar yields range from 15–25 wt. %. The remaining 10–15% is syngas. Slow pyrolysis is a term that refers to processes that use lower heating rates, and the primary product of these processes is typically biochar. Pyrolysis can be self-sustaining, as the syngas and a portion of biooil or biochar combustion provide all of the energy required to drive the reaction.

2.2.5 *Fermentation*

Microbes (mostly strains of the yeast Saccharomyces cerevisiae) convert sugar (glucose, fructose, or other monosaccharides) to ethanol during the fermentation process (Chandrasekaran and Sivamani 2018). Monosaccharides are formed directly from disaccharides that have been broken down by invertase enzymes or from starch that has been hydrolyzed by amylase enzymes. Apart from ethanol, water, and carbon dioxide are also produced. Equation (2.1) represents the glucose-to-ethanol conversion:

$$C_6H_{12}O_6 + 2ADP + 2Pi \rightarrow 2C_2H_5OH + 2CO_2 + 2H_2O + 2ATP \tag{2.1}$$

The most common processes produce a fermentation broth containing between 5 and 10% ethanol by volume, as ethanol is toxic to microorganisms. Due to the use of adapted and specialized yeasts, more advanced facilities can increase the concentration by up to 20%. The theoretical yield coefficient for the conversion of glucose to ethanol is 0.51 g ethanol/g glucose. The theoretical maximum yield of ethanol cannot be exceeded.

2.3 Applications of Nanomaterials in Biooil Production

Biooil is a dense complex mixture of organic compounds that have been oxygenated. It has a fuel value of about 50–70% that of petroleum-based fuels and can be used as boiler fuel or converted to renewable transportation fuels. Its density is greater than 1 kg L^{-1}, making it more cost-effective to transport than biomass. As such, we envision a distributed processing model in which numerous small-scale pyrolyzers (i.e., farm scale) convert biomass to biooil, which is then transported to a centralized location for refining. Our studies demonstrate that when distributed "farm scale" systems are used to feed a central gasification plant (for the production of Fisher–Tropsh liquids), the transportation cost savings alone are sufficient to offset the higher operational and biomass costs (Aswin et al. 2020).

Yasir et al. (2014) published a review article on the development of supported nanocatalysts over the last few decades, comparing the catalytic performance and deactivation rate of catalysts used to upgrade biooil to create a value-added and efficient transportation fuel. Upgrading biooil obtained through biomass pyrolysis is one of the most attractive technological and economic methods of producing fuel. Developing cost-effective, long-lived, and highly active catalysts is a significant challenge in this area of study. The addition of support material to the nanocatalyst not only extends its life but also provides additional active sites and reduces the cost of the catalyst by reducing the amount of active metal used. Additionally, proper support selection facilitates the active phase's efficient dispersion. Overall, significant progress has been made toward improving the hydrodeoxygenation reaction through the use of various nanosized rare earth metal support materials with increased catalytic efficiency. These materials will eventually be implemented in industries for the purpose of upgrading pyrolysis biooil.

Xu et al. (2021) developed an integrated process that incorporates catalytic co-pyrolysis of biomass/plastic wastes and in-line catalytic upgrading of pyrolysis gas to produce aromatics-rich biooil and carbon nanotubes simultaneously (CNTs). The effect of the ratio of feedstocks on the properties of biooil and carbon nanotubes was determined. Additionally, the carbon deposition reaction mechanism was investigated during the system's operation. The results indicated that co-feeding plastic to biomass increased the selectivity of monoaromatics (benzene, toluene, and xylene) from 5.6% for pure biomass to 44.4% for a 75.0% plastic ratio, while decreasing naphthalene and its derivatives from 85.9 to 41.7%. At a plastic ratio of 25%, the greatest synergistic effect on BTX selectivity was observed. The multi-walled carbon nanotubes were synthesized successfully on a nickel catalyst using pyrolysis gas as a feedstock. For pure biomass, the smallest yield of CNTs with ultrafine diameters of 3.9–8.5 nm was obtained via disproportionation of CO derived from the decarboxylation and decarbonylation of oxygenates on the ZSM-5 acid sites. With the increase in plastic ratio, sufficient hydrocarbons were produced to support the growth of CNTs, endowing them with long, straight tube walls and uniform diameters (16 nm). The yield of carbon nanotubes increased to 139 mg/g catalyst. Additionally, the decreased CO_2 inhibited dry reforming of C1–C4 hydrocarbons and deposited carbon, thereby avoiding excessive etching of CNTs. Thus, when the plastic ratio in the feedstock exceeded 50%, high-purity carbon nanotubes with fewer defects were fabricated. The strategy is expected to enhance biomass pyrolysis's sustainability and economic viability.

Saber et al. (2016) conducted a review of techniques for producing and upgrading biooil from algae (micro- and macro-algae). This article also discusses the chemistry and properties of biooil, as well as future research recommendations. Due to the depletion of oil reserves, renewable fuels must be developed to replace petroleum-based fuels. Fuels derived from biomass that emit no net CO_2 are a promising alternative to fossil fuels. Biooil has been considered as a possible alternative fuel source. Currently, two major processes for producing biooil from biomass are available: pyrolysis and hydrothermal liquefaction. Pyrolysis necessitates the drying of feedstock, increasing the process's energy consumption. Hydrothermal liquefaction, on the other hand, takes place in an aqueous environment, making it ideal for aquatic plants and wet biomass. One of the primary difficulties in applying biooil is the oil's low quality. Biooil has undesirable properties such as a high oxygen content and acidity, necessitating its upgrading prior to being used as a fuel.

Zahid et al. (2018) investigated recent advances in pyrolysis. The increasing importance of carbon-based nanomaterials for a variety of applications in the modern era has prompted efforts to develop cost-effective methods for facile synthesis from abundantly available wastes. By thermally treating organic waste, pyrolysis in a broad spectrum is frequently used to synthesize carbon nanostructures. The growth mechanism of nanoparticles dictates their functional distribution based on their size, medium, and physicochemical properties. The growth of carbon nanomaterials is a complex process that is significantly influenced by temperature, catalyst, and precursor type. Nowadays, significant progress has been made in improving the techniques for growing nanomaterials, thereby paving the way for commercial production of carbon-based nanomaterials. The methods that use hydrocarbon-rich organic waste as a feedstock are the most promising. This review discusses the different pyrolysis techniques used to synthesize carbon-based nanomaterials, specifically carbon nanotubes (CNTs), carbon nanofibers (CNFs), and graphene (G).

Using iron (Fe) and cobalt (Co) particles as susceptors, Debalina et al. (2017) synthesized nanostructured biochar from lignocellulosic biomass, sugarcane bagasse. The experiments were conducted in a bench-scale pyrolysis reactor at a temperature of 500 °C using 600 W microwave power. The effect of the bagasse: susceptor composition on the quality of biochar was investigated using a variety of characterization techniques, including scanning electron microscopy, porosimetry, X-ray diffraction, and Raman spectroscopy. When Fe was added as a susceptor, nanoparticles, and nanotubes with an average diameter of 30–120 nm and 20–50 nm, respectively, were formed. Due to the increased rate of localized heating caused by

Table 2.2 Summary of literature for biooil production

Raw material	Nanomaterial used	Process	Product	Reference
Biomass	Nanocatalysts	Pyrolysis	Biooil	Yasir et al. (2014)
Plastic waste	Carbon nanotube	Pyrolysis	Biooil	Xu et al. (2021)
Micro- and macro-algae	Nanoparticles	Liquefaction	Biooil	Saber et al. (2016)
Organic waste	Carbon-based nanomaterials	Pyrolysis	Biooil	Zahid et al. (2018)
Sugarcane bagasse	Fe and Co nanoparticles	Pyrolysis	Biooil	Debalina et al. (2017)

the addition of more Fe particles, the size of nanoparticles and nanotubes increased. In comparison to biochar obtained without the addition of susceptor, the biochar obtained with the addition of susceptor was higher in carbon content. By adding a mixture of Fe and Co as a susceptor, small graphitic flakes and fine nanotubes were formed. In addition to biochar, the compositions of biooil and non-condensable gases were determined. In biooil, the major organic functionalities were furan derivatives, simple phenols, guaiacol, syringol, and their derivatives. When 20 wt.% Fe and 33.3 wt.% Fe + Co were used as susceptors, the maximum energy recovery was 49%. Apart from CO_2 and CO_2, adding Fe resulted in a high yield of H_2 and CH_4 gases. Thus, microwave pyrolysis is demonstrated to be a promising technique for tailoring the morphological properties of biochar via metallic susceptors while also producing high-quality biooil and gases. The literature review for biooil production is summarized in Table 2.2.

2.4 Applications of Nanomaterials in Bioethanol Production

Bioethanol is generated through the fermentation of fermentable sugars derived from biomass. In bioethanol production, the use of nanomaterials is uncommon. Nanomaterials have the potential to be used to immobilize enzymes used in the production of bioethanol (Anushya et al. 2019). Bioethanol is an oxygenated biofuel that contributes to air pollution reduction.

Singhvi et al. (2021) used a greenly synthesized nanomaterial, functionalized few-layer graphene (FFG), to depolymerize cellulose moieties derived from corn cob (CC). Additionally, they investigated the combined effect of cellulase and a very low concentration of FFG (0.02 weight % of biomass) on the hydrolysis of CC-derived cellulose, demonstrating a 38% increase in hydrolysis when compared to the control (i.e., without FFG). In cellulose, the hydroxyl groups adjacent to the 1,4 glycosidic linkages form a crystalline structure, making depolymerization difficult. Notably, hydrolysis experiments were conducted under mild conditions, i.e., 50 °C, resulting in the formation of fermentable sugars without the formation of inhibitor compounds. Additionally, hydrolyzed CC cellulose was used to ferment bioethanol by Saccharomyces cerevisiae, which produced 1.53 times the amount of ethanol when FFG was added. Additionally, the solid residue was used to fabricate cellulose-based nanocomposite films. We attempted to create a zero-waste process here by utilizing each component, which is an attractive sustainable option. To our knowledge, this is the first report on the cellulase mimicking activity of FFG nanosheets and their application to the hydrolysis of CC cellulose for enhanced bioethanol production, which may eliminate the need for expensive cellulase enzymes required for biomass hydrolysis. Thus, the FFG-associated biomass to bioethanol conversion process may pave the way for the development of green, sustainable biofuel technology with a primary application as a transportation fuel.

Devi et al. (2021) discussed the role of nanotechnology and nanomaterials in pretreatment, enzymatic hydrolysis, and fermentation steps during bioethanol production, as well as in transesterification to improve the efficiency of the biodiesel manufacturing process. Due to the rapid depletion of fossil fuels and the world's growing population, demand for alternative and renewable energy sources is gaining global attention. At the moment, the majority of the world's energy needs are met by fossil fuels, which is rapidly depleting them. The combustion of fossil fuels, on the other hand, contributes to environmental problems such as global warming and pollution. Biofuels such as bioethanol and biodiesel are gaining popularity around the world due to their renewable nature, environmental friendliness, and safety. There is an abundance of lignocellulosic biomass that can be converted to bioethanol via pretreatment, hydrolysis, and fermentation. Numerous physical, chemical, and biological pretreatments are used to prepare biomass for enzymatic hydrolysis; however, these approaches have some limitations that can be overcome through the use of alternative, cost-effective, and environmentally friendly technologies such as nanotechnology. Magnetic nanoparticles can be used to immobilize enzymes during the enzymatic hydrolysis process, enhancing the efficiency of the process. Additionally, the microorganisms can be immobilized in a variety of matrixes, including magnetic nanoparticles, which can aid in the production of ethanol. Additionally, magnetic nanomaterials can be recycled and repurposed through the use of magnetic fields. Additionally, nanocatalysts can be used to improve the yields of the transesterification process used to produce biodiesel. By lowering processing costs and increasing productivity, nanotechnology can play a critical role in biofuel production.

Leo and Singh (2018) examined a variety of novel approaches and current trends in the retroactive application of nanotechnology to bioethanol production. The twenty-first century's technological advancements and global energy requirements have resulted in alarming global warming and the depletion of non-renewable fossil fuels. The search for alternative energy sources in order to reduce our reliance on fossil fuels has resulted in increased interest in biofuels such as bioethanol. Due to its eco-friendly and renewable characteristics, bioethanol is an extremely useful fuel additive. Bioethanol production utilizes fermentation technology to convert carbohydrate-rich biomass to biofuel, though high production costs and some technical difficulties have been identified as disadvantages. Nanotechnology has

Table 2.3 Summary of literature for bioethanol production

Raw material	Nanomaterial used	Product	Reference
Corncob	Functionalized graphene	Bioethanol	Singhvi et al. (2021)
Lignocellulosic feedstocks	Nanomaterials	Bioethanol	Devi et al. (2021)
Carbohydrate	Nanomaterials	Bioethanol	Leo and Singh (2018)
Cellulosic materials	Nanoparticles	Bioethanol	Kushwaha et al. (2018)

the potential to assist in overcoming these obstacles and ensuring the sustainable production of these biofuels. Numerous nanoparticles and nanomaterials have been shown to have an effect on the production of bioethanol and other biofuels.

Kushwaha et al. (2018) discussed the current state of the art in the application of nanoparticles in various stages of bioalcohol production (bioethanol/biobutanol). Only when biofuels are decoupled from food crops will they become globally competitive. All non-food feedstocks for the production of bioalcohols (bioethanol/biobutanol) present inherent difficulties in converting cellulose to simpler sugars that can be fermented to produce end products. Due to the high cost of the process currently used to convert complex cellulosic feedstocks to sugars, the biofuels obtained via these routes are not economically viable. Globally, enormous efforts are being made to convert second- and third-generation feedstocks to bioethanol/biobutanol, and several prominent global energy producers are investing significant sums of money to make this dream a reality. With the application of various nanoparticles, the complexity and cost of the various stages of bioalcohol production can be reduced. To understand the economics of bioethanol production, a three-pronged approach is required, beginning with improved crop production technology, improved feedstock processing, and the development of new biofuels such as biobutanol and renewable hydrocarbons. Numerous nanoparticles have been used in the production of biofuels, including iron oxide, nickel cobaltite, zinc oxide, and various nanocomposites. By incorporating these nanomaterials into various bioconversion processes, we can achieve sustainability by lowering raw biomass processing and production costs, as well as minimizing negative environmental impacts. The bioethanol production literature review is summarized in Table 2.3.

2.5 Applications of Nanomaterials in Biodiesel Production

Transesterification of vegetable or animal fat results in the production of biodiesel. Catalysts for biodiesel synthesis are classified as homogeneous, heterogeneous, or biological (Fig. 2.6). Catalysts that are homogeneous are in phase with the feed mixture. Catalysts that are heterogeneous are in a phase different from the feed mixture. Catalysts derived from biological sources are enzyme- or microorganism-based. Catalysts classified as heterogeneous are further classified into three categories: acid, base, and functional materials. Functional materials are those that exhibit one or more properties that can be altered significantly in response to external stimuli

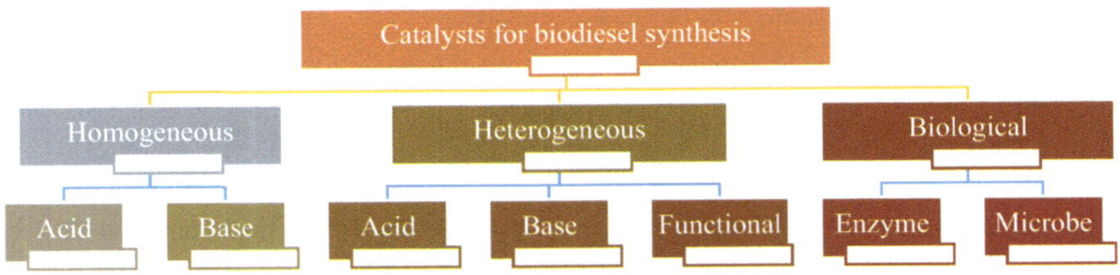

Fig. 2.6 Catalysts used for biodiesel synthesis

in a controlled manner. Nanomaterials are functional materials that contribute to the production of biodiesel (Sivamani et al. 2021a, b).

Xie et al. (2018) sought to develop magnetically recyclable solid catalysts for use in the production of biodiesel. To accomplish this, magnetically susceptible Fe_3O_4/MCM-41 composites with a core-shell structure were synthesized, and sodium silicate was then cross-linked onto the magnetic materials using epichlorohydrin. Numerous techniques were used to characterize the core-shell structured magnetic support and resulting solid catalyst. The results of the characterization indicated that the sodium silicate was chemically bonded to the Fe_3O_4/MCM-41 composites without causing obvious damage to the magnetic support structure. The solid catalyst exhibited a strong magnetic response and extraordinary catalytic activity in the transesterification of soybean oil to produce biodiesel, with an oil conversion of 99.2% after 8 h of reaction using a methanol-to-oil molar ratio of 25:1 and a catalyst loading of 3 wt. % at reflux of methanol. Additionally, simple magnetic decantation was used to separate the solid catalyst from the reaction mixture without significant mass loss, and the solid catalyst could be reused five times for heterogeneous transesterification.

Wang et al. (2015) synthesized crystalline Fe/Fe_3O_4 core/shell magnetic nanoparticles (MNPs) with sulphamic acid and sulphonic acid functionalized silica coatings that demonstrated excellent stability. They investigated their catalytic properties using oleic acid esterification and glyceryl trioleate transesterification as model reactions for biodiesel production. In the esterification of oleic acid, acid functionalized MNPs exhibited excellent catalytic activity. However, MNPs functionalized with sulphamic acid exhibited increased reactivity in the transesterification reaction of glyceryl trioleate. Magnetoprecipitation allows recovery of both acid-functionalized MNPs. The sulphamic acid functionalized MNPs maintained a high degree of reactivity (>95% conversion) over the course of five continuous reaction runs, indicating their potential for biodiesel production from low-grade feedstocks such as waste cooking oil containing a high concentration of free fatty acids.

De Medeiros et al. (2020) synthesized and investigated the use of bulk, fiber, acid- and thermal-treated carbon nitrides as heterogeneous catalysts for canola oil transesterification to biodiesel. While biodiesel remains one of the most promising sustainable alternatives to fossil fuel-derived energy, process limitations and high production costs associated with the use of homogeneous catalysts have limited global acceptance and adoption. While heterogeneous catalysts have been proposed as viable alternatives, they have a number of disadvantages, including a high metal content and a high price. As a result, a search for novel and affordable alternatives is underway. Although graphitic carbon nitride macrostructures have been extensively investigated for photocatalytic applications, their role in catalyzing chemical reactions remains relatively unexplored. A conversion of 96% was achieved with a catalyst loading of 1 wt. %, a low oil to methanol ratio of 1:24, and reaction temperatures and times of 150 °C and 3 h, respectively. Carbon nitrides provide a high-temperature stable, cost-effective, and metal-free alternative to the numerous metal-containing heterogeneous catalysts proposed for biodiesel production, while maintaining >90% biodiesel conversion.

Bharti et al. (2021) discussed nanoferrite heterogeneous catalysts for biodiesel production from soybean and canola oil. Transesterification is the primary route of synthesis because it is simple, cost-effective, allows for better process control, and yields a high conversion rate. The depletion of fossil fuels and the resulting pollution necessitate the development of alternative, renewable energy sources such as biofuels. Current challenges include developing efficient processes and catalysts for the conversion of a variety of feedstocks to biofuels. Catalysis with nanoferrites and composites enables yields greater than 95% conversion at 80 °C in 1–2 h with less than 5.0 wt.% catalyst loading. With a moderate alcohol/oil molar ratio, i.e., between 12:1 and 16:1, conversion yields of greater than 90% are achievable. Catalyst recovery is simplified by the magnetic properties of nanoferrite, which can be reused up to four times without losing more than 10% of its catalytic efficiency.

Ibrahim et al. (2019) synthesized solid acid catalysts based on ZrO_2 for use in environmentally friendly applications such as biodiesel production. Environmental pollution control is a difficult area for heterogeneous catalysts because it requires the avoidance of hazardous materials in chemical synthesis. ZrO_2 loading was carried out on a variety of supports (Al_2O_3, Fe_2O_3, TiO_2, and SiO_2) using a hybrid sol-gel auto combustion method. The properties of the structure, the surface, and the acidity are examined. Catalytic activity was determined by converting stearic acid to ester and then producing biodiesel. Because of its large surface area and abundance of Lewis acid sites, ZrO_2/SiO_2 has the highest conversion rate. The conversion rate reached 48.6% under optimal reaction conditions of 120:1 alcohol/acid molar ratio, 0.1 catalyst mass ratio to acid, 3 h reaction time, and 120 °C reaction temperature. Additionally, the catalysts were reused for at least five runs without experiencing significant activity loss.

Bet-Moushoul et al. (2016) synthesized five different types of calcium oxide-based catalysts supported on gold nanoparticles (AuNPs) (dubbed nanocatalysts), including commercial CaO, eggshell, mussel shell, calcite, and dolomite, and used them in the synthesis of biodiesel. The transesterification process was performed optimally at 65 °C, with a methanol-to-oil molar ratio of 9:1, a reaction time of 3 h, and a catalyst loading of 3%. Optimal conditions resulted in an oil conversion rate of 90–97% and a glycerol concentration of 3.9–4.3 (mg/kg) for all samples. By comparing supported AuNPs catalysts to conventional CaO catalysts, gas chromatographic analysis revealed that a high-quality biodiesel product was synthesized. The results of the reusability test indicated that all prepared nanocatalysts could be reused up to ten times without losing activity. The impregnation method was found to yield the highest catalytic activity for CaO–AuNPs nanocatalysts.

Hara (2009) discussed the environmentally friendly production of biodiesel using a variety of heterogeneous catalysts, including solid bases, acid catalysts, and immobilized enzymes. The production of esters of higher fatty acids from plant materials is highly desirable for biodiesel production. Heterogeneous catalysts may open up new avenues for producing biodiesel in an environmentally friendly manner. Particulate heterogeneous catalysts can be easily separated from products after the reaction, allowing the catalyst to be reused, resulting in less waste and energy consumption. Diesel engines are compact and powerful, offering numerous benefits in terms of energy efficiency and cost. As a result, the production of higher fatty acid esters from plant materials has gained interest in recent years as a means of producing biodiesel, a clean-burning alternative fuel. Biodiesel is primarily produced industrially using "soluble" catalysts such as alkali hydroxides and liquid acids. Purification of products and catalyst separation require a significant amount of energy, and these catalysts are also nonrecyclable. This process consumes a significant amount of energy and generates a significant amount of chemical waste. Following reaction, particulate heterogeneous catalysts can be easily separated from products, allowing the catalyst to be reused and consuming less energy.

Ambat et al. (2019) investigated the production of biodiesel from rapeseed oil using a Fe_3O_4-CeO_2 nanocatalyst that was potassium-impregnated. We investigated the catalytic conversion of rapeseed oil to triglyceride methyl ester using various concentrations of potassium-impregnated Fe_3O_4-CeO_2. The Fe_3O_4-CeO_2 nanocatalyst with a 25 weight % potassium impurity produced the most biodiesel. The nanocatalyst was characterized using FTIR, XRD, SEM, TEM, BET, and the Hammett indicator to determine its basicity. GC-MS, 1H, and 13C NMR were used to characterize the biodiesel. Additionally, the optimal reaction parameters for transesterification were determined using 1H NMR, including the catalyst concentration (wt. percent), the oil to methanol ratio, the reaction time, and the reaction temperature. The maximum yield of 96.13% was obtained at a catalyst concentration of 4.5 weight %, a ratio of 1:7 oil to methanol, and a temperature of 65 °C for 120 min. Biodiesel was found to have an acid value of 0.308 mg KOH/g and a kinematic viscosity of 4.37 mm^2/s. The flash point and density of the fuel were also determined. The reusability of the catalyst was investigated, and it was found to be stable for up to five cycles without significant activity loss. Utilizing a distillation process setup, excess methanol was recovered following the transesterification reaction.

Baskar et al. (2017) used manganese-doped zinc oxide as a heterogeneous catalyst to synthesize biodiesel from Mahua oil. XRD and SEM analysis were used to characterize the manganese-doped zinc oxide nanocatalyst. SEM and XRD analysis confirmed the catalyst's hexagonal structure and 24.18 nm particle size. Biodiesel production has accelerated in recent years as a result of the environmental benefits associated with its ability to reduce pollution. Heterogeneous catalysts are preferred for producing biodiesel due to their ease of recovery and lack of aqueous treatment requirements. The transesterification of fatty acids into biodiesel using transition metal oxide results in a rapid conversion of the fatty acids to their methyl esters. The optimum process conditions for a maximum biodiesel yield of 97% were determined to be a catalyst concentration of 8% (w/v), an oil to methanol ratio of 1: 7% (v/v), a reaction time of 50 min, and a reaction temperature of 50 °C. FTIR and GC-MS analysis both confirmed the presence of methyl esters in biodiesel.

Dantas et al. (2015) proposed to evaluate the performance of the nanomagnetic catalyst $Ni_{0.5}Zn_{0.5}Fe_2O_4$ in the transesterification of soybean oil to produce biodiesel by varying the processing conditions (temperature, molar ratio of oil to alcohol, and catalyst amount) during the catalytic reaction. Magnetic catalysts can be easily removed from the reaction process, which reduces the amount of wastewater generated. The catalyst was prepared via combustion and characterized via XRD, BET, magnetic measurements, and gas chromatography. The results indicated the formation of the inverse spinel phase $B(AB)_2O_4$, with an isotherm profile classified as type V and a hysteresis loop classified as type 3 (H3), and a surface area of 48.39 m^2g^{-1}. The magnetic hysteresis curve exhibited the typical behavior of soft magnetic materials with a saturation magnetization of 55 emu/g. Chromatographic analysis confirmed the catalytic activity of the magnetic nanoparticles and the direct effect of processing conditions on the conversion to esters.

Dantas et al. (2020) synthesized magnetic nanoparticles (MNP) of $Ni_{0.5}Zn_{0.5}Fe_2O_4$ using a combustion reaction, with a special differential in the production, starting at a scale of 10 g/production and reproducible up to 200 g/ production, and used and reused them as heterogeneous magnetic nanocatalysts in biodiesel production reactions. The nanocatalysts synthesized were extremely efficient in the production of biodiesel from soybean oil via the methyl and ethyl routes. It exhibited the highest activity in the esterification reaction, with conversions reaching 99.54 0.16% via the methyl route and 99.38 0.18% via the ethyl route. Maximum conversion of transesterification occurred at a rate of 14%. The viscosity, density, acidity, and iodine ratios of biodiesel were determined, and the obtained values indicate that the produced biodiesel complies with the specifications applicable to the quality standards for commercialization. The XRD, BET, TEM, AGM, and TPD techniques were used to characterize MNP. Additionally, the nanocatalyst was recovered using a simple external magnetic field (magnet) and reused three times without significant loss of catalytic activity, indicating high stability. Thus, the nanoferrite $Ni_{0.5}Zn_{0.5}Fe_2O_4$ can be validated as a novel environmentally responsible catalyst for heterogeneous catalysis in the field of biodiesel production.

Dantas et al. (2018) used a catalyst composed of $Ni_{0.5}Zn_{0.5}Fe_2O_4$ nanoferrite in esterification reactions. Biodiesel is a biodegradable fuel made from renewable biomass such as soybean and cottonseed oil. It can be made in a variety of ways and requires a catalyst to catalyze the reactions. Iron spinel is an excellent material for catalyzing the reactions involved in the production of biodiesel. The nanoferrite was synthesized on a large scale, 200 g/batch, where the reaction time and temperature were monitored and the material was characterized using XRD and surface area measurements. The biodiesel was produced by methyl and ethyl esterification of acidified soybean oil and cottonseed oil at a temperature of 160 °C for 1 h, with a molar ratio of 1:15, 3% catalyst, and chromatographic analysis. The nanoferrite exhibited enhanced peaks corresponding to the spinel's majority crystal phase and a surface area of 64.17 m^2/g. The methyl route converted 99.08% of the soybean oil, while the ethyl route converted 98.38%. The methyl route yielded 88.79% of cottonseed oil, and the ethyl route yielded 78.65%.

Dantas et al. (2017) investigated the effect of Cu^{2+} doping on the structure, morphology, and magnetic properties of $Ni_{0.5}Zn_{0.5}Fe_2O_4$ nanoferrites in preparation for their use in methyl transesterification of soybean oil to produce biodiesel. The nanoferrites with a composition of $Ni_{0.5}$ × Cu × $Zn_{0.5}Fe_2O_4$ (0.0 × 0.4) were synthesized via combustion reaction in a conical reactor with a batch size of 10 g and characterized via X-ray diffraction, textural analysis using N_2 adsorption, magnetic measurements, thermal analysis using temperature-programmed desorption, and biodiesel analysis using gas chromatography. All compositions produced a single phase of the inverse spinel type $B(AB)_2O_4$ and isotherms of adsorption/desorption with the same profile, classified as type V, with a hysteresis loop type 3 (H3). However, increasing the doping concentration of Cu ions from 0.0 to 0.4 mol resulted in a reduction of 37% in surface area and 36.4% in saturation magnetization. However, the ferrimagnetic property was retained, and the material is still strongly attracted by a magnet. Additionally, the presence of Cu ions increased the conversion values of methyl esters obtained through soybean oil transesterification by 5.5–85%. These findings suggest that NiZn nanoferrite compositions doped with Cu^{2+} could be used as heterogeneous nanocatalysts for biodiesel production.

Dantas et al. (2016) synthesized mixed nanoferrites of $Ni_{0.5}Zn_{0.5}Fe_2O_4$ and $Ni_{0.2}Cu_{0.3}Zn_{0.5}Fe_2O_4$ in batches of 10 g using urea as fuel and a conical reactor as a heating source, and then tested them as heterogeneous nanocatalysts in the transesterification and esterification reactions of soya bean oil via the methyl The time and temperature of the reactions were monitored throughout the synthesis, as well as the evolved gases and the color of the emitted flames. XRD, EDX, FTIR, SEM/EDS, BET, and gas chromatography analysis were used to characterize the samples. The reaction was carried out for 1 h with 10 g of oil, a molar ratio of 1: 12 oil:alcohol, and a catalyst concentration of 2% (w/w), all at 180 °C. The XRD patterns and FTIR spectra revealed the presence of the $B(AB)_2O_4$ inverse spinel phase type. The morphology revealed the formation of fragile agglomerates with a high surface area. Chromatographic analysis revealed excellent results in the esterification reactions for both samples under the tested conditions, with particular emphasis on $Ni_{0.5}Zn_{0.5}Fe_2O_4$, which converted 91.4% of methyl esters and 77.8% of ethyl esters, respectively, while the $Ni_{0.2}Cu_{0.3}Zn_{0.5}Fe_2O_4$ sample converted 75.1% and 65.1%, respectively. The methyl and ethyl transesterification conversions were 14% and 2% for the $Ni_{0.5}Zn_{0.5}Fe_2O_4$ sample, respectively, and 11% and 3% for the $Ni_{0.2}Cu_{0.3}Zn_{0.5}Fe_2O_4$ sample, respectively.

Dantas et al. (2013) investigated the performance of $Ni_{0.5}Zn_{0.5}Fe_2O_4$ ferrite doped with 0.1 and 0.4 mol of Cu as a catalyst for the transesterification of soybean oil to biodiesel in the presence of methanol. SEM, nitrogen adsorption, and X-ray diffraction were used to characterize the samples. The reaction was carried out for

2 h at a temperature of 160 °C with 10 g of soybean oil, a molar ratio of oil to alcohol of 1:10, and a catalyst concentration of 4% (w/w). Gas chromatography was used to characterize the reaction product, confirming that it was converted to methyl esters. Only a $Ni_{0.5}Zn_{0.5}Fe_2O_4$ ferrite phase with a crystallite size of 29 nm was observed in the diffraction patterns. Surface area and particle size of samples doped with 0.1 and 0.4 mol of Cu were 22.17 m^2g^{-1} and 50.47 nm, respectively, and 23.49 m^2g^{-1} and 47.64 nm, respectively. Both samples had a brittle block-shaped agglomerate morphology and a broad particle size distribution. A comparison of the two catalysts revealed that the catalyst doped with 0.4 mol of Cu performed better, achieving a conversion rate of 50.25%, compared to the catalyst doped with 0.1 mol of Cu, which achieved a conversion rate of 42.71%.

Dantas et al. (2014) evaluated the performance of the new catalysts $Ni_{0,5}Zn_{0,5}Fe_2O_4$ (pure) and $Ni_{0,1}Cu_{0,4}Zn_{0,5}Fe_2O_4$ (doped with 0.4 mol of Cu^{2+}) as nanoferrites in transesterification reactions of soybean oil methyl ester. Both samples were synthesized using urea as a fuel source and a resistance heating coil in the stoichiometry of the combustion reaction. Time and combustion temperature were recorded during synthesis. They were later characterized using X-ray diffraction (XRD), thermogravimetric analysis (TGA), textural analysis (BET), and catalytic tests on a bench. The transesterification reaction conditions were as follows: 10 g oil, 2 h reaction time, 1:20 molar ratio of oil to alcohol, 4% (w/w) catalyst, and a reaction temperature of 160 °C. Finally, gas chromatography was used to characterize the reaction product for its conversion to methyl esters. Only the inverse spinel phase, typical of Ni-Zn ferrite, was detected in both samples, with crystallite sizes of 26 and 29 nm, respectively. Thermogravimetric analysis revealed that the samples are thermally stable, losing only 4.9 and 3.7% of their weight, respectively. The surface area and particle size were 48.89 m^2g^{-1} and 23 nm, respectively, while the surface area and particle size were 18.06 m^2g^{-1} and 62 nm. The conversion rates of $Ni_{0,5}Zn_{0,5}Fe_2O_4$ and $Ni_{0,1}Cu_{0,4}Zn_{0,5}Fe_2O_4$ were 13% and 50%, respectively, indicating that the sample containing copper converted 26% more effectively, indicating that copper is a promising catalyst for the transesterification reaction, which aims to produce biodiesel. Research focusing on the global energy transition to renewable energy sources undoubtedly indicates that the use of biodiesel would be a viable option for increasing rural income, reducing spending on oil derivatives, and creating new job opportunities.

Erdem et al. (2018) demonstrated the fabrication of an inorganic shell composed of silica on the surface of magnetic iron oxide nanoparticles through the deposition of preformed colloids and functionalization of these particles. Chlorosulphonic acid is used to functionalize magnetic nanoparticles that are uncoated and coated with a silica layer via the Stöber method. Magnetic nanoparticles (MNPs) with a diameter of 10–13 nm have the potential to be used as an acid catalyst in the production of biodiesel due to their superparamagnetic properties. Various methods such as XRD, EDX, FTIR, and VSM were used to characterize the prepared nanoparticles. The catalytic activity of coated and uncoated solid acids was investigated in the palmitic acid-methanol esterification reaction, which is used in industry to synthesize biodiesel. Although the thin silica layer presents only a minor impediment to magnetism, it can significantly accelerate mass transportation due to its relatively porous structure, and the magnetic core may be more stable in the acidic reaction medium due to the covering process. As a result, coating strategies may be an effective means of enabling MNPs to be used in acid-catalyzed esterification.

Kelarijani et al. (2019) investigated the effect of nanomagnetic catalysts on rapeseed oil-based biodiesel production. The catalysts produced were characterized using X-ray diffraction (XRD), transmission electron microscopy (TEM), a BET surface area analyzer, and a vibrating sample magnetometer (VSM). Ultrasonic irradiation was used for 35 min at various lithium/Fe_3O_4 and lithium/ZnO–Fe_3O_4 molar ratios. The highest yield of 99.8% was obtained with a Li/Fe_3O_4 molar ratio of 3/1 and a Li/ZnO–Fe_3O_4 molar ratio of 3/1. The vibrating sample magnetometer demonstrated that the catalysts were easily separated due to their ferromagnetic properties, even after they were regenerated and used for esterification again. After three times of regeneration and application, the superparamagnetic behavior of the particles demonstrated a high yield of catalyst recovery. Additionally, the results indicated that when compared to the conventional mechanical stirring method, using the basic nano magnetic catalyst and ultrasonic waves at a frequency of 37 kHz could reduce the reaction temperature and time, while increasing the yield of biodiesel. The viscosity and flash point of the fatty acid methyl esters produced met the ASTM D6751 requirements.

Firouzjaee and Taghizadeh (2017) synthesized a nanomagnetic catalyst composed of CaO/NaY-Fe_3O_4 and used it to produce biodiesel from canola oil. X-ray diffraction, field emission scanning electron microscopy, the Brunauer-Emmett-Teller method, Fourier transform infrared spectroscopy, and the vibrating sample magnetometer method were used to characterize the structure of the catalysts. To optimize the effect of operating variables on the yield of the transesterification reaction, such as the methanol/canola oil molar ratio, the amount of catalyst, and the reaction time, a Box-Behnken design was used. The cubic model predicted the optimal values for these variables and they were found to be extremely close to the experimental values. Due to diminishing oil supplies, alternative fuels such as biodiesel are necessary.

Ghalandari et al. (2019) synthesized a KOH/Fe_3O_4@-Al_2O_3 nanocatalyst using the Fe_3O_4@-Al_2O_3 core-shell structure as a support and KOH as the active component. X-ray diffraction (XRD), field emission scanning electron microscopy (FE-SEM), energy-dispersive X-ray spectroscopy (EDS), Fourier transform infrared (FTIR), Brunauer-Emmett-Teller (BET), and vibrating sample magnetometry (VSM) were used to characterize the prepared samples. We investigated the transesterification of canola oil to methyl esters (biodiesel) in the presence of a KOH/Fe_3O_4@-Al_2O_3 nanocatalyst with a magnetic core and a mesoporous shell. To optimize the effect of critical operating variables on biodiesel yield, a response surface methodology (RSM) based on the Box-Behnken design (BBD) was used. Under optimal reaction conditions, a yield of 97.4% biodiesel was achieved. Between experimental and predicted results, there was excellent agreement.

Gurunathan and Ravi (2015) investigated the optimal transesterification reaction and its kinetics for the production of biodiesel from neem oil using a copper doped

Table 2.4 Summary of literature for biodiesel production

Raw material	Nanomaterial used	Product	References
Soybean oil	Fe_3O_4/MCM-41	Biodiesel	Xie et al. (2018)
Glyceryl trioleate	Functionalized Fe/Fe_3O_4	Biodiesel	Wang et al. (2015)
Soybean and canola oil	Nanoferrites	Biodiesel	Bharti et al. (2021)
Stearic acid	ZrO_2 nanoparticles	Biodiesel	Ibrahim et al. (2019)
Rapeseed oil	Impregnated Fe_3O_4-CeO_2	Biodiesel	Ambat et al. (2019)
Mahua oil	Doped ZnO	Biodiesel	Baskar et al. (2017)
Soybean oil	Nanomagnetic $Ni_{0.5}Zn_{0.5}Fe_2O_4$	Biodiesel	Dantas et al. (2015)
Soybean oil	Magnetic nanoparticles	Biodiesel	Dantas et al. (2020)
Soybean oil	Nanoferrite	Biodiesel	Dantas et al. (2018)
Soybean oil	Mixed nanoferrites	Biodiesel	Dantas et al. (2017)
Soybean oil	Nanoferrite doped with Cu	Biodiesel	Dantas et al. (2016)
Soybean oil	Nanoferrite doped with divalent copper	Biodiesel	Dantas et al. (2013)
Soybean oil	Nanomagnetic particles	Biodiesel	Dantas et al. (2014)
Canola oil	Carbon nitrides	Biodiesel	de Medeiros et al. (2020)
Vegetable oil	Gold nanoparticles	Biodiesel	Bet-Moushoul et al. (2016)
Vegetable oil	Magnetic nanoparticles	Biodiesel	Erdem et al. (2018)
Rapeseed oil	Nanomagnetic catalyst	Biodiesel	Kelarijani et al. (2019)
Canola oil	Nanomagnetic catalyst	Biodiesel	Firouzjaee and Taghizadeh (2017)
Canola oil	Magnetic mesoporous nanocatalyst	Biodiesel	Ghalandari et al. (2019)
Neem oil	Copper doped ZnO	Biodiesel	Gurunathan and Ravi (2015)

ZnO (CZO) nanocatalyst. Researchers have chosen heterogeneous nanocatalysts to improve vegetable oil transesterification to biodiesel. AFM analysis confirmed the CZO nanocatalyst's highly porous and non-uniform surface, which results in the aggregation of CZO nanoparticles into multi-layered nanostructures. The yield of 97.18% biodiesel was obtained in 60 min at a temperature of 55 °C using a 10% (w/w) CZO nanocatalyst and a 1:10 (v:v) oil:methanol ratio. Using recycled nanocatalyst in the sixth cycle, a yield of 73.95% biodiesel was obtained. GC–MS and 1H NMR analysis were used to confirm the biodiesel obtained. On the basis of testing reaction kinetic models for biodiesel production, a first-order kinetic model with $R^2 = 0.9452$ was found to be the most closely fitting to the experimental data. The transesterification of neem oil into biodiesel using CZO nanocatalyst required an activation energy of 233.88 kJ/mol. Table 2.4 summarizes the biodiesel production literature review.

2.6 Conclusion

In this review article, the applications of nanomaterials in the production of liquid biofuels were discussed. Liquid biofuels include vegetable and animal oils, biodiesel, biooil, and bioethanol. The biomass conversion processes used to produce liquid biofuels are solvent extraction, melting, transesterification, liquefaction, pyrolysis, and fermentation. The feedstocks for liquid biofuels include woody biomass, agricultural- and agro-industrial-based residues, aquatic materials, animal, domestic, and industrial residues. Nanomaterials are gaining more significance today mainly because of their large surface area and reusability. Nanomaterials find limited application in the production of vegetable and animal oils. Biooil and bioethanol. But nanomaterials are widely used heterogenous catalysts for biodiesel production.

References

Akbar E, Yaakob Z, Kamarudin SK, Ismail M, Salimon J (2009) Characteristic and composition of Jatropha curcas oil seed from Malaysia and its potential as biodiesel feedstock feedstock. Eur J Sci Res 29(3):396–403

Alonge AF, Jackson NI (2018) Extraction of vegetable oils from agricultural materials: a review. In: Nigeria: proceedings of the 12th CIGR section VI international symposium, held at the International Institute of Tropical Agriculture, Ibadan, Oyo State, Nigeria

Ambat I, Srivastava V, Haapaniemi E, Sillanpää M (2019) Nano-magnetic potassium impregnated ceria as catalyst for the biodiesel production. Renew Energy 139:1428–1436

Anushya A, Swathika M, Sivamani S, Sivakumar N (2019) Bioprocessing of cassava stem to bioethanol using soaking in aqueous ammonia pretreatment. In: Bioprocessing for biomolecules production, pp 429–441

Aswin S, Ranjithkumar SP, Sivamani S (2020) Modelling and simulation of pyrolysis of teak (Tectona Grandis) sawdust. In: Biofuel production technologies: critical analysis for sustainability. Springer, Singapore, pp 325–342

Baskar G, Gurugulladevi A, Nishanthini T, Aiswarya R, Tamilarasan K (2017) Optimization and kinetics of biodiesel production from Mahua oil using manganese doped zinc oxide nanocatalyst. Renew Energy 103:641–646

Behrendt F, Neubauer Y, Oevermann M, Wilmes B, Zobel N (2008) Direct liquefaction of biomass. Chemical Engineering & Technology: Industrial Chemistry-Plant Equipment-Process Engineering-Biotechnology, 31(5), 667–677

Bet-Moushoul E, Farhadi K, Mansourpanah Y, Nikbakht AM, Molaei R, Forough M (2016) Application of CaO-based/Au nanoparticles as heterogeneous nanocatalysts in biodiesel production. Fuel 164:119–127

Bharti MK, Chalia S, Thakur P, Sridhara SN, Thakur A, Sharma PB (2021) Nanoferrites heterogeneous catalysts for biodiesel production from soybean and canola oil: a review. Environ Chem Lett 19:1–20

Chakrabarty MM (2003) Chemistry and technology of oils & fats, vol. 1. Allied Publishers

Chandrasekaran AP, Sivamani S (2018) Statistical modeling and optimization of pretreatment for fermentable sugars production from cotton gin waste. Energy Sources Part A Recov Util Environ Eff 40(4):400–405

Chornet E, Overend RP (1985) Biomass liquefaction: an overview. In: Fundamentals of thermochemical biomass conversion, pp 967–1002

Dantas J, Santos JRD, Cunha RBL, Kiminami RHG, Costa ACF (2013) Use of Ni-Zn ferrites doped with Cu as catalyst in the transesterification of soybean oil to methyl esters. Mater Res 16(3): 625–627

Dantas J, Santos JRD, Silva FN, Silva AS, Costa ACFM (2014) Study of nanoferrites Ni0, 5Zn0, 5Fe2O4 and Ni0, 1Cu0, 4Zn0, 5Fe2O4 as catalysts in the methyl transesterification of soybean oil. In: Materials science forum, vol. 775, pp 705–711. Trans Tech Publications Ltd.

Dantas J, da Silva FN, de Oliveira Pereira KR, Silva AS, Costa ADM (2015) Evaluation of the processing conditions in the transesterification for biodiesel production using the nanomagnetic catalyst Ni0. 5Zn0. 5Fe2O4. In: Materials science forum, vol. 820, pp 113–118. Trans Tech Publications Ltd

Dantas J, Leal E, Mapossa AB, Silva ASA, Costa ACFDM (2016) Synthesis, characterization and catalytic performance of mixed nanoferrites submitted to transesterification and esterification reaction using methyl and ethyl route for biodiesel production. Revista Materia (Online) 21(4)

Dantas J, Leal E, Mapossa AB, Cornejo DR, Costa ACFM (2017) Magnetic nanocatalysts of Ni0. 5Zn0. 5Fe2O4 doped with Cu and performance evaluation in transesterification reaction for biodiesel production. Fuel 191:463–471

Dantas J, Leal E, Feitosa AC, Vasconcelos EV, Costa ACFDM (2018) Biodiesel from fatty acids found in Brazilian native cultures as soybean and cotton using the nanocatalyst Ni0. 5Zn0. 5Fe2O4. In: Materials science forum, vol. 912, pp 274–279. Trans Tech Publications Ltd

Dantas J, Leal E, Cornejo DR, Kiminami RHGA, Costa ACFM (2020) Biodiesel production evaluating the use and reuse of magnetic nanocatalysts Ni0. 5Zn0. 5Fe2O4 synthesized in pilot-scale. Arab J Chem 13(1):3026–3042

das Neves MA, Kimura T, Shimizu N, Nakajima M (2007) State of the art and future trends of bioethanol production. Dyn Biochem Process Biotechnol Mol Biol 1(1):1–14

de Medeiros TV, Macina A, Naccache R (2020) Graphitic carbon nitrides: efficient heterogeneous catalysts for biodiesel production. Nano Energy 78:105306

Debalina B, Reddy RB, Vinu R (2017) Production of carbon nanostructures in biochar, biooil and gases from bagasse via microwave assisted pyrolysis using Fe and Co as susceptors. J Anal Appl Pyrolysis 124:310–318

Demirbas A (2011) Competitive liquid biofuels from biomass. Appl Energy 88(1):17–28

Devi A, Singh A, Bajar S, Owamah HI (2021) Nanomaterial in liquid biofuel production: applications and current status. Environ Sustain 4:1–11

Erdem S, Erdem B, Öksüzoğlu RM (2018) Magnetic nano-sized solid acid catalyst bearing sulfonic acid groups for biodiesel synthesis. Open Chem 16(1):923–929

Firouzjaee MH, Taghizadeh M (2017) Optimization of process variables for biodiesel production using the Nanomagnetic catalyst CaO/NaY-Fe3O4. Chem Eng Technol 40(6):1140–1148

Ghalandari A, Taghizadeh M, Rahmani M (2019) Statistical optimization of the biodiesel production process using a magnetic core-mesoporous shell KOH/Fe3O4@ γ-Al2O3 nanocatalyst. Chem Eng Technol 42(1):89–99

Gurunathan B, Ravi A (2015) Process optimization and kinetics of biodiesel production from neem oil using copper doped zinc oxide heterogeneous nanocatalyst. Bioresour Technol 190:424–428

Hara M (2009) Environmentally benign production of biodiesel using heterogeneous catalysts. ChemSusChem 2(2):129–135

Ibrahim MM, Mahmoud HR, El-Molla SA (2019) Influence of support on physicochemical properties of ZrO2 based solid acid heterogeneous catalysts for biodiesel production. Catal Commun 122:10–15

Kelarijani AF, Zanjani NG, Pirzaman AK (2019) Ultrasonic assisted transesterification of rapeseed oil to biodiesel using nano magnetic catalysts. Waste Biomass Valori 1–9

Kushwaha D, Upadhyay SN, Mishra PK (2018) Nanotechnology in bioethanol/biobutanol production. In: Green nanotechnology for biofuel production. Springer, Cham, pp 115–127

Leo VV, Singh BP (2018) Prospectus of nanotechnology in bioethanol productions. In: Green nanotechnology for biofuel production. Springer, Cham, pp 129–139

Nde DB, Foncha AC (2020) Optimization methods for the extraction of vegetable oils: a review. Processes 8(2):209

Saber M, Nakhshiniev B, Yoshikawa K (2016) A review of production and upgrading of algal biooil. Renew Sustain Energy Rev 58:918–930

Salvi BL, Subramanian KA, Panwar NL (2013) Alternative fuels for transportation vehicles: a technical review. Renew Sustain Energy Rev 25:404–419

Shahidi F (ed) (2005) Bailey's industrial oil and fat products, edible oil and fat products: processing technologies, vol 5. Wiley, Hoboken

Sharma H, Giriprasad R, Goswami M (2013) Animal fat-processing and its quality control. J Food Process Technol 4(8):252

Sheikh SM, Kazi ZS (2016) Technologies for oil extraction: a review. Int J Environ Agric Biotechnol 1(2):238506

Singhvi MS, Deshmukh AR, Kim BS (2021) Cellulase mimicking nanomaterial-assisted cellulose hydrolysis for enhanced bioethanol fermentation: an emerging sustainable approach. Green Chem 23:5064

Sivamani S, Cuento A, Kaaf ASAA, Jadad ARSA, Al-Shanfari AAN, Wazeer KKAB, Kashoub MSM, Marjan ANSB (2021a) Experimental investigations and concise review on biodiesel production from crude sunflower oil using lime-based catalysts. In: Bioenergy research: integrative solution for existing roadblock. Springer Nature, Singapore

Sivamani S, Manickam A, Karthiban S, Karthikeyan S, Balajii M (2021b) Process modelling and simulation of biodiesel synthesis reaction for non-edible yellow oleander (yellow bells) oil and waste chicken fat. In: Bioenergy research: revisiting latest development. Springer, Singapore, pp 129–160

Srinivasan GR, Jambulingam R (2018) Comprehensive study on biodiesel produced from waste animal fats—a review. J Environ Sci Technol 11(3):157–166

Subroto E, Manurung R, Heeres HJ, Broekhuis AA (2015) Optimization of mechanical oil extraction from Jatropha curcas L. kernel using response surface method. Ind Crop Prod 63: 294–302

Tirtiaux A (1983) Tirtiaux fractionation: industrial applications

Wang H, Covarrubias J, Prock H, Wu X, Wang D, Bossmann SH (2015) Acid-functionalized magnetic nanoparticle as heterogeneous catalysts for biodiesel synthesis. J Phys Chem C 119(46):26020–26028

Xie W, Han Y, Wang H (2018) Magnetic Fe3O4/MCM-41 composite-supported sodium silicate as heterogeneous catalysts for biodiesel production. Renew Energy 125:675–681

Xu D, Yang S, Su Y, Shi L, Zhang S, Xiong Y (2021) Simultaneous production of aromatics-rich biooil and carbon nanomaterials from catalytic co-pyrolysis of biomass/plastic wastes and in-line catalytic upgrading of pyrolysis gas. Waste Manag 121:95–104

Yasir M, Chowdhury S, Mansor N, Mohamed NM, Uemura Y (2014) Upgrading of pyrolysis biooil to fuel over supported nanomaterials—A review. In: Applied mechanics and materials, vol. 625, pp 357–360. Trans Tech Publications Ltd

Yusuf AK (2018) A review of methods used for seed oil extraction. Int J Sci Res 7(12):233–223

Zahid MU, Pervaiz E, Hussain A, Shahzad MI, Niazi MBK (2018) Synthesis of carbon nanomaterials from different pyrolysis techniques: a review. Mater Res Express 5(5):052002

Zhang L, Xu CC, Champagne P (2010) Overview of recent advances in thermo-chemical conversion of biomass. Energy Convers Manag 51(5):969–982

Chapter 3
Applications of Nanomaterials in Gaseous Biofuels Production

B. S. Naveen Prasad, B. Saikat, and S. Sivamani

Abstract Nanotechnology is being studied in several sectors, including medical, engineering, the environment, electronics, military, etc. Many studies are being conducted to advance knowledge in terms of size, capacity, and cost. The nanomaterials are synthesized and characterized by different methods, and it is used for the different applications. The review presented here is on the role of nanomaterials in biogas and biohydrogen production. The addition of NP increased the yield of biogas and biohydrogen, the influence of size and shape of metal oxide NP and metallic NP on biogas and biohydrogen production.

Keywords Biogas · Biohydrogen · Nanoparticles · Microorganisms

3.1 Introduction

Throughout the preceding century, nanotechnology has grown in popularity. A flurry of research initiatives is now being conducted on nanotechnology. Nanotechnology is well-defined as the "evolving, synthesizing, characterization, and application of materials and electronics at the nanoscale." The prefix "nano" is used as a phrase in every field, including commercial advertising. True, the word "nano" comes from either the Greek nanos or the Latin nanus, which both imply "dwarf." It includes chemistry, physics, solid state physics, materials science, and biosciences. As a result, mastery of a single discipline will no longer suffice; a thorough understanding of physics, chemistry, material science, solid state physics, and biosciences will be needed. Nanotechnology is the science of altering matter at the atomic level to create unique nanomaterials with a variety of features, whereas nanoscience is the study of the atom arrangement and nanoscale properties (Singh and Gupta 2016). Nanotechnology is gaining popularity in practically every engineering field, yet the general public is unaware of its practical applications. Despite

B. S. Naveen Prasad (✉) · B. Saikat · S. Sivamani
Chemical Engineering Section, Engineering Department, University of Technology and Applied Sciences (Salalah College of Technology), Salalah, Oman

© The Author(s), under exclusive license to Springer Nature Singapore Pte Ltd. 2023
M. Srivastava, P. K. Mishra (eds.), *NanoBioenergy: Application and Sustainability Assessment*, Clean Energy Production Technologies,
https://doi.org/10.1007/978-981-19-6234-9_3

this, it is being more widely used in medical, engineering, the environment, electronics, the military, and security. Despite the fact that this technology has applied to a wide application, new unique nanomaterials have the potential to be developed in a variety of industries for the advancement of mankind. The growth of information in terms of size, capability, and cost fascinates and inspires researchers. As a result, there is a lot of desire in the medical and scientific industries to miniaturize products to save money. In the future, nanotechnology will rule all aspects of human life, including living, working, and communicating. As a result, interest in the topic develops, prompting a discussion of nanotechnology's fundamental and major characteristics. Nanomaterials are the foundational and indispensable elements of nanotechnology. Nanomaterials are with a size of fewer than 100 nm in at least one dimension. In other words, they are far smaller than microscale. Nanomaterials have a size of about 10^{-9} m or one billionth of a meter. Nanoparticles differ from bulk materials in terms of their physicochemical qualities, which are mostly governed by their size and shape. Surprisingly, by altering their form and size at the nanoscale level, nanoparticles develop a distinct personality with new features and capabilities. Nanomaterials come in a variety of morphologies, including nanorods, nanoparticles, and nanosheets, and are defined by their dimensionality. Nanoparticles are zero-dimensional nanomaterials, nanorods, or nanotubes are one-dimensional nanomaterials, and films/layers of type one or two-dimensional nanomaterials. The physical properties of two or more particles will change as a result of their interaction. Bulk nanomaterials, also known as three-dimensional nanomaterials, are made up of a range of compounds. Nanotechnology has the potential to improve human living standards when combined with other interconnected technologies (cognitive sciences, biotechnology, and information technology), with applications in areas like pharmaceutical development, electronics, life sciences, etc., (Demetzos 2016; Wolf and Medikonda 2012). Because of its dynamic physicochemical features, nanotechnology has the potential for industrial use. Nanoparticles are expected to grow in economic relevance greatly in the foreseeable future, likely from 2000 tons in 2004 to 58,000 tons recently. (Shi and Ma 2010; Shi and Ma 2011). Nanoparticles used in a various industry, biofuel production and environmental cleanup (Srivastava et al. 2014). Nanoparticles can boost the efficiency of green fuels like bioethanol and biohydrogen, hence nanomaterials could help boost biofuel production (Srivastava et al. 2015). According to the findings, nanomaterials could be employed to improve cellulase production, strength, and catalytic activity. Furthermore, the inclusion of nanoparticles increases the processing of lignocellulosic biomass (Srivastava et al. 2015). Depending on their stoichiometry, goethite, ferrihydrite, magnetite, wüstite, hematite, maghemite, and other iron oxide nanoparticles have a range of crystalline phases. Fe_3O_4 and g-Fe_2O_3 are the most studied FeO-NPs. (Cavas et al. 2013; Haddad et al. 2015), including the pretreatment and hydrolysis of biomass to form sugars (Cavas et al. 2013; Haddad et al. 2015; Ubale and Belkhedkar 2015; Srivastava et al. 2017a, b).

3.2 Nanomaterial Synthesis and Characterization

Methods for making nanomaterials Nanoparticles can be made in three different methods. The specifics are (1) Biological methods, (2) Physical methods (3) Chemical methods. The biological process is simple and fast, often involving only one step, and it is also environmentally friendly. Microbes, as well as diverse plant parts, could be employed to create nanomaterials in this situation, as shown in Fig. 3.1.

3.2.1 Microorganisms Are Used to Make Nanomaterials

Microorganisms like bacteria, fungi, and algae could be used to generate a variety of nanomaterials from aqueous metal salt solutions. Living animals will produce nanoparticles utilizing a protein as part of the biomineralization process. Under anaerobic conditions, magnetotactic bacteria, for example, use magnetosomes, are used for the synthesis of magnetic FeO nano crystals, to prepare magnetic particles (Krishnan 2016; Bazylinski et al. 1994); Faivre and Schuler (2008); Schüler and Frankel (1999); Timko et al. (2012) reported that homogeneous particles with a core diameter of 20–45 nm can be produced in vitro. Magnetosomes, despite this, show good magnetic properties in medical applications such as hyperthermia (Hergt et al. 2005; Molcan et al. 2016). Photosynthetic bacteria like *Rhodopseudomonas capsulata* were used to make gold nanoparticles with a size of 10–20 nm. The bacterial enzyme NADH-dependent reductase is important in the transformation of Au ions into Au-NPs (He et al. 2007). *Pseudomonas* microbes from alpine habitats were used to make extracellular palladium nanoparticles (Schlüter et al. 2014).

Fig. 3.1 Green synthesis approach for nanoparticles

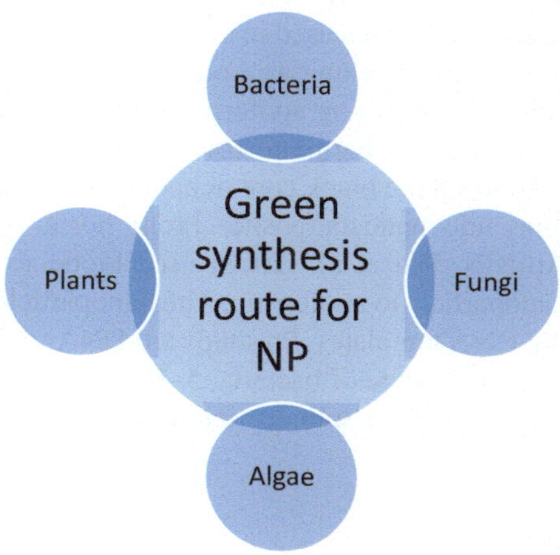

3.2.2 Fungi Are Used

Extracellular silver nanoparticles were made using the *Fusarium oxysporum fungus*. The enzymatic activity of NADH-reductase produces long-term stable nanoparticles. In fungal cells, the amount of protein released is higher than in bacterial cells (Ahmad et al. 2003). Animal feed, food, medications, textile sector, and paper sector all use it these days.

3.2.3 Algae Usage

In 2007, Singaravelu et al. studied the Au-NPs production using *Sargassum wightii*. Ninety-five percent of the product was ready after only 12 h of incubation. They investigated how algae can be utilized to manufacture nanoparticles, something that has not been done before. Some bacteria, fungi, and algae are pathogenic, necessitating the creation of safety procedures.

3.2.4 Nanomaterials Are Being Synthesized for Use as Biological Templates

A biological mechanism could be used to create nanoparticles within the organism. The following are the key tools used to create this biological template. To make distinctive and technologically advanced nanostructures, scientists use biological templates like proteins and DNA. Biosensors (Zhang et al. 2019), bioNEMS, and bioelectronics systems can all benefit from this nanoparticle (Zhang et al. 2017a, b). Nanocomposite materials rely heavily on proteins. Prokaryotes and eukaryotes, for example, have ferritin as an intracellular iron-storage protein. Iron oxide is accumulated and then released in a controlled manner. It balances iron deficiency and excess in humans by acting as a buffer. It is made up of a protein shell with an iron oxide core. For obtaining apoferritin, the core containing FeO can be the selective dissolution without harming surrounding proteins. FeO or any other nanoparticle can be used to fill the blank core of apoferritin. It has now been developed a protein with an inorganic nanocomposite. From horse spleen ferritin, Fan et al. (2010) made Au-NPs. Wu et al. (2008) used biotin to bind a yttrium phosphate radionuclide nanoparticle to apoferritin. The nanoparticles can also be assembled with the help of the DNA templates. Plasmids are DNA molecules with a closed circular form found in a wide range of bacteria. 5–10 nanometer CdS DNA nanoparticle compound can be produced by spin coating a mixture of plasmid DNA and cadmium perchlorate.

3.2.5 *Plant Components Are Used to Prepare Nanomaterials*

The nanoparticles were also made with plants and plant extracts. Metal NPs are reduced by phytochemicals found in plants. In the production of nanoparticles, phytochemicals such as flavones, organic acids, and quinones act as natural reducing agents. Au-NPs of different forms are generated from the biomass of the *Medicago sativa* (*alfalfa*) plant (Gardea-Torresdey et al. 2003) and the leaves of the *Pelargonium graveolens* (Geranium) plant (Pandian et al. 2013). Au core-Ag shell bimetallic nanoparticles and bimetallic Au, Ag and are made via *Azadirachta indica* (neem) leaves. The Plant contains sugars/terpenoids that act as reducing agents (Shankar et al. 2004). Au nanotriangles are made from leaf extract of *aloe vera* (Chandran et al. 2006). Cobalt, zinc, Silver, nickel, and copper nanoparticles are made from plants such as *Helianthus annuus* (sunflower) (Banerjee et al. 2014) and *Brassica juncea* (Indian mustard) (Banerjee et al. 2014).

3.2.6 *Physical Methods Can Be Used to Prepare Nanomaterials*

Physical methods are split into two categories: top-down and bottom-up approaches. Mechanical milling is a method for crushing bigger materials into smaller particles that works from the top down. The difficulty in achieving the proper particle size and shape is the method's principal disadvantage (Thakore et al. 2014). Due to flaws in lattice features formed during the milling process, the magnetic properties of the samples generated by the milling operation differ from normal particles of the same size (Dutz et al. 2007). The "bottom-up" technique is used to condense nanoparticles in a liquid or gaseous phase, with the larger materials emerging from the chemical combination of the smaller ions.

3.3 Characterizations of Nanomaterials

The nanoparticles have one-of-a-kind physicochemical properties. Nanoparticles have a wide range of qualities depending on their size and form. A variety of devices must be employed to determine the properties of nanoparticles. UV Spectrophotometer, FT-IR Spectroscopy, Atomic Force Microscopy (AFM), Transmission Electron Microscopy (TEM), Scanning Electron Microscopy (SEM), Vibrating Sample Magnetometer (VSM), SQUID, Energy Dispersive X-ray Spectroscopy (EDS), X-ray Photoelectron Spectroscopy (XPS), Magnetic Force Microscopy (MF), Vibrating Sample Magnetometer (Galloway et al. 2015).

3.3.1 Nanoparticle Surface Morphology, Surface Area, Size, and Shape

The size and form of nanoparticles are most important factors in defining their characteristics. TEM, FESEM, and AFM can all be used to investigate surface morphology. These approaches will produce images that show whether nanoparticles are spherical, rod-shaped, or porous, as well as their morphology. TEM will provide information on the composition, shape, and crystallinity of nanoparticles, as opposed to SEM. The signals will reveal the surface topography and sample composition. The surface topography and composition of the sample will be revealed by these signals. As a result, the samples' surfaces should be electrically conductive at the very least. SEM and TEM can accomplish the same tasks as AFM, but only on dry samples. HRTEM, FESEM, and XRD for assessing the size of nanoparticles. In comparison to light microscopes, TEM can collect high-resolution images. The TEM can be used to determine crystal orientation, lattice gap, electron structure, aggregation state, and electron phase shift. Using the TEM, it is simple to find the nanoparticle size (Gabbasov et al. 2015). The size is calculated using the Scherrer equation and XRD spectroscopy. The size of nanoparticles is determined by the strong XRD peaks. Because non-crystalline nanoparticles contain big XRD peaks, size determination is more difficult than with a TEM as seen in Fig. 3.2. The surface area of nanoparticles was calculated using the Brunauer–Emmet–Teller method.

3.3.2 Mineral and Element Content Determination

Elemental composition and surface morphology of a sample can be determined by EDS in conjunction with TEM and SEM. Inductively Coupled Plasma Mass Spectroscopy and Atomic Absorption Spectroscopy are two techniques that can be used for elemental computations. AAS will not be used to apply solid nanoparticles directly. Before using, they must first be dissolved in the appropriate acids or bases. XRD may be used to figure out what minerals are in aggregated crystalline nanoparticles (Kubickova et al. 2013). XPS can also provide information on the composition of elements.

3.3.3 Nanoparticle Structure and Bonding Types

To obtain the structure and bonding qualities, a variety of methods can be applied. Techniques such as XPS, FT-IR, Raman Spectroscopy, TGA, and XAS are helpful. The metal–oxygen connections will be confirmed using XPS and FTIR. It is possible to preserve data on the binding energy and oxidation state of materials with various

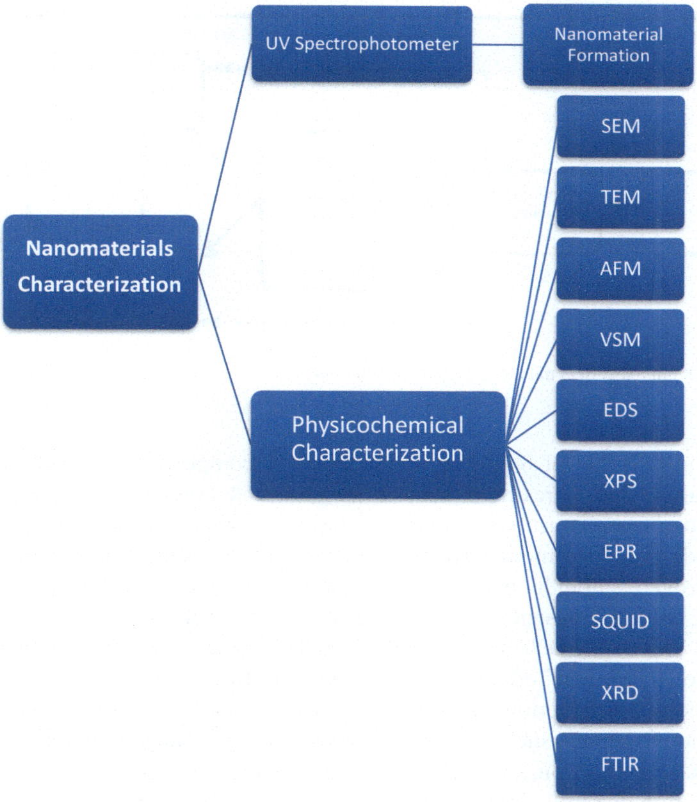

Fig. 3.2 Characterization of nanomaterials

elemental compositions. The structure and spinel lattice of a substance are determined via Raman spectroscopy. XAS will provide the neighboring atoms, oxidation states, bond length, coordination number, and any other necessary information (Gräfe et al. 2014; Reddy et al. 2012).

3.4 Nanomaterial Applications in Biogas Production

Anaerobic digestion is regarded as a complex biological procedure that combines hydrogen sulfide (H_2S), methane (CH_4), water vapor, carbon dioxide (CO_2), hydrogen, and ammonia with complex organic wastes. Anaerobic digestion consists of several steps process including four types of microbiological phases: hydrolysis, acetogenesis, acidogenesis, and methanogenesis, all of which are driven by a microbial community, as shown in Fig. 3.3 (Abdelsalam et al. 2016).

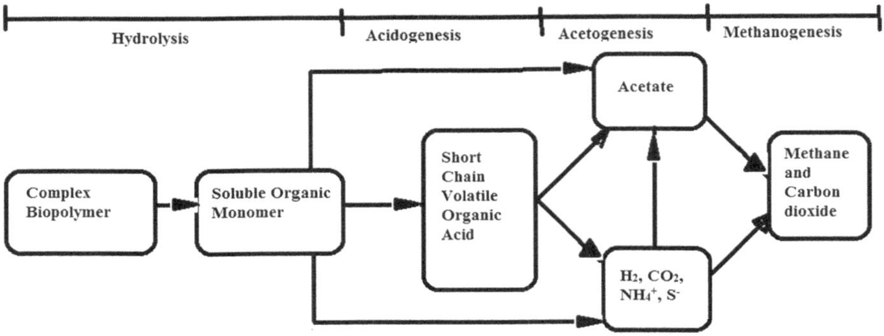

Fig. 3.3 Mechanism of Anaerobic Digestion Process

The sluggish rate of biological degradation of composite biological layers and lignocellulosic layers restricts the effectiveness and efficiency of AD processes. Furthermore, toxic constituents such as relatively higher molecular weight organic compounds and ammonium derivatives in sewage water, high demand for ammonia and Hydrogen sulphide intensity throughout and biochemical oxygen demand (BOD), COD total basicity, and salinity at certain levels may necessitate initiate treatment processes such as heat and acid/alkali because toxic components may reduce the AD process' effectiveness (Ahmed and Rodrigues 2013).

The recent introduction of various innovations under the Integrated Biofuels Production Scheme, such as biogas production, bioaugmentation, anaerobic co-digestion, and so on, as well as the use of nanoparticles (NPs), are attracting more attention due to their small size and various notable features.

Furthermore, nanoparticles have higher field of volume ratios, which increases the number of active sites, which is a requirement for increasing various reactions. NPs, for example, might make hydropower processes easier by offering large surface areas per volume ratio for microbes to adhere to organic molecules' active sites (Hsieh et al. 2016). Otherwise, NPs can boost yield by acting as the donors of electron and cofactors of essential number of enzymes in diverse biological processes (Ali et al. 2017).

Richard Feynman invented the term nanotechnology at the first meeting of the American Physical Society. A rolled sheet of carbon known as fullness/sentence force was discovered in the tubular construction in subsequent years. Eric Drexler produced the first scientific publication on nuclear engineering 22 years after Feynman's famous presentation on the topic of nanotechnology.

Introducing the Basics of Nanotechnology, which he published in 1986, was a seminal work. Dresler was the first batch of student to get a doctorate degree in the course of molecular-based nanotechnology in the year 1991, despite the fact that he delivered a recognized course on the arena of nanotechnology and nanomaterial formation in the year 1987. They work with their own bulk materials, which have a variety of chemical, physical, and biological properties.

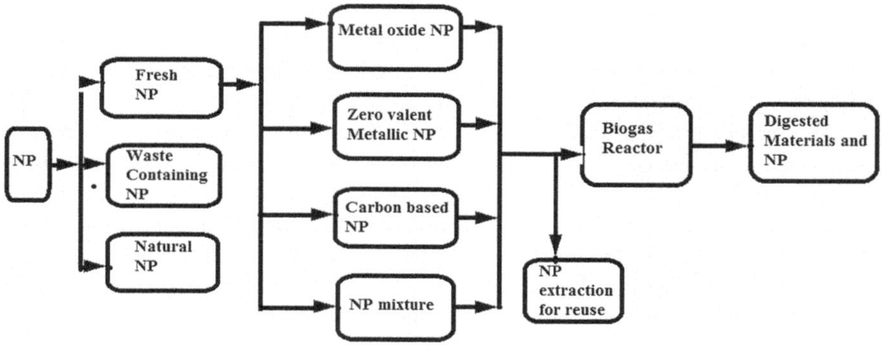

Fig. 3.4 Classification of NPs for biogas production

In biogas, the potential for nanoparticles to be used in a range of disciplines, including the treatment of pollutants from soil and water, is virtually unaltered. Scott (2006) addressed the overall prospect of improving biogas production from animal manure of observed NPs in one of the first experiments (Ambuchi et al. 2017).

However, the effect of particle size of a few metal oxides (CuO as well as ZnO) on the biogas production as well as CH_4 in the anaerobic digestion process, which was only discovered in 2011. The researchers noted an accumulation of NPs in municipal wastes in the following years, due to the increased use of these chemicals in numerous businesses. In the same year, it began looking into the impacts of numerous metallic oxide and metallic NPs on the AD of Municipal Slides (Amen et al. 2017).

Preliminary research has found that certain NPs at particular concentrations (e.g., Ag and TiO_2) may not have a negative impact on AD processes in methane in addition to biogas production, as well as the number of microbiological organisms participating in their variety (Au, CeO_2). This can have a negative impact on the procedure. The effects of NPs on AD were continuing to be explored in the subsequent years, and certain additional NPs (such as copper oxide, nZVI, zinc oxide, selenium oxide and titanium dioxide) were also researched.

In light of the foregoing, this study focuses on new research on a sophisticated application of different nanomaterials utilized to increase/progress biogas as well as generation using AD processes (Amen et al. 2018). The NPs in this study are separated into three different categories: (1) ZVMNPs, (2) carbon-based NPs & (3) metallic and metal oxide nanoparticles as shown in Fig. 3.4.

3.4.1 Role of Biological Sensors for the Monitoring of Biogas

Anaerobic digestion evaluation, uninterrupted checking of volatile fatty acids and organic matter is used to develop mediators for the situation of unstable progression. The depletion of fossil fuels has sparked a surge in public interest in biogas production. AD provides the benefit of absorbing energy production from industrial waste, so addressing a contemporary issue (Zhou et al. 2019). Improvements in several economic and technological areas have emerged from the long-term sustainability of efficient methane generation and process stability.

These include the right feedstock mix, biogas purification technology, and appropriate biogas furnace conditions depending on a variety of biochemical as well as physical and factors like pH of the solution, alkalinity of the solution and quality of gas evolved from the process. Reactor acidification occurs as a result of the manufacture and synthesis of alcohol and unstable fatty acids (e.g., propionate, acetate, butrate). Gas chromatography, spectroscopy as well as high-performance liquid chromatography are the most popular procedures for determining acid composition, and they are normally performed by an external source as part of a high-cost analysis (Gonzalez-Estrella et al. 2013).

3.4.2 Nanomaterial

Both ethnographic and natural resources are used to make nanoparticles. NP concentrations can reach dangerously high levels in waste mud. The toxicity and impacts of NPs in the sludge treatment stream, on the other hand, are currently being studied. The hazardous potential of plants and bacteria, as well as the sludge dehydration process, are determined by the effects of the sludge AD process on NP and CO_2 nanoparticles. Nanoparticle concentration is crucial in identifying their significance in the creation of methane and biogas.

Although not all nanoparticles activate the microbial digestion system, as compared to controls, certain nanoparticles dramatically limit production rates. The exposure density of 1000 mg/L of ZnO resulted in 64.3% biogas volume and 47.7% methane. The barrier effect can be overcome after 14 days of exposure to zinc oxide at the concentration used for permanent exposure.

3.4.3 Role of Oxides of Metal Nano Particles in Anaerobic Ingestion Process

Oxides of metal NPs offer distinct chemical and physical features, including smaller size, superior surface structure, higher volume ratio, solubility, and strong catalytic action. The effect of chemical and physical properties of oxides of metal-based NPs on the AD process can vary according to size, density, type of NP, HRT, and layering type. This emphases on the influences of concentration and size of metallic oxide NPs on biogas manufacturing; AD procedure durability; production of H_2S and other contaminants in biogas manufacture from the standpoint of different levels and HRT (Baniamerian et al. 2019).

3.4.4 Influences on the Yield of Biogas

The AD process produces biogas, which is a useful and significant byproduct. Biogas yield has critical index to assess AD process' efficacy. The consequence of Fe_3O_4 NP on methane output of metropolitan compact waste was studied by certain researchers. Fe_3O_4 NP was tested at concentrations of 50, 75, 100, and 125 mg/L. The addition of 76 mg/L Fe_3O_4 NP boosted methane generation by 54.3%, according to the findings. Adding high quantities of Fe_3O_4 NP, on the other hand, resulted in limited CH4 generation.

The effects of Fe_3O_4 NP at three types of diverse concentrations on biogas manufacture from poultry waste were studied: 16, 60, and 150 mg/L. In comparison to controls, the addition of 120 mg/L Fe_3O_4 NP is caused by a 28% rise in CH4 manufacture. The effect of Fe_3O_4 NP at 25 mg/L on methane manufacture from cattle manure was studied by certain researcher.

When Fe_3O_4 NP was added, biogas and methane manufacture is increased by 1.8 as well as 2.5 times, respectively. According to some researchers, adding 120 mg/g-TSS iron oxide NP increased methane manufacture by 12%. Insolubility of Fe_2O_3 NPS and electron transfer may cause this increase in methane manufacture to be disrupted. Farghali et al. (2019) looked at the iron oxide and titanium dioxide effects of NP on the making of biogas from cattle dung.

Biogas and CH_4 generation rates were 1.1, 1.16, and 1.19, 1.21 times for Fe_2O_3 NP, 30 & 150 mg/L concentrations respectively. The authors assume that this discrepancy is because to changes in NPS size, experimental settings, and levels, as these outcomes are inferior to those reported by way of Abdelsalam et al. (2016). Furthermore, TiO_2 NPs in the concentration of 600 mg/L produced 339.25 mg/g VS of biogas and CH_4 of 194.31 mg/g VS.

Low-density (12 mg/L) CeO_2 NPs methane as well as biogas manufacture increased by 12%, according to Gonzalez-Estrella et al. (2013); however, adding concentrations beyond 100 mg/L CO_2 NP lowers biogas output compared to the control technique. The drop in biogas manufacture could be related to the high hardness of sludge and the low diffusion characteristics of CO_2 NP in the media, according to the scientists. Mu et al. (2011) looked at how varying concentrations of four types of NPs (TiO_2, Al_2O_3, SiO_2, and ZnO) affected CH_4 production during Sludge AD (Gottschalk et al. 2013).

In case of quantities used up to 160 mg/g-total TSS, TIO_2, SiO_2, and Al_2O_3 NP had no inhibitory impact; however, ZnO NPs had an inhibitory impact when the dose was increased. When TiO_2, CO_2, Al_2O_3 NP up to dosage of 160 mg/g-Total Suspended Solids lowered the control reaction by 40 and 160 mg/g-TSS compared to the nitrogen and CH_4 generation control response, CH_4 generation was equivalent to control, respectively, 100% by a whopping 18.9% (Pieta et al. 2018).

3.4.5 Influences on the Stability of Oxygen Limiting Ingestion Method

COD, volatile fatty acids (VFS), volatile solids (VS), and Total solids (TS) are essential pointers; so, observing these values is adequate to assure AD process and metabolic stability. The influence of CO, Nickel, iron, and Fe_3O_4 NP on the breakdown of TS and VS was explored by Abdelsalam et al. (2016) and found to be 4.99% and 4.36%, respectively, with the largest loss of TS and VS recorded. These findings are consistent with those of Ali et al., respectively. Ali et al. (2017), who found that adding 78 mg/L Fe_3O_4 NP reduced total solid and volatile solid by 32.15 and 42.15%, correspondingly, as linked to controls (Ramesha et al. 2011).

The effects of Fe_2O_3 NP on VS reduction were also examined by some researchers. When compared to the control technique, the VC reduction for 30% and 150 mg/L Fe_2O_3 NPs was 1.03 and 1.15 times, respectively, after 30 days. The use of metallic oxide NP in AD has an impact on TS and VS digestion, as well as microbial activation at various stages of the anaerobic digestion process. Some researcher also found methanogenic bacteria in anaerobic granular sludge with eight NPs (CuO, ZnO, CeO_2, Al_2O_3, Fe_2O_3, TiO_2, SiO_2, and MnO_3). There are a few activity's effects.

ZnO NPs had an IC50 value of 52–222 mg/L, whereas CuO, CeO_2, Al_2O_3, Fe_2O_3, TiO_2, SiO_2, and MnO_3 have no effect on methane-producing bacteria. According to Zhang et al. (2017a, b), increases to ZnO NP have a detrimental impact on hydrolysis-acidification, VFA, and biogas generation (Suanon et al. 2017).

According to the authors, a rise in ZnO NP may effectively block hydrolysis-acidification, particularly in sludge-containing proteins. This finding is in line with Mu et al. (2011), who found that TiO_2, Al_2O_3, CuO, and ZnO only had no effect on the formation of soluble proteins and sugars during acidification and high levels of ZnO in methane NP hydrolysis. NP had an impact on the layer (Güngör-Demirci and Demirer 2004).

3.4.6 Influences on the Contaminants Like Hydrogen Sulphide for the Production of Biogas

H_2S and different sulfate is mostly created during the AD process by bacteria scattering H2S sulfate or by proteinaceous material depletion. The presence of hydrogen sulfide (H_2S) in biogas can impair the life of pipes and other structures. To reduce the H_2S level in biogas and the toxicity of sulfide of methanogenic bacteria, a combination of oxygen or sulfide precipitation or anaerobic bioreactors could be used. Endogenous chemical drive oxidants, like H_2O_2, ferrous salts, nitrates, and ferrite (VI), be unsuccessful to manage H_2S in drainage sludge processes over time and can harm AD (Batstone and Jensen 2011).

The use of metallic oxide NPs in AD will affect biogas production and future phases, as well as the reduction of H_2S in biogas mixes. Several studies have been undertaken on the effects of metallic oxide nanoparticles on hydrogen sulfide and other pollutants present metal organic synthesis. Farghali et al. (2019) looked at the significance of hydrogen sulfide mitigation as well as TiO_2 and Fe_2O_3 NP on cattle manure biogas generation. H_2S Fe_2O_3 nanoparticles decreased 2.13 and 2.38 times when compared to 20 and 100 mg/L supplementary controls. Hassanein et al. (2019) investigated the effects of three different Fe_2O_3 nanoparticles concentrations on H_2S generation in poultry litter. Adding 15, 50, and 100 mg/L iron oxide nanoparticles to the control system reduced hydrogen sulfide by 8.3%, 26.1%, and 9.27% (Thanh et al. (2016).

3.4.7 Influence of Metal Nanomaterial (Zero-Valance) for the Production of Biogas

The pure metals like Cu, Fe, Ni, Ag, and Co are zero-valent metallic NPs. In comparison to the bigger part, zero-valent metallic NPs have exhibits physical and chemical properties because of the effects of quantum surface and small object (Hassanein et al. 2019).

This section of the review discusses the effects of various sizes, densities, and types of zero-valent metallic nanoparticles on production of biogas, AD method strength, Hydrogen Sulfide, as well as other contaminants to digest and produce biogas at various levels of HRT.

3.4.8 Influences on the Yield of Biogas

Amen et al. (2018) investigated the impact of 5 types of different concentrations of zero-valent Fe-NPs (nZVI) on municipal sludge statistics AD to CH_4 content (100, 250, 500, 1500, and 3000 mg/L). Growing biogas output decreased by 19% and 9.5% when zero-valent Fe-NPs of 50 and 100 mg/L were added, respectively; however, when 260, 600, 1600, and 3500 mg/L of nZVI were added, amount of biogas increased by 1.29, 1.41, 1.83, and 2.15 times, respectively. The effects of different concentrations of nZVI (600, 1600, 1900, and 2300 mg/L) on CH_4 as well as biogas manufactured from sludge were also investigated by the researcher (Han and Yan 2016).

The nZVI addition of 1200 mg/L increased biogas generation cumulatively by 18.11% over the control sample. These are matched with Suanon et al. (2017) and Zhou et al. (2019). Microorganism activity was inhibited, and aggregate biogas production was lowered by 28.30% and 46.45%, respectively, when 1200 and

Table 3.1 Summary of application of NP in AD

NP type	Size in nm	Main effect
Fe_3O_4	94–400	H2S yield increased 8%
Fe_2O_3	20–40	Methane yield increased 2 times
ZnO	<100	Methane yield increased 1.18 times
CeO_2	50	Methane yield decreased 35%
TiO_2	25	Methane yield increased 2 times
SiO_2	20	No change
Fe	200	Catalyst reactivity enhanced
Ni	30	Methane yield increased 1.19 times
Co	30	Methane yield increased 1.17 times
Ag	100	Methane yield decreased 74%
Carbon-based NPs	10	Methane yield increased 51%

300 mg/L nZVI were added. Due to changes in NP size, as well as residue type and experiment, this could be the case (Rokaya et al. 2019).

Hassanein et al. (2019) studied the impacts of three types of NPs (Ni, Fe, and Co) in poultry litter biogas generation at three different concentrations. Ni NPs is found to be 11 mg/L, Co NPs is found to be 5.6 mg/L, and Fe NPs is found to be 120 mg/L which enhanced CH_4 generation by 37.48%, 25.7%, and 28.1%, respectively, as associated to control. This is matched with Abdelsalam et al. (2016), who found that nanoparticles produced methane and biogas of the three NP additions (Carpenter et al. 2015) (Table 3.1).

Fe NPS additions, according to Abdelsalam et al. (2016), can boost CH_4 generation in two ways.

For starters, Fe aids in the production of NPS acetate. Second, Fe NPs reduce CH_4 by acting as direct electronic donors. In addition, the size and form of Fe NP are critical for improving biogas and CH_4 generation, according to the findings. By introducing 2 mg/L Co between the three different types of Co nanoparticle concentrations (0.6, 1.2, and 2.5 mg/L). Abdelsalam et al. (2017a) found that production of biogas reduced by 0.95 times in comparison with reference sample (Su et al. 2013). AG NPs & Fe effects on CH_4 production AD were examined by Wang et al. (2016). Substrate CH_4 production increased by 120% when treated by 10 mg/g-TSS Iron nanoparticles, while methane production fall by 73.52% as compared to the 500 mg/g-TSS Iron nanoparticles control. Because Iron nanoparticles increase quantity of bacteria and activate critical enzymes, but 500 Ag NPs suppress them, the scientists believe that adding 10 mg/g TSS may boost CH_4 production. CuO and AgO NPs were also discovered to impede methanogenesis by Gonzalez-Estrella et al. (2013).

3.4.9 Influences on the Stability of Oxygen Limiting Ingestion Method

The impacts of nano zero-valent ion of 160 nm and 0.2 mm of powder iron on AD methodologies in sewage sludge were investigated by Sanon et al. Adding powder iron and nZVI raises the pH for the first week, then reduces it after that, but not below the under-year level.

In this scenario, the water may operate as an oxidant, and the oxidation was carried out from FeO to Fe^{2+}, which may raise the pH for the first week, but after that, the creation of VFA causes a fall in pH, which is consistent with Zhou et al. (2019) and Zia has discovered that adding 1000 mg/L nZVI resulted in a function of pH that was well-regulated. Furthermore, when compared to controls, COD removal efficiency rose by 53.4 percent and 65.2%, respectively, for powder iron and nZVI (Hao et al. 2019).

In addition, the addition of Fe enhanced the CH4 concentration of iron powder and nZVI by 11.6% and 3.1%, respectively. Zhou et al. (2019) discovered that adding nZVI to VFA, coenzyme F420, TS, VS, and COD resulted in a downward trend in 30 days, which matched Abdelsalam et al. (2016), discovered that VS and TS increased pancreatitis by 24.47% and 16%, when compared to a 30 mg/L NP. The morphology of the sludge in the control test was rod-like, whereas the morphology of the sludge with nZVI additions was spherical, according to SEM pictures (Sreekrishnan et al. 2004).

The cobalt and nickel nanoparticles effect on methanogenic activity was examined by Abdelsalam et al. (2017a, b). At dosages of 0.6 mg/L and 1.2 mg/L, Co NP has a beneficial on methanogenic activity. Nickel nanoparticles have influence on methanogenic activity at all concentrations. All dosages of cobalt & nickel increased the breakdown of total solids and poly volatile solids, with the exception of 2.2 mg/l cobalt nanoparticles.

3.4.10 Influences on the Contaminants like Hydrogen Sulphide for the Production of Biogas

Hassanein et al. (2019) investigated how three different types of Ni, Ko, and Fe NPs influenced hydrogen sulfide concentration phases in poultry litter. Although the rising H_2S fell by 6.7% when compared to 5.4 mg/L nanoparticles coupling furnaces, there was no significant difference between 120 mg/L Fe nanoparticles and 7 mg/L Ni nanoparticles in comparison with the controls. Su et al. (2013) investigated the nano zero-valent ion effects on the amount of hydrogen sulfide in the production of biogas by sludge through aerobic digestion.

According to the findings, adding 0.1% nZVI to biogas resulted in a 98% drop in H_2S content. The zero-valent metal nanoparticles addition to biogas affects not only the level of hydrogen sulfide but also the levels of other pollutants such as carbon dioxide and ammonia. Carpenter et al. (2015) looked the effect of nano zero-valent ion on the carbon dioxide levels of biogas. Two forms of nZVI were used for the two concentrations. When using the equal concentration of 1.25 g/L and nZVI, synthesized, the carbon dioxide level of the biogas was reduced by 48 and 57% (Hassanein et al. 2019).

Zhou et al. (2019) nZVI studied the effects of NH_3. Ammonia is produced when amino acids are low (NH_3). The level of NH_3 inside the reactor is critical. When NH_3 concentrations are high, bacteria, on the other hand, block the utilization of NH_3 in their nitrogen sources with suitable flexibility. The amount of NH_3 in the plain is related to its pH. Jia et al. (2017) looked at how different ammonium nitrogen nZVI concentrations (0, 600, 1200, 1800, and 2400 mg/L) affected the results. When 1200 mg/L nZVI was added, ammonia nitrogen was considerably reduced, and the authors assume that ammonia nitrogen is reduced.

Overall, the negative effects of Fe^{+} and Fe^{3+} metallic ions in establishment of anaerobic microbial populations, as good as the time reduction behind AD processes, may be to blame for these biogas generation advances. During syntrophic methanogenesis, however, the use of Fe_3O_4 NP can aid in the transfer of directly intersected electrons (Tian et al. 2017).

It should be emphasized, however, that large quantities of these metallic oxides can hinder methanogenesis by acting on acetoclastic methanogens. CuO, Fe_2O_3 NPs, for example, have been demonstrated to restrict acetoclastic methanogen and H_2-consuming methanogen, respectively, reducing methane synthesis. In both groups, high quantities of CeO_2, Mn_2O_3 NP were linked to methane inhibition.

NPs with Zn, Cd and Cd have been linked to interference with production of biogas, in contrast to the metallic and metallic oxide NPs mentioned above, which have shown favorable benefits in specific quantities. These metallic species can block or kill the microbes that cause Alzheimer's disease, effectively stopping the disease from progressing. Changes in the membrane characteristics of microorganisms and/or the promotion of oxidative stress might affect cellular processes (Kaushal and Baitha 2019).

Because metal oxides require O_2 or sunlight to form reactive oxygen species, the northern side is rarely observed in AD. Temizel et al. studied the nano ZnO effect on municipal solid waste digestion using both traditional and bioreactor methods. Biogas production has been reduced by up to 15% in both conventional and bioreactor mode, presumably due to ZNO's inhibitory effect on methanogenic microbes (Wang et al. 2016).

Furthermore, over 98% of ZNOs can survive for a long time in the waste matrix, with an unknown fate and environmental impact. Mu et al. (2011) found the impact of ZnO NP on generation of biogas has no influence on methane production at a concentration of coenzymes F420 dependent on 6 mg/g-TSS 150, 30 and 1 mg zinc oxide reduced NPs/g to 66.2%, 89.8%, and 99.3%, respectively, of total stagnant water present.

The presence of mutant-producing CuO nanoparticles, a similar mechanism was found while reducing biogas by ZnO NP, mostly due to the existence of acetoclastic methane rather than hydrogen-consuming methane. Lunadelisisk et al. reported on

Table 3.2 Summary of application of carbon-based NP in AD

Carbon-based NP type	Size in nm	Main effect
G	6	Methane yield increased
GO	10	Methane yield decreased 7%
MWCNT	200	Methane yield increased 46.9%
CNT	2	VFA yield increased 49.2%
C60	200	Methane yield increased 33%

the deleterious impacts of both of the above-mentioned NPs on methane generation (e.g., ZnO, CuO), stressing their harmful effects on aerial species engaged in the AD process (Cuéllar and Webber 2008).

In their research, using 120 mg/L copper (II) oxide or zinc oxide to minimize biogas production at 36 °C was found to be effective. Despite the fact that studies have shown that microorganism adaptation to these nanoparticles from the medium, such as via bioremediation improves slightly in the last days of testing.

The size of NPs has an effect on Alzheimer's disease. Smaller NPs, to be more precise, are preferable. The addition of metallic and metallic oxide nanoparticles, like Knight, Fe, Co, Fe_3O_4, Fe_2O_3, MgO, and Al_2O_3, identified to improve the production of biogas and methane content in the literature, however, papers highlighting the inhibitory effects have also been published. When utilized in high concentrations, the production process comprises ZnO and TiO_2 (Table 3.2).

3.5 Role of Carbon Nano Particles in Anaerobic Ingestion Process

Carbon-based NPs' broad use in electronic chemistry and biochemistry eventually expands to environmental and wastewater treatment plants, causing environmental problems. As a result, the influence of biogas generation, carbon-based NP, AD, H_2S, and other biogas pollutants must be evaluated at different levels.

3.5.1 Influences on the Yield of Biogas

Dakshin and Bautha evaluated the effects of a mixture of biogas and CH_4 levels of GO NPs on food waste, wheat straw and fertilizer. According to the findings, introducing GO NPs increased methane output by 2.2 times in comparison with the control. Hao et al. (2019) looked at how fullerenes & multiwall carbon nanotubes at two concentrations of 60 g and 600 mg/kg affected biogas and methane production in sheep dung. When compared to the high concentration (600 mg/kg) control, multiwall carbon nanotubes and fullerenes yields increased by 47.4% and 34.5%, respectively (Kulkarni and Ghanegaonkar 2019).

The Brazilian sugar sector has also looked at the effects of MWCNTTs on production of biogas from wastewater. The addition of 1500 mg/g-Volatile Suspended Solids MWCNTS increased generation of biogas by 8.9%. The graphene NPs impact on the SLAD genogenesis process was proposed by Tian et al. (2017). This indicates that 30 mg/L and 120 mg/L graphene NP supplementation higher production of methane by 17.0% and 51.4%.

Direct intersecting electron transfer (DIET) can be facilitated by SWCNTS during the AD process. The addition of SWCNTS enhanced feedstock consumption and methane production rates compared to controls, while methane production yields were unchanged (14 million after 180 hours). DIET, on the other hand, permits microbes to exchange electrons without requiring direct configuration or hydrogen transfer.

As a result, materials such as electric vehicles (such as SWCNT) can act as ducts between electron-receiving and electron-contributing throughout the DIET process. Worryingly, in the presence of conductive material, methane synthesis is more efficient than hydrogen, DIET, or electron carriers recognized as linked electron transfer. A multiwall carbon nanotube (MWCNT) is another electrically conductive component that can boost production of biogas and methane by converting electrons to methanogen (Dang et al. 2016).

CNTs can considerably enrich electronic microorganisms like calorimeter SP in general. Geobacter SP in combination with other bacteria or methanogenic archaea. (methanosarcina, methenocyta, etc.). MWCNTs, for example, which include the top two phylloxera, bacterioiditis (11.2%) and pharmacites (0.4.4%) (Ambuchi et al. 2017), can enhance the bacterial community and enhance the environment. Ambuchi and colleagues found under mesophilic circumstance investigated the effects of MWCNTs on the AD of granular sludge from the 96-h beer business (Xu et al. 2014).

The findings showed that 1500 mg/L of MMWCNTS, the CH_4 generation increase was larger than in the reference sample (151.8 vs. 106 mL/g of VSS) at the conclusion of the experimental period. Due to probable interactions between MWCNT, bacteria, and microbial EPS, SEM scans showed that granules of that sludge from bioreactors treated with multiwall carbon nano tubes were reduced to microorganisms. Wang et al. looked the impact of MMWCNTS on the microbial activities and the structure in AGS for 110° in his recent study (Li et al. 2007).

MWCNTs had no influence on biogas output or COD reduction, but they did boost overall phosphorus removal by 29.34%. 16 S RNA gene sequencing suggests that MWCNTs have affected the structure and content of microorganisms but have had little effect on microbial diversity. The presence of MWCNTs, for example, relates to the potential for MWCNTs to have negative effects on specific microbial populations by substituting saccharobacteria protobacteria.

A variety of carbon-based additives can be used to boost the generation of ash-based nanomaterial to form biogas. Low and others (2012) got micro/nano bottom ash and micro/nano fly ash from solid waste incinerators and used them to boost biogas production. A significant rise in biogas generation can be noticed with varying concentrations of these ashes at 35 °C for 90 days' time. This observation can be linked to the expanded habitat of inorganic bacteria due to the rising nature of these nanocompounds. Long-term use of fullerene nanoparticles was found with no effect on biogas generation (Mihranyan et al. 2012).

3.5.2 Influences on the Stability of Oxygen Limiting Ingestion Method

The impacts of MWCNTS on the AD process sustainability of sugar beets effluent were also investigated by Ambuchi et al. (2017). Biogas production was improved with adding the 1500 mg/g-VSS MWCNTS, with no barrier impact, no significant difference in hydrogen generation, and faster use and removal efficiency of VFA when compared to greater COD control experiments. Tian et al. (2017) was looking at significance of graphene NPs on the sludge formed from the methanogenesis process. When compared to controls, introducing 30 mg/L and 120 mg/L graphene nanoparticles resulted in a rise in coenzyme F420 content. As coenzyme F420 levels rise, the rate at which acetate is used rises, and acetate is quickly converted to CH_4, increasing CH_4 synthesis.

Dong et al. (2019) studied the nanographene oxide effect on carbon conversion & methane production at 0.05 and 0.108 g Ng/g volatile solids. It has no influence on the nitrogen levels in the environment. Furthermore, the presence of coenzyme F420 was lowered when two concentrations of nanographene oxide were used. The authors believe that the reduction in methane formation caused by the addition of NGOs is due to the NGOs' strong organic layer synthesis, which was confirmed by Fourier spectroscopy (Demirel and Scherer 2011).

3.5.3 Influences on the Contaminants Like Hydrogen Sulphide for the Production of Biogas

Because of inhibition during AD, some animal manures and slurries include significant levels of ammonium (NH_4). In Alzheimer's disease, the use of carbon-based NPs can result in considerable improvements in function. ADB explored the impact of carbon nanotubes (CNTs) on resistant ammonia in Ian et al. The results demonstrate that methane yield is equivalent to all concentrations of CNT before ammonia addition and is better than VFAS service with VFAS concentration 3 g/L and 5 g/L CNT with 0 g/L and 1 g/L CNT. In comparison to pneumonia bodies, the rate of specific CH_4 generation reduced and the cumulative concentration of VFAs increased with the addition of ammonia.

3.5.4 Role of Mixture of Nano Particles in Anaerobic Ingestion Process

Combining NPSs is a new field in the AD process that takes advantage of their distinct characteristics. The effects of NPS mixes for AD performance on H2S content in specially produced biogas are the subject of this section of the review. Since AD, Hassanein et al. (2019) have looked at the concentration of H2S in chicken litter as well as the effect of NP mixes on biogas production. High to low NPS blends reduced H2S production by 100%, 71.9%, 40.9%, and 11.9%, respectively.

The inclusion of G1 and G2 reduced H2S generation by 150.9% compared to the control and increased it by 163.6%. It also cuts down on the time it takes to reach maximum biogas production. After 30 days, however, the VS decrease is 1.14 times greater than the G1 and G2 controls (Yadvika et al. 2004).

Eduok et al. (2017) investigated the impacts of experienced engineer nanoparticles on hydrogen sulphide reduction and methane generation from WAS AD. The alloys employed were Ag_2O (20 nm), TiO_2 (20 nm), and ZnO (21 nm) alloys with dry weight concentrations of 250, 2000, and 2800 mg/kg. It showed that applying tailored NP mixes did not result in marginal rise in methane emission when compared to the control system. The application of selected NP mixes, however, lowered H2S by two times in comparison with the reference, demonstrating that sulfide-reducing microorganisms have an inhibiting effect. The designed NP mixture also increased the quantity of VFAs by 1.1 times (Nzila 2017).

Amen et al. (2017) used nZVI and zolite mixes and nZVI-coated zolite to investigate overall eddy performance of household sludge. ICZs were separated into two groups based on iron concentration: 500 mg/L and 1000 mg/L. Biogas output increased by 130.8%, 149.9%, and 286.7%, respectively, when compared to IMZ, ICZ500, and ICZ1000. After 14 days, all bioreactors showed lower ammonia levels and a higher tendency for COD and total alkalinity throughout digestion than the control.

3.5.5 Impact on Environment on Biogas Generation and Nanoparticle Recovery Technologies Within the AD System

The usage of nanoparticles in a various industry has resulted in significant environmental issues. Many studies have looked into ways to decrease the environmental risk of nanoparticles in biogas production, as well as approaches to recover nanoparticles from AD. As a result, this section of the review concentrates on techniques for lowering the environmental risk of nanoparticles in biogas production as well as strategies for recovering nanoparticles from AD (Dong et al. 2019).

3.5.6 Limit the Effects of Nanomaterials on Biogas Generation on the Atmosphere

The measures to mitigate the environmental risk of nanoparticles in production of biogas are mentioned below. (1) Because NPC-containing wastes used as an alternative to using NP in AD process (Wang et al. 2016); (2) ZNS Digest can restore replacement using agricultural fertilizers by treating few nanoparticles such as Q and ZN through AD process; (3) From natural sources nanoparticles are prepared.

3.5.7 Recovery of Nanomaterial in the Anaerobic Digestion Reactor

Several studies have described methods for recovering NPs within the AD reactor and identifying nanoparticles when the AD process has finished. Hassanein et al. (2019) looked into different methods for recovering NPs from an AD reactor.

The techniques of recovery were for the test:

(1) Liquid and hard samples were taken from the four reactors mentioned after AD, dried at 35 °C for 20 min, (2) liquid and hard samples were taken from the four reactors after AD, dried at 35 °C for 20 min, (3) Magnetic stray bars covered in plastic parafilm were used to collect fluid and hard samples from the interiors of the reactors. The results demonstrate that of the three methods tried, recovering metallic NPs with a magnetic string bar wrapped in plastic parafilm was the only one that worked.

Thousand mg/L of concentration, iron nanoparticles were identified in all NMSPs, and their size was calculated. Low nanoparticles were not identified in any of the treatments because to low concentrations, while neon NPs were only found at a concentration of 120 mg/L. The third strategy is preferred because it combines NP recovery with post-AD tracking, whereas the first and second techniques can only track nanoparticles after AD (Eduok et al. 2017).

3.5.8 Analysis of Energy and Costs Based on NPs Supply

The cost analysis and energy balance were used to explore the effects of various nanoparticles on biogas output. It was calculated by using the quantity of biogas produced from 1 m^3 of fertilizer. Biogas is a renewable source of energy. It demonstrates that the maximum energy level of biogas is 1000 mg/L Iron nanoparticles & 403.0 kWh, with a profit of 676.5 American dollar, but the lowest is 5 mg/L Iron nanoparticles, when compared to the 192.6 kW reference. Nanoparticles, on the other hand, not gained any for high concentrations of Iron and Iron oxide nanoparticles (>20 mg/L), but fared well when compared to the density range (0.5–2.0 mg/L) reference.

The three types of NPs employed in the AD process were metallic oxide NP, ZVNP, and carbon-based nanoparticle. Adding metal oxide nanoparticles has a variety of effects on biogas production due to the density, kind, and size of NPs. The combination of ZnO and CeO_2 has a considerable inhibitory effect on methanogenic bacteria.

The Fe_3O_4 and TiO_2 addition to biogas rates has demonstrated to be beneficial. Furthermore, TiO_2 demonstrated a 2.5-fold reduction in H_2S generation compared to the control approach. Adding zero-valent nanoparticles boosted biogas generation, particularly in non- nanoparticles, although lower cobalt (0.5 mg/L) & gold (10 mg/g-TSS) concentrations enhanced methane, whereas Co (2 mg/L) increased. High quantities of NPs in agriculture (500 mg/g-TsS) in CH_4 generation. Furthermore, zero-valent NPs decompose Total Solids, Volatile Solids, VFA, and COD, resulting in a beneficial effect on ammonia concentration, VFA uptake, and chemical oxygen demand using carbon-based nanoparticles as in comparison with the reference (Yan et al. 2019).

Because of its unique characteristics, the nanoparticles combination provides advantages, and it is a new research topic in the AD approach. H_2S generation is completely inhibited depending on the concentration of the nanoparticle's mixture. The use of waste nanoparticles generated by natural nanoparticles, as well as improved nanoparticle recovery techniques inside AD reactors, can reduce the environmental risk of nanoparticles in other industries. For high density, Fe_3O_4 NP (20 mg/L) and Fe provided little benefit, and their inclusion was judged a high economic burden for AD systems. When compared to the reference, NPs with a wide range of concentrations (0.5–2 mg/L) produced significant net benefits.

Some possible future consequences of the foregoing analysis are as follows: (1) the application of NPS mixtures to understand their behavior in relation to impurities concentrations in biogas; (2) the method of recovering nanoparticles after digestion can be used to assess the environmental risk in AD process; (3) the method of recovering nanoparticles after digestion, can be used to assess the environmental risk in AD process (Yang et al. 2012).

Overall, the nanoparticle cost determines the economic viability of a large-scale nanoparticles enhanced AD process. Furthermore, there are some environmental dangers associated with using NP in any business (including the biogas business) that must be addressed. As a result, it is prudent to investigate AD as a bio mediating approach for the treatment of contaminated mud in NPs in order to be safe. In this context, future study should focus on the impact of various NPs on the Alzheimer's disease process. Two or more varieties of NP were used simultaneously, for example, and potential interactions and beneficial effects should be investigated. In addition, the presence of NPS supposition and residue of chemicals formed from NP or NP during the AD process should be recognized in the digest to avoid pollution of agricultural land (Farghali et al. 2020).

The use of digests from NPS-augmented AD (with specific NPs) to boost the growth of specific agricultural crops is another area of future research. Furthermore, determining the rate of absorption of various NPs in the form of intact NPs or their derivatives by plants and their deposit sites from polluted soil or water would be interesting (e.g., stalks, roots, fruits, leaves).

In this case, NPs-contaminated plant parts could be used as biogas boosting additives in the AD process to figure out the exact dosage.

3.6 Applications of Nanomaterials in Biohydrogen Production

Because of their unique properties, nanomaterials can assist improve the overall biomass to biohydrogen conversion method (Kumar et al. 2019; Pugazhendhi et al. 2019). Every phase of cellulosic biohydrogen production technology can be catalyzed by strong electroconductivity, increased surface area, and a high surface to volume ratio (Taherdanak et al. 2015). Biohydrogen production has been shown to be boosted by a number of nanomaterials, including Copper, Gold, Silver, Iron, Nickel, and Titanium, via a variety of biological processes (Taherdanak et al. 2015; Lin et al. 2016a, b). Nanomaterials, particularly Fe and Ni, act as cofactors in the active areas of nitrogenase and hydrogenase enzymes, boosting biohydrogen generation (Taherdanak et al. 2015; Lin et al. 2016a, b).

In the fermentation process, nanoparticles operate as oxygen scavengers, absorbing excess oxygen and reducing the oxidation-reduction potential. This provides the hydrogenase enzyme with the ideal anaerobic environment, resulting in increased biohydrogen generation. Using nanoparticles to pre-treat lignocellulosic biomass can improve lignin removal while also boosting sugar output (Wei et al. 2015). NPs can also increase synthesis of cellulase enzymes and their heat and pH stability, allowing for more efficient cellulose hydrolysis and sugar conversion (Srivastava et al. 2017a, b, c; Ladole et al. 2017; Bilal et al. 2018).

Due to their unique physicochemical qualities, nanoparticles are getting a clear view to enhance the production process of biohydrogen, and are thus capable of having a substantial impact on biohydrogen generation (Beckers et al. 2013). Nanoparticles have a considerable impact on the metabolic activity of microorganisms by increasing effective electron transfer to acceptors, resulting in improved productivity and biohydrogen productivity (Patel et al. 2018). Because of its small size and large surface area, the enzyme interacts with the substrate and catalyzes the biohydrogen production (Vaghari et al. 2016; Sekoai et al. 2019).

Furthermore, nanomaterials with enhanced surface area and quantum size have ability to absorb and increase the rate of transfer between nanoparticles and enzymes, resulting in higher H_2 production (Patel et al. 2018). Furthermore, in biological processes for biohydrogen generation, two enzymes are essential for structural and functional integrity: hydrogenase and nitrogenase, both of which have Fe and Ni as metallic cofactors at their active sites (Engliman et al. 2017). Once delivered to bioreactor, nanomaterials, particularly magnetic nanomaterials such as Fe and/or Ni, goes to the active sites of these enzymes, enhancing biohydrogen generation (Mohanraj et al. 2014; Patel et al. 2018).

As a photo-fermentation substrate, organic acids formed as a byproduct of dark fermentative biohydrogen generation can be used (Zagrodnik and Łaniecki 2017; Nikolaidis and Poullikkas 2017; da Silva Veras et al. 2017). The nanoparticles can increase the yield of photo-fermentative biohydrogen synthesis (Dolly et al. 2015). In addition to fermentative biohydrogen synthesis, nanomaterials used as catalyst in biohydrogen generation (Nagarajan et al. 2017; da Silva Veras et al. 2017). In microalgae and cyanobacteria, photocatalytic nanomaterials like TiO_2, SiC nanomaterials to enhance photosystem light absorption, which is essential for breaking water molecules and biohydrogen generation (Giannelli and Torzillo 2012). NPs bind to these enzymes' active sites and participate in the ETC, which speeds up reaction and boosts biohydrogen production. Giannelli and Torzillo (2012) reported that using *Chlamydomonas reinhardtii* in a 50 L tubular photobioreactor in the presence of light nanoparticle (silica) solution, biohydrogen production was increased to over 850 mL, about 1.7 times higher than control, at a production rate of 0.17 mL/L/h. Nanomaterials improve not only biohydrogen output and productivity, but also enzyme reusability, allowing enzymes to maintain their activity after numerous cycles, completing speed up the reactions, and more cost-effective.

3.6.1 Biohydrogen Production

Due to its low cost and environmentally friendly character, biohydrogen production via biological processes has long been believed to be a particularly promising technology (Engliman et al. 2017; Srivastava et al. 2018). Biohydrogen made from cheap organic waste found in nature, which is abundant and renewable (Gasparatos et al. 2011; Thakur et al. 2014; Colmenares et al. 2017). All Biohydrogen generation's functional feasibility can solve various concerns, environmental sustainability and waste management (Nigam and Singh 2011). Two ways for creating biological hydrogen are bacterial fermentation and cyanobacteria-based bio-photolysis (Harish et al. 2015). Dark fermentation has more popularity and is currently recognized as an effective technique for biohydrogen production (Azman et al. 2016).

Though dark fermentation allows for improved biohydrogen synthesis, the method's main drawbacks have been recognized as low yield and insufficient energy recovery from organic substrates. Nonetheless, cellulose-rich organic substrate is the preferred feedstock for dark fermentation biohydrogen production; the waste's complex structure of hemicellulose, lignin, and cellulose must be broken for bioconversion into fermentable sugars for further biohydrogen generation (Bajaj et al. 2014; Balat and Balat 2009). According to Fontes and Gilbert (2010), microbial processing can hydrolyze 1011 tons of plant biomass and generate enough energy to power. However, biomass must be processed to separate lignin and hemi cellulosic barriers and release cellulose for enzymatic hydrolysis, which is one of the rate-limiting processes of biohydrogen synthesis (Gupta et al. 2016). Peng et al. 2009; Srivastava et al. 2017a, b, c). Chemical pretreatment is widely used method to eliminate lignin from biomass, cost of doing so limits the pilot size reaction (Dawson 2005). Furthermore, inefficiency and cellulase enzyme high cost have a significant impact on the economic viability of biohydrogen generation (Wahono et al. 2014). According to present discoveries, successfully convert cellulosic biomass into fermentable sugars for biohydrogen production, a low-cost technique for pre-treating it as well as its enzymatic hydrolysis is required.

Fermentation biohydrogen synthesis requires boosting enzymes that affect the dark fermentation response in addition to the substrate (Ansari and Husain 2012). Because fermentative biohydrogen generation is a microbially mediated process, hydrogenase and nitrogenase, which govern dark fermentation directly (Sompong et al. 2008; Bastidas-Oyanedel et al. 2015). Improvements in these enzymes, as well as the addition of metal elements as a co-factor, are required to address the current limits of biohydrogen-generating technology (Chong et al. 2009). Biohydrogen synthesis is likewise reliant on bacterial multiplication because biohydrogen enzymes are microbial enzymes. (Sompong et al. 2008; Bastidas-Oyanedel et al. 2015).

3.6.2 Hydrogen Production Via Dark Fermentation

To explore if Au/Fe–Zn–Mg–Al–O hydrotalcites-supported Au catalysts could change biohydrogen generation, Wimonsong et al. (2013) created and tested a number of them. When the catalyst was used, the Au/Zn–Mg–Al hydrotalcite produced hydrogen yield of 2.74 mol hydrogen per mol of sucrose. When compared to a control group that did not utilize nanoparticles, the use of ferric oxide nanoparticles increased H_2 generation from *Enterobacter aerogenes* by 17%. According to their findings, adding 200 mg/L FeO nanoparticles boosted H_2 synthesis via an acetate-mediated route while reducing ethanol production. Additionally, images taken with SEM showed the existence of bacterial nanowires, which are responsible an effective electron exchanges with neighboring microbial cells (Lin et al. 2016a, b).

According to Gadhe et al. (2015), the addition of Ni and hematite NPs process for the production of hydrogen from dairy effluent increased the process efficiency by over 27% when compared to the individual nanobiocatalysts. Adding ZVFe-activated carbon micro-electrolysis to a mixed consortia also encouraged *Clostridium sp.* enrichment and showed 50% increase in H_2 output by Zhang and Shen (2007). Wang and Wan (2009) improved the hematite nanoparticle concentration (0–1600 mg/L) on production of hydrogen from sucrose with the pH (4–10.0), and at optimum pH of 6, 3.57 mol H_2/mol of sucrose. Furthermore, TEM study of the gradual discharge of hematite nanoparticles indicated changes in the morphology of bacteria as the length rose from 2.0–3.6 m to around 2.6–5.6 m. Furthermore, Zhang and Shen (2007) discovered that the size of the biocatalysts used had a substantial impact on the mixed consortia's ability to produce hydrogen from sucrose. The inclusion of 5 nm gold particles improved the bioactivity of hydrogen-producing bacteria with an HY of 4.47 mol H_2/mol sucrose. Mullai et al. found that adding a small amount of nickel nanoparticles (5.67 mg/L) boosted hydrogen generation efficiency by 22.7%, resulting in an HY of 2.54 mol H2/mol glucose. Hsieh et al. (2016) used RTPCR to monitor hydrogenase enzyme activity on hydrogen production by *Clostridium pasteurianum* after adding the of Iron-mediated nanoparticles in the medium. The addition of Iron-mediated nanoparticles to the medium stimulated H_2 production, but only marginally. As a result, the capacity to track hydrogen producer activity following nanoparticle addition is critical for a better understanding of the stimulatory impacts of nano-mediated biocatalysts on biohydrogen production.

3.6.3 Fe and FeO -NPs in Biohydrogen Generation

In dark fermentation, hydrogenase is an enzyme that creates hydrogen under ideal conditions. [Fe-Fe] hydrogenase is one example of a cluster form (Mullai et al. 2013). Only a few bacteria species have been shown to produce [Fe-Fe]- hydrogenase when exposed to high Fe NP concentrations (Bao et al. 2013). When 200 mg/L hematite nanoparticles added, the amount of hydrogen produced has risen by 32.6% (Han et al. 2011). According to studies, the ZV Iron Oxide nanoparticles can lower DO content in the medium, increasing the oxygen-sensitive hydrogenase enzyme efficiency. When compared to control samples, 150 mg/L concentration of $FeSO_4$ improved the hydrogen generation yield as much as 163%. Starch was also fermented in a batch fermenter utilizing anaerobic sludge in dark fermentation conditions (Yang and Shen 2006). *Enterobacter cloacae* was used in glucose fed fermentation systems in the absence of light and with 125 mg/L FeO NPs. Biohydrogen was created at a rate of 258 mL/g VS during the test (Mohanraj et al. 2014). In the cassava starch dark fermentation with *Enterobacter aeruginos*, FeO NPs were found to be effective, generating 192.4 $\pm$ 1.14 mL/g H_2. The conductivity was high for ferric oxide nanoparticles would boost electron transfer rates, effective in H_2 production, according to the study (Lin et al. 2016b). A recent study combined 200 mg/L FeO NPs/carbon nanoparticles with glucose and mixed bacteria under anaerobic fermentation conditions to produce 218.63 mL hydrogen/g glucose (Zhang et al. 2018).

3.6.4 Synthesis of Biohydrogen by Nickel Nanoparticles

[NieFe]-hydrogenase is a cluster that prohibits dark fermentation of substrate by hydrogenase enzymes, resulting in biohydrogen production. Bacteria generating [NieFe] hydrogenase may successfully produce biohydrogen by nickel nanoparticles in dark fermentation (Bao et al. 2013). Nickel nanoparticles reported for the rise in hydrogen production with hydrogenases in general. Mullai et al. (2013) discovered that adding 5.67 mg/L of Ni NPs to dark hydrogen fermentation with glucose boosted the yield by 22.7%. At 200 mg/L concentration of nickel nanoparticles were employed to form molasses into 1.30 mol hydrogen per mol hexose by 4.8% (Engliman et al. 2017).

3.6.5 TiO_2 Nanoparticles Are Used to Produce Biohydrogen

According to current research, using 100 mg/L concentration of titanium dioxide nanoparticles increased H_2 generation by up to 46.1%. Titanium dioxide nanoparticles boosted biohydrogen generation by breaking down proteins and polysaccharides into smaller organic compounds bits that hydrogen-producing bacteria could easily consume. TiO_2 also improved nitrogenase activity while decreasing hydrogen uptake by the hydrogenase enzyme (Zhao and Chen 2011).

3.6.6 Biohydrogen Can Be Produced Utilizing a Variety of Nanoparticles

Dark fermentation produced around 104.75 ± 12.39 mL of hydrogen per gram in anaerobic baffled fermenter injected with immobilized sludge by maghemite nanoparticles. 200 mg/L concentration, Fe_2O_3/C nanoparticles increased H_2 production by 33.7%, implying that FOC nanoparticles can enhance the activity of hydrogenase enzyme and speed up the dark fermentation (Zhang et al. 2018). Co-precipitation was employed in a study to make M^{2+-}Mg-Al HT materials. A maximum hydrogen output of 2.30 ± 0.37 mol hydrogen per mol was found in concentration range of 0.833 mg/L of Fe-Zn-Mg-Al HTs, which was 44% greater than the control sample by Wimonsong et al. (2013). To increase biohydrogen production, Le and Nitisoravut (2015) employed Ni-Mg-Al HT at concentrations ranging from 83.417 mg/L. Using sucrose as the substrate, approximately 3.37 ± 0.17 of hydrogen per mol of H_2 was produced at 250 mg/L concentration of Ni-Mg-Al and HT. This is due to the simplicity with which magnesium and nickel can be used to aid electron transport. Using 60 mg/L Ni-Ni graphene nanocomposite, researchers were able to produce 41.28 ± 1.69 mL/g of Chemical Oxygen Demand biohydrogen in dark fermentation (Ni-NiGr NC). In this work, industrial wastewater containing monoethylene glycol is used to generate about 105% biohydrogen. In addition to Fe, Ti, and Ni NPs, other nanocomposites could be capped or coated to allow for high biohydrogen synthesis (Elreedy et al. 2017).

3.6.7 Enhancement of Biohydrogen Production in Dark Fermentation with Nano Zero-Valent Iron Implanted on Chitosan

The nZVI/CS used as a catalyst improved production of biohydrogen at Sewwandi and Nitisoravut 2020. *Enterobacter aerogenes* carried conducted dark fermentation utilizing glucose at 37 °C in a mesophilic environment. The presence of nZVI/CS was confirmed by TEM examination. 243 mL H_2/g glucose was the greatest biohydrogen yield of 30% greater than reference. With a starting pH of 6.8, the optimum Chitosan-supported nano zero-valent iron concentration was 150 mg/L. The acetic route was used to make biohydrogen based on the distribution of volatile fatty acids, resulting in a favorable relationship between H_2 yield and CH_3COOH content (VFA) (Table 3.3).

3.7 Conclusion

Nanomaterials are being manufactured and employed in a variety of industries, which was grown in recent years. Different techniques that it could be utilized to characterize the required size, shape, and property of nanomaterials, which can then be used for a variety of applications, including biomedicine, electronic storage

Table 3.3 Summary of literature for production of biohydrogen by using nanomaterial

Sl.	Nanomaterial used	Biocatalyst	Biohydrogen yield	References
1	Hematite NPs	*Clostridium butyricum*	3.21 mole of hydrogen per mole of sucrose	Han et al. (2011)
2	Fe_2O_3-Fe_3O_4/ carbon nanocomposite	Anaerobic mixed bacteria	Percentage yield of hydrogen approximately 33.7% more	Zhang et al. (2018)
3	Hematite NPs and NiO- NPs	Thermophilic mixed bacteria	Yield percentage of hydrogen was approximately 34.38% more and for α-Fe_2O_3 NPs was approximately 5.47% more for NiO NP	Engliman et al. (2017)
4	Maghemite NPs	Hydrogenase enzyme	Hydrogen production upto 33.7%	Zhang et al. (2018)
5	γ-Fe_2O_3 NPs	*Enterobacter aerogenes*	Yield percentage of hydrogen approximately 21% more	Mohanraj et al. (2014)
6	Ni NPs	*Anaerobic microflora*	Hydrogen yield 22.71% more than reference	Mullai et al. (2013)
7	Nano zero-valent iron	*Enterobacter aerogenes*	Biohydrogen yield was 243 mL H_2/g glucose which was 30% higher	Sewwandi and Nitisoravut (2020)

devices, and sensors. The use of nanomaterials in the production of biogas and biohydrogen was reviewed in this review paper. In this article, the effects of metal oxide and metallic NP on the production of biogas and biohydrogen were explored, as well as the influence of metal oxide and metallic NP on the production of biogas and biohydrogen. The potential for employing NPs in real-world industrial settings to improve biohydrogen production output and quality.

References

Abdelsalam E, Samer M, Attia YA, Abdel-Hadi MA, Hassan HE, Badr Y (2016) Comparison of nanoparticles effects on biogas and methane production from anaerobic digestion of cattle dung slurry. Renew Energy 87:592–598

Abdelsalam E, Samer M, Attia YA, Abdel-Hadi MA, Hassan HE, Badr Y (2017a) Effects of Co and Ni nanoparticles on biogas and methane production from anaerobic digestion of slurry. Energy Convers Manag 141:108–119

Abdelsalam E, Samer M, Attia YA, Abdel-Hadi MA, Hassan HE, Badr Y (2017b) Influence of zero valent iron nanoparticles and magnetic iron oxide nanoparticles on biogas and methane production from anaerobic digestion of manure. Energy 120:842–853

Ahmad A, Mukherjee P, Senapati S, Mandal D, Khan MI, Kumar R, Sastry M (2003) Extracellular biosynthesis of silver nanoparticles using the fungus Fusarium oxysporum. Colloids Surf B: Biointerfaces 28(4):313–318

Ahmed F, Rodrigues DF (2013) Investigation of acute effects of graphene oxide on wastewater microbial community: a case study. J Hazard Mater 256:33–39

Ali A, Mahar RB, Soomro RA, Sherazi STH (2017) Fe3O4 nanoparticles facilitated anaerobic digestion of organic fraction of municipal solid waste for enhancement of methane production. Energy Sources Part A: Recov Util Environ Eff 39(16):1815–1822

Ambuchi JJ, Zhang Z, Shan L, Liang D, Zhang P, Feng Y (2017) Response of anaerobic granular sludge to iron oxide nanoparticles and multi-wall carbon nanotubes during beet sugar industrial wastewater treatment. Water Res 117:87–94

Amen TW, Eljamal O, Khalil AM, Matsunaga N (2017) Biochemical methane potential enhancement of domestic sludge digestion by adding pristine iron nanoparticles and iron nanoparticles coated zeolite compositions. J Environ Chem Eng 5(5):5002–5013

Amen TW, Eljamal O, Khalil AM, Sugihara Y, Matsunaga N (2018) Methane yield enhancement by the addition of new novel of iron and copper-iron bimetallic nanoparticles. Chem Eng Process 130:253–261

Ansari SA, Husain Q (2012) Potential applications of enzymes immobilized on/in nano materials: a review. Biotechnol Adv 30(3):512–523

Azman, N. F., Abdeshahian, P., Kadier, A., Al-Shorgani, N. K. N., Salih, N. K., Lananan, I., . . . & Kalil, M. S. (2016). Biohydrogen production from de-oiled rice bran as sustainable feedstock in fermentative process. Int J Hydrog Energy, 41(1), 145–156.Ding, J., Wang, X,

Bajaj BK, Sharma M, Rao RS (2014) Agricultural residues for production of cellulase from Sporotrichum thermophile LAR5 and its application for saccharification of rice straw. J Mater Environ Sci 5(5):1454–1460

Balat M, Balat M (2009) Political, economic and environmental impacts of biomass-based hydrogen. Int J Hydrog Energy 34(9):3589–3603

Banerjee M, Dey B, Talukdar J, Kalita MC (2014) Production of biodiesel from sunflower oil using highly catalytic bimetallic gold–silver core–shell nanoparticle. Energy 69:695–699

Baniamerian H, Isfahani PG, Tsapekos P, Alvarado-Morales M, Shahrokhi M, Vossoughi M, Angelidaki I (2019) Application of nano-structured materials in anaerobic digestion: current status and perspectives. Chemosphere 299:188–199

Bao MD, Su HJ, Tan TW (2013) Dark fermentative bio-hydrogen production: effects of substrate pre-treatment and addition of metal ions or L-cysteine. Fuel 112:38–44

Bastidas-Oyanedel JR, Bonk F, Thomsen MH, Schmidt JE (2015) Dark fermentation biorefinery in the present and future (bio) chemical industry. Rev Environ Sci Biotechnol 14(3):473–498

Batstone DJ, Jensen PD (2011) Anaerobic processes, treatise on water science. Academic, Oxford, pp 615–640

Bazylinski DA, Garratt-Reed AJ, Frankel RB (1994) Electron microscopic studies of magnetosomes in magnetotactic bacteria. Microsc Res Tech 27(5):389–401

Beckers L, Hiligsmann S, Lambert SD, Heinrichs B, Thonart P (2013) Improving effect of metal and oxide nanoparticles encapsulated in porous silica on fermentative biohydrogen production by Clostridium butyricum. Bioresour Technol 133:109–117

Bilal S, Ali L, Khan AL, Shahzad R, Asaf S, Imran M et al (2018) Endophytic fungus Paecilomyces formosus LHL10 produces sester-terpenoid YW3548 and cyclic peptide that inhibit urease and α-glucosidase enzyme activities. Arch Microbiol 200(10):1493–1502

Carpenter AW, Laughton SN, Wiesner MR (2015) Enhanced biogas production from nanoscale zero valent iron-amended anaerobic bioreactors. Environ Eng Sci 32(8):647–655

Cavas M, Gupta RK, Al-Ghamdi AA, Gafer ZH, El-Tantawy F, Yakuphanoglu F (2013) Preparation and characterization of dye sensitized solar cell based on nanostructured Fe2O3. Mater Lett 105:106–109

Chandran SP, Chaudhary M, Pasricha R, Ahmad A, Sastry M (2006) Synthesis of gold nanotriangles and silver nanoparticles using Aloevera plant extract. Biotechnol Prog 22(2): 577–583

Chong ML, Sabaratnam V, Shirai Y, Hassan MA (2009) Biohydrogen production from biomass and industrial wastes by dark fermentation. Int J Hydrog Energy 34(8):3277–3287

Colmenares JC, Varma RS, Nair V (2017) Selective photocatalysis of lignin-inspired chemicals by integrating hybrid nanocatalysis in microfluidic reactors. Chem Soc Rev 46(22):6675–6686

Cuéllar AD, Webber ME (2008) Cow power: the energy and emissions benefits of converting manure to biogas. Environ Res Lett 3(3):034002

da Silva Veras T, Mozer TS, da Silva César A (2017) Hydrogen: trends, production and characterization of the main process worldwide. Int J Hydrog Energy 42(4):2018–2033

Dang Y, Holmes DE, Zhao Z, Woodard TL, Zhang Y, Sun D, Wang LY, Nevin KP, Lovley DR (2016) Enhancing anaerobic digestion of complex organic waste with carbon-based conductive materials. Bioresour Technol 220:516–522

Dawson L (2005) Optimization of pretreatment conditions for high efficiency ethanol production from post-harvest sugarcane residue (Doctoral dissertation, MS thesis submitted to Nicholls State University, Thibodaux, LA)

Demetzos C (2016) Pharmaceutical nanotechnology: fundamentals and practical applications. Springer, Puchong

Demirel B, Scherer P (2011) Trace element requirements of agricultural biogas digesters during biological conversion of renewable biomass to methane. Biomass Bioenergy 35(3):992–998

Dolly S, Pandey A, Pandey BK, Gopal R (2015) Process parameter optimization and enhancement of photo-biohydrogen production by mixed culture of Rhodobacter sphaeroides NMBL-02 and Escherichia coli NMBL-04 using Fe-nanoparticle. Int J Hydrog Energy 40(46):16010–16020

Dong B, Xia Z, Sun J, Dai X, Chen X, Ni BJ (2019) The inhibitory impacts of nano-graphene oxide on methane production from waste activated sludge in anaerobic digestion. Sci Total Environ 646:1376–1384

Dutz S, Hergt R, Mürbe J, Müller R, Zeisberger M, Andrä W, Töpfer J, Bellemann ME (2007) J Magn Magn Mater 308:305–312

Eduok S, Ferguson R, Jefferson B, Villa R, Coulon F (2017) Aged-engineered nanoparticles effect on sludge anaerobic digestion performance and associated microbial communities. Sci Total Environ 609:232–241

Elreedy A, Ibrahim E, Hassan N, El-Dissouky A, Fujii M, Yoshimura C, Tawfik A (2017) Nickel-graphene nanocomposite as a novel supplement for enhancement of biohydrogen production from industrial wastewater containing mono-ethylene glycol. Energy Convers Manag 140:133–144

Engliman NS, Abdul PM, Wu SY, Jahim JM (2017) Influence of iron (II) oxide nanoparticle on biohydrogen production in thermophilic mixed fermentation. Int J Hydrog Energy 42(45): 27482–27493

Faivre D, Schuler D (2008) Magnetotactic bacteria and magnetosomes. Chem Rev 108(11): 4875–4898

Fan R, Chew SW, Cheong VV, Orner BP (2010) Fabrication of gold nanoparticles inside unmodified horse spleen apoferritin. Small 6(14):1483–1487

Farghali M, Andriamanohiarisoamanana FJ, Ahmed MM, Kotb S, Yamashiro T, Iwasaki M, Umetsu K (2019) Impacts of iron oxide and titanium dioxide nanoparticles on biogas production: hydrogen sulfide mitigation, process stability, and prospective challenges. J Environ Manag 240:160–167

Farghali M, Andriamanohiarisoamanana FJ, Ahmed MM, Kotb S, Yamamoto Y, Iwasaki M, Yamashiro T, Umetsu K (2020) Prospects for biogas production and H2S control from the anaerobic digestion of cattle manure: the influence of microscale waste iron powder and iron oxide nanoparticles. Waste Manag 101:141–149

Fontes CM, Gilbert HJ (2010) Cellulosomes: highly efficient nanomachines designed to deconstruct plant cell wall complex carbohydrates. Annu Rev Biochem 79:655–681

Gabbasov R, Polikarpov M, Cherepanov V, Chuev M, Mischenko I, Lomov A et al (2015) Mössbauer, magnetization and X-ray diffraction characterization methods for iron oxide nanoparticles. J Magn Magn Mater 380:111–116

Gadhe A, Sonawane SS, Varma MN (2015) Enhancement effect of hematite and nickel nanoparticles on biohydrogen production from dairy wastewater. Int J Hydrog Energy 40(13): 4502–4511

Galloway JM, Talbot JE, Critchley K, Miles JJ, Bramble JP (2015) Developing biotemplated data storage: room temperature biomineralization of L10 CoPt magnetic nanoparticles. Adv Funct Mater 25(29):4590–4600

Gardea-Torresdey JL, Gomez E, Peralta-Videa JR, Parsons JG, Troiani H, Jose-Yacaman M (2003) Alfalfa sprouts: a natural source for the synthesis of silver nanoparticles. Langmuir 19(4): 1357–1361

Gasparatos A, Stromberg P, Takeuchi K (2011) Biofuels, ecosystem services and human wellbeing: putting biofuels in the ecosystem services narrative. Agric Ecosyst Environ 142(3–4):111–128

Giannelli L, Torzillo G (2012) Hydrogen production with the microalga Chlamydomonas reinhardtii grown in a compact tubular photobioreactor immersed in a scattering light nanoparticle suspension. Int J Hydrog Energy 37(22):16951–16961

Gonzalez-Estrella J, Sierra-Alvarez R, Field JA (2013) Toxicity assessment of inorganic nanoparticles to acetoclastic and hydrogenotrophic methanogenic activity in anaerobic granular sludge. J Hazard Mater 260:278–285

Gottschalk F, Sun T, Nowack B (2013) Environmental concentrations of engineered nanomaterials: review of modeling and analytical studies. Environ Pollut 181:287–300

Gräfe M, Donner E, Collins RN, Lombi E (2014) Speciation of metal (loid) s in environmental samples by X-ray absorption spectroscopy: a critical review. Anal Chim Acta 822:1–22

Güngör-Demirci G, Demirer GN (2004) Effect of initial COD concentration, nutrient addition, temperature and microbial acclimation on anaerobic treatability of broiler and cattle manure. Bioresour Technol 93(2):109–117

Gupta VK, Kubicek CP, Berrin JG, Wilson DW, Couturier M, Berlin A et al (2016) Fungal enzymes for bio-products from sustainable and waste biomass. Trends Biochem Sci 41(7): 633–645

Haddad PS, Britos TN, Li LM, Li LDS (2015) Preparation, characterization and tests of incorporation in stem cells of superparamagnetic iron oxide. J Phys Conf Series 617(1):012002. IOP Publishing

Han Y, Yan W (2016) Reductive dechlorination of trichloroethene by zero-valent iron nanoparticles: reactivity enhancement through sulfidation treatment. Environ Sci Technol 50(23):12992–13001

Han H, Cui M, Wei L, Yang H, Shen J (2011) Enhancement effect of hematite nanoparticles on fermentative hydrogen production. Bioresour Technol 102(17):7903–7909

Hao Y, Wang Y, Ma C, White JC, Zhao Z, Duan C, Zhang Y, Adeel M, Rui Y, Li G et al (2019) Carbon nanomaterials induce residue degradation and increase methane production from livestock manure in an anaerobic digestion system. J Clean Prod 240:118257

Harish BS, Ramaiah MJ, Uppuluri KB (2015) Bioengineering strategies on catalysis for the effective production of renewable and sustainable energy. Renew Sustain Energy Rev 51: 533–547

Hassanein A, Lansing S, Tikekar R (2019) Impact of metal nanoparticles on biogas production from poultry litter. Bioresour Technol 275:200–206

He S, Guo Z, Zhang Y, Zhang S, Wang J, Gu N (2007) Biosynthesis of gold nanoparticles using the bacteria Rhodopseudomonas capsulata. Mater Lett 61(18):3984–3987

Hergt R, Hiergeist R, Zeisberger M, Schüler D, Heyen U, Hilger I, Kaiser WA (2005) Magnetic properties of bacterial magnetosomes as potential diagnostic and therapeutic tools. J Magn Magn Mater 293(1):80–86

Hsieh PH, Lai YC, Chen KY, Hung CH (2016) Explore the possible effect of TiO2 and magnetic hematite nanoparticle addition on biohydrogen production by Clostridium pasteurianum based on gene expression measurements. Int J Hydrog Energy 41(46):21685–21691

Jia T, Wang Z, Shan H, Liu Y, Gong L (2017) Effect of nanoscale zero-valent iron on sludge anaerobic digestion. Resour Conserv Recycl 127:190–195

Kaushal R, Baitha R (2019) Biogas and methane yield enhancement using graphene oxide nanoparticles and Ca (OH) 2 pre-treatment in anaerobic digestion. Int J Ambient Energy 5:1–8

Krishnan KM (2016) Fundamentals and applications of magnetic materials. Oxford University Press

Kubickova, S., Vejpravova, J., Holec, P., & Niznansky, D. (2013). Correlation of crystal structure and magnetic properties of Co (1− x) NixFe2O4/SiO2 nanocomposites. J Magn Magn Mater, 334, 102–106

Kulkarni MB, Ghanegaonkar PM (2019) Methane enrichment of biogas produced from floral waste: a potential energy source for rural India. Energy Sources Part A: Recov Util Environ Eff 41(22): 2757–2768

Kumar AN, Bandarapu AK, Mohan SV (2019) Microbial electro-hydrolysis of sewage sludge for acidogenic production of biohydrogen and volatile fatty acids along with struvite. Chem Eng J 374:1264–1274

Ladole MR, Mevada JS, Pandit AB (2017) Ultrasonic hyperactivation of cellulase immobilized on magnetic nanoparticles. Bioresour Technol 239:117–126

Le DTH, Nitisoravut R (2015) Ni-Mg-Al Hydrotalcite for improvement of dark fermentative hydrogen production. Energy Procedia 79:301–306

Li XQ, Brown DG, Zhang WX (2007) Stabilization of biosolids with nanoscale zero-valent iron (nZVI). J Nanopart Res 9(2):233–243

Lin HN, Hu BB, Zhu MJ (2016a) Enhanced hydrogen production and sugar accumulation from spent mushroom compost by Clostridium thermocellum supplemented with PEG8000 and JFC-E. Int J Hydrog Energy 41(4):2383–2390

Lin R, Cheng J, Ding L, Song W, Liu M, Zhou J, Cen K (2016b) Enhanced dark hydrogen fermentation by addition of ferric oxide nanoparticles using Enterobacter aerogenes. Bioresour Technol 207:213–219

Mihranyan A, Ferraz N, Strømme M (2012) Current status and future prospects of nanotechnology in cosmetics. Prog Mater Sci 57(5):875–910

Mohanraj S, Kodhaiyolii S, Rengasamy M, Pugalenthi V (2014) Phytosynthesized iron oxide nanoparticles and ferrous iron on fermentative hydrogen production using Enterobacter cloacae: evaluation and comparison of the effects. Int J Hydrog Energy 39(23):11920–11929

Molcan M, Gojzewski H, Skumiel A, Dutz S, Kovac J, Kubovcikova M et al (2016) Energy losses in mechanically modified bacterial magnetosomes. J Phys D Appl Phys 49(36):365002

Mu H, Chen Y, Xiao N (2011) Effects of metal oxide nanoparticles (TiO2, Al2O3, SiO2 and ZnO) on waste activated sludge anaerobic digestion. Bioresour Technol 102(22):10305–10311

Mullai P, Yogeswari MK, Sridevi KJBT (2013) Optimisation and enhancement of biohydrogen production using nickel nanoparticles—a novel approach. Bioresour Technol 141:212–219

Nagarajan D, Lee DJ, Kondo A, Chang JS (2017) Recent insights into biohydrogen production by microalgae–from biophotolysis to dark fermentation. Bioresour Technol 227:373–387

Nigam PS, Singh A (2011) Production of liquid biofuels from renewable resources. Prog Energy Combust Sci 37(1):52–68

Nikolaidis P, Poullikkas A (2017) A comparative overview of hydrogen production processes. Renew Sustain Energy Rev 67:597–611

Nzila A (2017) Mini review: update on bioaugmentation in anaerobic processes for biogas production. Anaerobe 46:3–12

Pandian M, Marimuthu R, Natesan G, Rajagopal RE, Justin JS, Mohideen AJAH (2013) Development of biogenic silver nano particle from Pelargonium graveolens leaf extract and their antibacterial activity. Am J Nanosci Nanotechnol 1(2):57–64

Patel SKS, Lee J-K, Kalia VC (2018) Beyond the theoretical yields of dark-fermentative biohydrogen. Indian J Microbiol 58(4):529–530

Peng F, Ren JL, Xu F, Bian J, Peng P, Sun RC (2009) Comparative study of hemicelluloses obtained by graded ethanol precipitation from sugarcane bagasse. J Agric Food Chem 57(14): 6305–6317

Pieta IS, Epling WS, Kazmierczuk A, Lisowski P, Nowakowski R, Serwicka EM (2018) Waste into fuel-catalyst and process development for MSW valorisation. Catalysts 8(3):113

Pugazhendhi A, Shobana S, Nguyen DD, Banu JR, Sivagurunathan P, Chang SW et al (2019) Application of nanotechnology (nanoparticles) in dark fermentative hydrogen production. Int J Hydrog Energy 44(3):1431–1440

Ramesha GK, Kumara AV, Muralidhara HB, Sampath S (2011) Graphene and graphene oxide as effective adsorbents toward anionic and cationic dyes. J Colloid Interface Sci 361(1):270–277

Reddy LH, Arias JL, Nicolas J, Couvreur P (2012) Magnetic nanoparticles: design and characterization, toxicity and biocompatibility, pharmaceutical and biomedical applications. Chem Rev 112(11):5818–5878

Rokaya B, Kerroum D, Hayat Z, Panico A, Ouafa A, Pirozzi F (2019) Biogas production by an anaerobic digestion process from orange peel waste and its improvement by limonene leaching: investigation of H2O2 pre-treatment effect. Energy Sources Part A: Recov Util Environ Eff 1–9

Schlüter M, Hentzel T, Suarez C, Koch M, Lorenz WG, Böhm L et al (2014) Synthesis of novel palladium (0) nanocatalysts by microorganisms from heavy-metal-influenced high-alpine sites for dehalogenation of polychlorinated dioxins. Chemosphere 117:462–470

Schüler D, Frankel RB (1999) Bacterial magnetosomes: microbiology, biomineralization and biotechnological applications. Appl Microbiol Biotechnol 52(4):464–473

Scott NR (2006) Impact of nanoscale technologies in animal management. In: Rosati A, Tewolde A, Mosconi C (eds) Animal production and animal science worldwide. Wageningen Academic Publishers, Wageningen, pp 283–291

Sekoai PT, Ouma CNM, Du Preez SP, Modisha P, Engelbrecht N, Bessarabov DG, Ghimire A (2019) Application of nanoparticles in biofuels: an overview. Fuel 237:380–397

Sewwandi KAHS, Nitisoravut R (2020) Nano zero valent iron embedded on chitosan for enhancement of biohydrogen production in dark fermentation. Energy Rep 6:392–396

Shankar SS, Rai A, Ahmad A, Sastry M (2004) Rapid synthesis of Au, Ag, and bimetallic Au core–Ag shell nanoparticles using Neem (Azadirachta indica) leaf broth. J Colloid Interface Sci 275(2):496–502

Shi W, Ma Z (2010) Amperometric glucose biosensor based on a triangular silver nanoprisms/chitosan composite film as immobilization matrix. Biosens Bioelectron 26(3):1098–1103

Shi W, Ma Z (2011) A novel label-free amperometric immunosensor for carcinoembryonic antigen based on redox membrane. Biosens Bioelectron 26(6):3068–3071

Risal Singh, ShipraMital Gupta Introduction to nanotechnology (2016)

Sompong O, Prasertsan P, Karakashev D, Angelidaki I (2008) Thermophilic fermentative hydrogen production by the newly isolated Thermoanaerobacterium thermosaccharolyticum PSU-2. Int J Hydrog Energy 33(4):1204–1214

Sreekrishnan TR, Kohli S, Rana V (2004) Enhancement of biogas production from solid substrates using different techniques—a review. Bioresour Technol 95(1):1–10

Srivastava N, Rawat R, Oberoi HS (2014) Application of thermostable cellulases in biofuels production. In: Recent advances in bio-energy research, vol. III, pp 121–129. isbn:978–81–927097-2-7

Srivastava N, Srivastava M, Mishra PK, Singh P, Ramteke PW (2015) Application of cellulases in biofuels industries: an overview. J Biofuels Bioenergy 1(1):55–63

Srivastava N, Srivastava M, Kushwaha D, Gupta VK, Manikanta A, Ramteke PW, Mishra PK (2017a) Efficient dark fermentative hydrogen production from enzyme hydrolyzed rice straw by Clostridium pasteurianum (MTCC116). Bioresour Technol 238:552–558

Srivastava N, Srivastava M, Manikanta A, Singh P, Ramteke PW, Mishra PK (2017b) Nanomaterials for biofuel production using lignocellulosic waste. Environ Chem Lett 15(2): 179–184

Srivastava N, Srivastava M, Mishra PK, Singh P, Pandey H, Ramteke PW (2017c) Nanoparticles for biofuels production from lignocellulosic waste. Nanosci Food Agric 4:263–278

Srivastava N, Srivastava M, Gupta VK, Ramteke PW, Mishra PK (2018) A novel strategy to enhance biohydrogen production using graphene oxide treated thermostable crude cellulase and sugarcane bagasse hydrolyzate under co-culture system. Bioresour Technol 270:337–345

Su L, Shi X, Guo G, Zhao A, Zhao Y (2013) Stabilization of sewage sludge in the presence of nanoscale zero-valent iron (nZVI): abatement of odor and improvement of biogas production. J Mater Cycles Waste Manag 15(4):461–468

Suanon F, Sun Q, Li M, Cai X, Zhang Y, Yan Y, Yu CP (2017) Application of nanoscale zero valent iron and iron powder during sludge anaerobic digestion: impact on methane yield and pharmaceutical and personal care products degradation. J Hazard Mater 321:47–53

Taherdanak M, Zilouei H, Karimi K (2015) Investigating the effects of iron and nickel nanoparticles on dark hydrogen fermentation from starch using central composite design. Int J Hydrog Energy 40(38):12956–12963

Thakore S, Rathore PS, Jadeja RN, Thounaojam M, Devkar RV (2014) Sunflower oil mediated biomimetic synthesis and cytotoxicity of monodisperse hexagonal silver nanoparticles. Mater Sci Eng C 44:209–215

Thakur VK, Thakur MK, Raghavan P, Kessler MR (2014) Progress in green polymer composites from lignin for multifunctional applications: a review. ACS Sustain Chem Eng 2(5):1072–1092

Thanh NT, Watari T, Thao TP, Hatamoto M, Tanikawa D, Syutsubo K, Fukuda M, Tan NM, Anh TK, Yamaguchi T et al (2016) Impact of aluminum chloride on process performance and microbial community structure of granular sludge in an upflow anaerobic sludge blanket reactor for natural rubber processing wastewater treatment. Water Sci Technol 74(2):500–507

Tian T, Qiao S, Li X, Zhang M, Zhou J (2017) Nano-graphene induced positive effects on methanogenesis in anaerobic digestion. Bioresour Technol 224:41–47

Timko M, Molcan M, Hashim A, Skumiel A, Muller M, Gojzewski H et al (2012) Hyperthermic effect in suspension of magnetosomes prepared by various methods. IEEE Trans Magn 49(1): 250–254

Ubale AU, Belkhedkar MR (2015) Size dependent physical properties of nanostructured α-Fe2O3 thin films grown by successive ionic layer adsorption and reaction method for antibacterial application. J Mater Sci Technol 31(1):1–9

Vaghari H, Jafarizadeh-Malmiri H, Mohammadlou M, Berenjian A, Anarjan N, Jafari N, Nasiri S (2016) Application of magnetic nanoparticles in smart enzyme immobilization. Biotechnol Lett 38(2):223–233

Wahono SK, Darsih C, Rosyida VT, Maryana R, Pratiwi D (2014) Optimization of cellulose enzyme in the simultaneous saccharification and fermentation of sugarcane bagasse on the second-generation bioethanol production technology. Energy Procedia 47:268–272

Wang J, Wan W (2009) Factors influencing fermentative hydrogen production: a review. Int J Hydrog Energy 34(2):799–811

Wang T, Zhang D, Dai L, Chen Y, Dai X (2016) Effects of metal nanoparticles on methane production from waste-activated sludge and microorganism community shift in anaerobic granular sludge. Sci Rep 6:25857

Wei Z, Zeng G, Huang F, Kosa M, Huang D, Ragauskas AJ (2015) Bioconversion of oxygen-pretreated Kraft lignin to microbial lipid with oleaginous Rhodococcus opacus DSM 1069. Green Chem 17(5):2784–2789

Wimonsong P, Llorca J, Nitisoravut R (2013) Catalytic activity and characterization of Fe–Zn–Mg–Al hydrotalcites in biohydrogen production. Int J Hydrog Energy 38(25):10284–10292

Wolf EL, Medikonda M (2012) Understanding the nanotechnology revolution. Wiley, Hoboken

Wu H, Wang J, Wang Z, Fisher DR, Lin Y (2008) Apoferritin-templated yttrium phosphate nanoparticle conjugates for radioimmunotherapy of cancers. J Nanosci Nanotechnol 8(5): 2316–2322

Xu J, Yuan H, Lin J (2014) Evaluation of thermal, thermal-alkaline, alkaline and electrochemical pretreatment on sludge to enhance anaerobic biogas production. J Taiwan Inst Chem Eng 45: 2531–2536

Yadvika S, Sreekrishnan TR, Kohli S, Rana V (2004) Enhancement of biogas production from solid substrates using different techniques—a review. Bioresour Technol 95(1):1–10

Yan W, Lu D, Liu J, Zhou Y (2019) The interactive effects of ammonia and carbon nanotube on anaerobic digestion. Chem Eng J 372:332–340

Yang H, Shen J (2006) Effect of ferrous iron concentration on anaerobic bio-hydrogen production from soluble starch. Int J Hydrog Energy 31(15):2137–2146

Yang Y, Zhang C, Hu Z (2012) Impact of metallic and metal oxide nanoparticles on wastewater treatment and anaerobic digestion. Environ Sci: Processes Impacts 15(1):39–48

Zagrodnik R, Łaniecki M (2017) Hydrogen production from starch by co-culture of Clostridium acetobutylicum and Rhodobacter sphaeroides in one step hybrid dark-and photofermentation in repeated fed-batch reactor. Bioresour Technol 224:298–306

Zhang Y, Shen J (2007) Enhancement effect of gold nanoparticles on biohydrogen production from artificial wastewater. Int J Hydrog Energy 32(1):17–23

Zhang L, He X, Zhang Z, Cang D, Nwe KA, Zheng L, Li Z, Cheng S (2017a) Evaluating the influences of ZnO engineering nanomaterials on VFA accumulation in sludge anaerobic digestion. Biochem Eng J 125:206–211

Zhang T, Liu J, Wang C, Leng X, Xiao Y, Fu L (2017b) Synthesis of graphene and related two-dimensional materials for bioelectronics devices. Biosens Bioelectron 89:28–42

Zhang J, Fan C, Zhang H, Wang Z, Zhang J, Song M (2018) Ferric oxide/carbon nanoparticles enhanced bio-hydrogen production from glucose. Int J Hydrog Energy 43(18):8729–8738

Zhang NMY, Qi M, Wang Z, Wang Z, Chen M, Li K et al (2019) One-step synthesis of cyclodextrin-capped gold nanoparticles for ultra-sensitive and highly-integrated plasmonic biosensors. Sensors Actuators B Chem 286:429–436

Zhao Y, Chen Y (2011) Nano-TiO_2 enhanced photofermentative hydrogen produced from the dark fermentation liquid of waste activated sludge. Environ Sci Technol 45(19):8589–8595

Zhou J, You X, Jia T, Niu B, Gong L, Yang X, Zhou Y (2019) Effect of nanoscale zero-valent iron on the change of sludge anaerobic digestion process. Environ Technol 6:1–11

Chapter 4
Microbial Assisted Synthesis of Nanophotocatalysts for Dark Fermentative Hydrogen Generation

Shanthipriya Ajmera, Souvik Roy, M. Noorjahan, J. Immanuel Suresh, and Debashree Borthakur

Abstract The generation of hydrogen by using nanophotocatalyst in dark fermentation is playing a major role for decades. Nanoparticles recently gained a lot of interest due to their feasibility to enhance the effect of metabolic engineering by improving performance or product yield. Biohydrogen produced by the microbes shows enhanced production in anaerobic conditions using NPs. Whereas, these NPs improve reaction kinetics with an increased transfer of electrons. Though several NPs are reported to enhance dark fermentation for hydrogen production. Nanomaterials have the advantages of cost-effective recovery and recyclability; therefore, features such as strong electro-conductivity, bigger surface area, and high surface-to-volume ratio have been researched in silver, gold, palladium, iron, nickel, copper, and other metals. The renewable capacity of hydrogen is also very useful for developing the biological production of hydrogen. The production of hydrogen is done anaerobically both facultative and obligate, microorganisms include *E. coli*, *Sporolactobacillus* spp., *Citrobacter intermedius*, *Enterobacter aerogenes,* and *E. cloacae*. The sources used in hydrogen production are molasses, organic wastes, and sludges as substrates. The production of biohydrogen includes two steps; the Formate pathway and the NADH pathway. H_2 is employed for

S. Ajmera (✉)
Department of Microbiology, Palamuru University, Mahabubnagar, Telangana, India
e-mail: drspajmera@palamuruuniversity.ac.in

S. Roy
Post-Graduate Department of Biotechnology, St. Xavier's College (Autonomous), Kolkata, West Bengal, India

M. Noorjahan
Department of Chemistry, Palamuru University, Mahabubnagar, Telangana, India

J. I. Suresh
Department of Microbiology, The American College, Madurai, Tamil Nadu, India

D. Borthakur
Assam Don Bosco University, Guwahati, Assam, India

© The Author(s), under exclusive license to Springer Nature Singapore Pte Ltd. 2023
M. Srivastava, P. K. Mishra (eds.), *NanoBioenergy: Application and Sustainability Assessment*, Clean Energy Production Technologies,
https://doi.org/10.1007/978-981-19-6234-9_4

hydration of considerable oils for fuel production, food hydration, diminishing nitrate, perchlorate, and selenite. This chapter also includes the merits of hydrogen generation by nanophotocatalyst like viable process, effective method, and demerits like compromised reliability, low substrate conversion efficiency, a consequent low yield of H_2, etc. Some of the safety measures are also followed during the generation of hydrogen by nanophotocatalyst. Unsafe acts are due to the human elements which include; physical and mental characteristics, knowledge, skill, and attitudes of the individuals.

Keywords Nanophotocatalyst · Hydrogen · Dark fermentation · Biofuel · Anaerobic fermentation · Sugars

4.1 Introduction

The increasing gap created between the global energy demands and an inadequate supply of the same has caused a mammoth increase in the usage of fossil fuels. Presently, exhaustive energy generation by the traditional fossil fuel combustion methods has lead to a huge depletion of natural resources like combustible geologic deposits buried underground, along with a significant negative impact on the global climate following emissions of greenhouse gases (GHGs) in large quantities. Currently, the CO_2 concentration in the atmosphere has crossed a whooping 350 parts per million (ppm) (Chandrasekhar et al. 2015). This huge rise in CO_2 levels in the atmosphere potentially aggravates the greenhouse effect, resulting in a steep increase in global temperatures. This gradual, but steep decrease in the global fossil fuel reserves, and sincere concerns about climatic changes resulting from huge bulks of GHGs emitted have led new-age scientists to investigate and develop eco-friendly, nonpolluting, and renewable energy sources (Ayodele et al. 2020).

Production of hydrogen gas by microorganisms has gained significant attention over the past few decades due to its eco-friendly and recyclable nature, which involves a highly efficient conversion of substrates in the process. Cyanobacteria, algae, and bacteria are nowadays widely employed for the large-scale efficient production of hydrogen gas from organic wastes (Goyal et al. 2013). Improvement in microbial hydrogen gas production is recently accomplished by introducing new cutting-edge technologies like bioaugmentation (Marone et al. 2012), cell immobilization (Kumar et al. 2016), microbial electrolysis cell (Kadier et al. 2016), and genetic engineering (Srirangan et al. 2011). Bio-based energy is gradually showing its tip as a sustainable replacement of fossil fuel-based energy sources, as it can not only reduce the prevailing crises in the global energy supplies but also can reduce the natural calamities soon to befall the world due to global warming. To name, one such potential alternative is biohydrogen (H_2). The high-energy yield/unit mass of this fuel is 122 kJ/g, which is around 2.75 times more than the yield/unit mass of conventional hydrocarbon fuels (Chandrasekhar et al. 2015). Combustion of H_2 forms H_2O as the only major by-product thus reducing GHG emissions. As it is

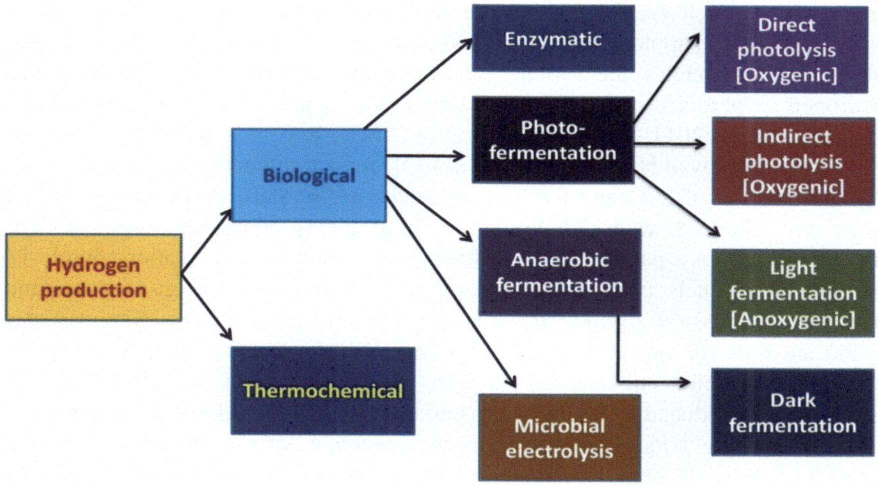

Fig. 4.1 The different routes of biohydrogen production

much similar to electricity with regard to the framework of energy systems rather than fossil fuels, biohydrogen is a much sought-after energy carrier.

Recently, a variety of biological routes for the production of H_2 have been developed (Khabirul Islam et al. 2021). Based on the systems that lead to H_2 generation, these routes are grouped into four categories: (1) water-splitting photosynthesis (photolysis), (2) photofermentation, (3) dark fermentation, and (4) microbial electrolysis process (electro-fermentation). Figure 4.1 shows the different routes of biohydrogen production (Hallenbeck et al. 2018; Chandrasekhar et al. 2015).

Each of the above processes has its own characteristic advantages and disadvantages with regard to energy production efficiency and practicality. Hence, the choice of a suitable biocatalyst, as well as inoculum concerned with H_2 production, will be different for different processes.

Despite the eco-friendly advantages of biohydrogen production, low substrate conversion efficiencies leading to low production rates, and the production of acid-rich intermediates from acidogenic processes are some of the demerits that the biological production of H_2 suffers from (Chandrasekhar et al. 2015). To ensure this, a number of sophisticated techniques for ensuring very high yields of hydrogen are nowadays being applied, which employ tools of metabolic engineering to reroute metabolic pathways to improve the efficiency of utilization of substrates through the microbial expression of heterologous proteins, and to increase the electron flux required for H^+ reduction.

Dark fermentation (DF) is a highly efficient concept for biohydrogen production (Preethi et al. 2019). DF is a comparatively low-tech, cheap process, producing moderate rates of H_2 and organic removal (Sun et al. 2019a, b). In this method, complex forms of organic substrates can be utilized by anaerobic microorganisms or microalgae (Preethi et al. 2019). In this regard, the different avenues still open for continued research are in the pretreatment of feedstock, process optimization, co-fermentation, supplementation of additives such as metal ions (for example, nickel that accelerates the action of hydrogenase, contributing to an increased hydrogen evolution) and improving inoculum specificity for H_2 production (Srirangan et al. 2011; Khabirul Islam et al. 2021). Recently, it has been demonstrated that improved H_2 yields, as well as COD reduction rates, are achievable with the introduction of NiO and CoO nanoparticles to the dark fermentation process using rice mill wastewater (Khabirul Islam et al. 2021). Nanotechnology can definitely bring about a path-breaking difference in advanced fermentation, pyrolysis, biofuel cell, jet fuels, carbon capture storage, catalytic conversion gasification, and nano-based precision forming technologies (Nizami et al. 2018). Nanoparticles (NPs) are now the prime areas of focus for scientists because of their ability to enhance metabolic engineering through the improvement of product yield. Biohydrogen produced by the microbes under anaerobic conditions shows enhanced production using NPs. These NPs improve reaction kinetics with an increased transfer of electrons. Dark fermentation for hydrogen production has been improved using gold NPs, which shows a stimulatory effect on substrate utilization by 56% and increases yield by 46%. Their enhanced surface area to volume ratio provides better accessibility of the binding site with bacteria as well as enzyme (Zhang and Shen 2007).

4.2 Biohydrogen Production by Dark Fermentation

4.2.1 The Process

DF or heterotrophic fermentation is an alternative route of biohydrogen generation by anaerobic bacteria and microalgae from substrates, including carbohydrate type of organic wastes (feedstock) or wastewaters, in the absence of sunlight. It is an anaerobic process carried out in the dark, directly related to the acidogenesis stage of the anaerobic sludge digestion process (Antonopoulou et al. 2011).

DF can be represented by the general biochemical reaction:

$$\begin{aligned}\alpha - \text{biomass} + \beta - H_2O \longrightarrow \gamma &- \text{acetic acid} + \delta - \text{propionic acid} + \varepsilon \\ &- \text{butyric acid} + \zeta - \text{valeric acid} + \theta \\ &- \text{hexanoic acid} + \kappa - H_2 + \lambda CO_2 + \mu \\ &- \text{microbial biomass} + \pi - \text{others(e.g., ethanol)}\end{aligned}$$

4.2.1.1 Microbiology

Anaerobic degradation of organic wastes to produce hydrogen is carried out principally in two main successive stages, acidification and methanogenesis, performed by specific microorganisms through different complex syntrophic interactions (Sun et al. 2019a).

In this dark fermentation route of biological production of H_2, pure microbial cultures or a mixture of anaerobic microorganisms can be employed (Antonopoulou et al. 2011). Microorganisms involved in the process include the facultatively anaerobic bacteria *Escherichia coli, Sporolactobacillus spp., Citrobacter intermedius, Enterobacter aerogenes,* and *E. cloacae,* and the obligately anaerobic bacteria *Ruminococcus albus, Clostridium beijerinckii,* and *C. paraputrificum* (Fig. 4.2) (Chandrasekhar et al. 2015).

Biomass from microalgae, rich in their carbohydrate content, is considered an efficient feedstock for DF, promising elevated H_2 productivities. The most promising microalgal species for this purpose have been found to be *Chlorella* spp., *Scenedesmus* spp., and *Saccharina* spp. (Koutra et al. 2020). In them, the heterotrophic bacterial population present in the algal biomass slurries produced biohydrogen.

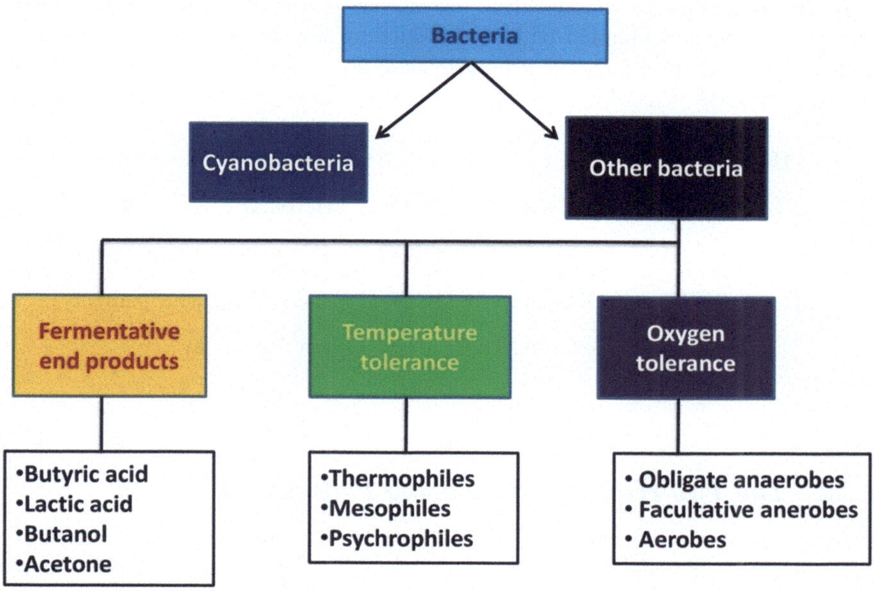

Fig. 4.2 The different bacteria (including cyanobacteria) involved in the biological production of H_2 by dark fermentation

4.2.1.2 Biochemistry

DF technologies for biohydrogen production have been developed in the laboratory with substrates like molasses, organic wastes, and sludges, the latter two being rich in complex carbohydrates. Carbohydrates, lipids, and proteins are hydrolyzed by the hydrogen-producing enzymes (hydrogenases) in DF. Since no oxygen is produced or consumed in this reaction, oxygen-labile hydrogenase is less likely to be inactivated (Antonopoulou et al. 2011).

In the DF process, the anaerobic microbes generate reducing equivalents like NAD(P)H and $FADH_2$, which are finally reoxidized via the electron transport chain (ETC) to form energy-rich molecules like ATP. Unlike in aerobic respiration, anaerobic respiration uses CO_2, NO_3^-, S^0, and SO_4^{2-} as terminal electron acceptors (TEAs), which are finally reduced with the regeneration of reducing powers (Chandrasekhar et al. 2015).

Glycolysis or EMP pathway is an anaerobic metabolic pathway in which the chief substrate glucose ($C_6H_{12}O_6$) is catabolized to pyruvate (Show et al. 2019):

$$C_6H_{12}O_6 + 2NAD^+ \rightarrow 2CH_3COCOOH + 2NADH + 2H^+.$$

Under anaerobic conditions, the pyruvate so produced goes to the different acidogenic pathways coupled with H_2 production, forming different volatile fatty acids (VFAs) like propionic acid, butyric acid, acetic acid, malic acid, and others (Show et al. 2019; Chandrasekhar et al. 2015). The corresponding reactions are:

$$C_6H_{12}O_6 + 2H_2 \rightarrow 2CH_3CH_2COOH + 2H_2O(\text{propionic acid pathway})$$

$$C_6H_{12}O_6 \rightarrow CH_3CH_2CH_2COOH + 2CO_2 + 2H_2(\text{butyric acid pathway})$$

$$C_6H_{12}O_6 + 2H_2O \rightarrow 2CH_3COOH + 2CO_2 + 4H_2(\text{acetic acid pathway})$$

$$C_6H_{12}O_6 + 2H_2 \rightarrow COOHCH_2CH_2OCOOH + CO_2(\text{malic acid pathway})$$

$$C_6H_{12}O_6 \rightarrow CH_3CH_2OH + CO_2(\text{ethanol pathway})$$

4.2.2 The Two Hydrogen-Producing Pathways

Interconversion of metabolic intermediates takes place during substrate breakdown in anaerobic fermentation, which increases the availability of reducing equivalents in bacterial cells.

The two main hydrogen-producing pathways in DF are the formate pathway and the NADH pathway (Fig. 4.3) (Chandrasekhar et al. 2015). Recent scientific studies have shown that these two hydrogen-generating pathways might be linked.

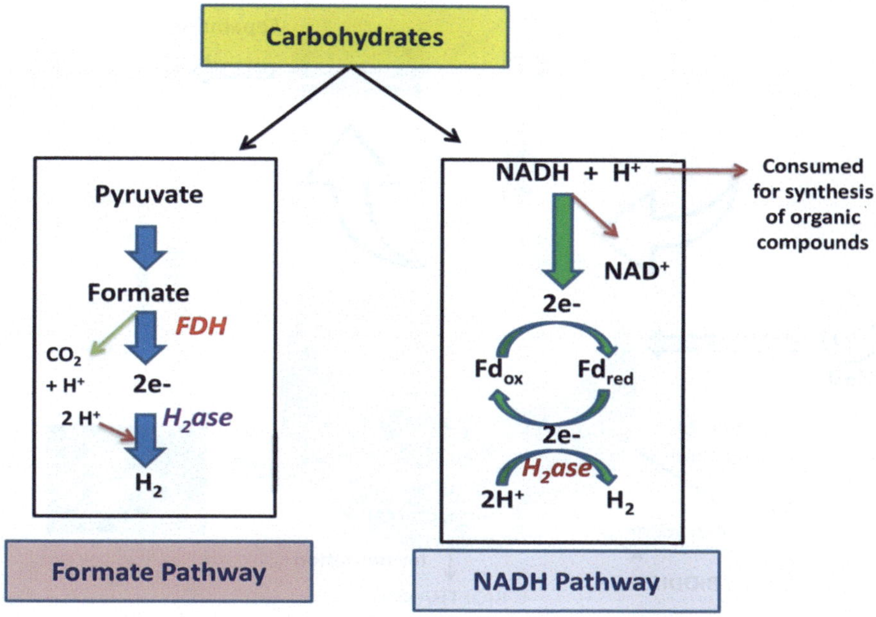

Fig. 4.3 The two hydrogen-producing pathways in Dark Fermentation

1. **The Formate pathway**: The main enzyme of the formate pathway is formate hydrogen lyase (FHL) complex. The core FHL complex includes formate dehydrogenase (FDH) and hydrogenase (H_2ase). This FHL complex catalyzes the oxidation of formate and reduction of proton to produce molecular H_2 and CO_2 in a 1:1 molar ratio. Facultatively anaerobic microorganisms first transform pyruvate to acetyl-CoA and formate by the enzyme pyruvate formate lyase and then produce H_2 by the use of the FHL complex. Many genetic engineering studies have been conducted on the FHL-related genes for regulating the metabolic reactions of the formate pathway to enhance biohydrogen production (Sinha et al. 2015).
2. **The NADH pathway**: This pathway, first reported in the 1980s, uses NADH as the precursor involved in biohydrogen production. This process is carried out principally by obligately anaerobic microorganisms. This route of H_2 production via reoxidization of NADH requires ferredoxin (Fd) reduction by NADH-Fd-oxidoreductase, and the participation of Fd-hydrogenase (Show et al. 2019):

$$NADH + H^+ + 2Fd^{2+} \xrightarrow{\text{NADH - Fd - oxidoreductase}} 2H^+ + NAD^+ + 2Fd^+$$

$$2Fd^{2+} + 2H^+ \xrightarrow{\text{Fd - hydrogenase}} 2Fd^+ + H_2$$

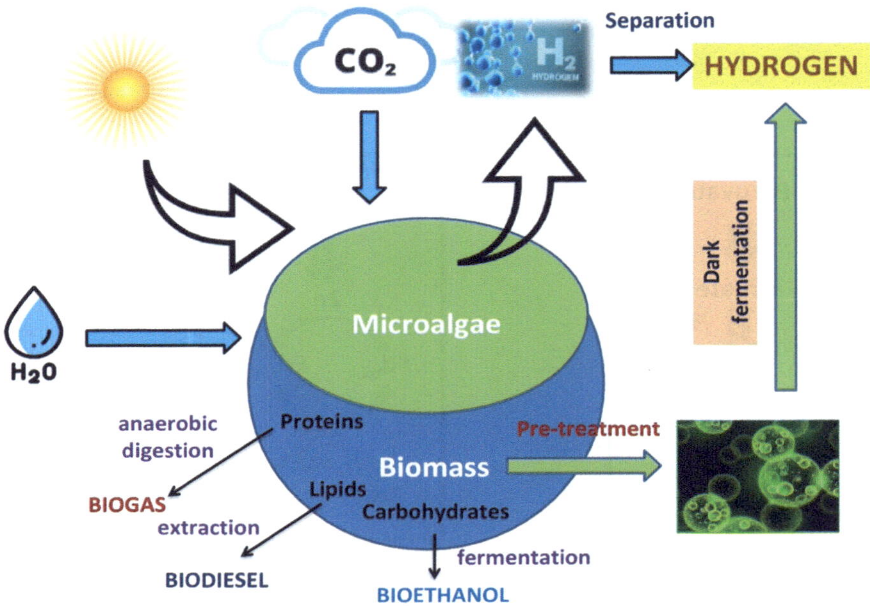

Fig. 4.4 Hydrogen Production from renewable sources

During fermentation, the H^+ released from NADH/FADH with NADH dehydrogenase is reduced to H_2 by hydrogenase with the reduced Fd. On the other hand, the membrane-bound NADH dehydrogenase and cytochrome bc_1 (cyt bc_1) protein complexes, and quinone and cytochrome c, the mobile electron carriers, participate in the electron transfer through the quinone pool. The redox interconversion between reduced Q (ubiquinol/QH_2) and H^+ transfers the electrons to the cyt bc_1 and to the cytochrome aa_3 (cyt aa_3). The reduced cyt aa_3 now reduces Fd, which donates an electron to the hydrogenase to produce H_2 (Chandrasekhar et al. 2015). These electron transfers are very much affected by the prevailing environmental conditions (Show et al. 2019). Hence, several detailed scientific studies were conducted to regulate the metabolism of NADH in order to enhance H_2 production. Figure 4.4 shows the H_2 production from renewable sources.

4.3 Mechanism of Synthesis of Microorganism-Assisted Nanoparticles

The processes of both intracellular and extracellular synthesis of nanoparticles (NP) through microorganisms from metals, metal oxides, or metalloids were properly documented in the literature (Lahiri et al. 2021). The extracellular method includes reduction of metal ions for NPs synthesis through microbial enzymes and proteins, bacterial or fungal cell wall components, or organic molecules present in the culture medium, whereas the intracellular system involves initial electrostatic enchantment of metal ions through carboxyl groups of the microbial cell wall, resulting in the passage of metal ions via the cells and reduction by intracellular proteins and cofactors to yield NPs (Lahiri et al. 2021; Siddiqi et al. 2018). Biochemical mechanisms concerning microorganism-mediated nanoparticle synthesis could also be seen as a part of microbial resistance mechanisms for cellular detoxification. This involves changes in the solubility of inorganic and toxic ions through enzymatic reduction and/or precipitation in the shape of nanostructures (Lahiri et al. 2021). Both extracellular and intracellular biocatalytic synthesis mechanisms were proposed, which specifically include oxidoreductase enzymes (e.g., NADH-based nitrate reductase, NADPH-based sulfite reductase flavoprotein subunit α, and cysteine desulfhydrase) and cellular transporters (Lahiri et al. 2021; Grasso et al. 2019). Nano-size substances are biosynthesized in the microorganisms with the aid of using binding target ions from the surrounding environment and changing those toxic metal ions into the corresponding element metal by cellular enzymes. Based on the region of synthesis of nanoparticles, it may be categorized into intracellular or extracellular (Lahiri et al. 2021). The intracellular approach includes transporting ions into the microbial cell to form nanoparticles in the presence of enzymes. The extracellular mode involves trapping the metal ions on the cell surface and reducing ions in the presence of enzymes (Lahiri et al. 2021; Li et al. 2011). Nanoparticles have recently attracted a lot of attention due to their potential to improve metabolic engineering by enhancing performance or yield. Advanced fermentation, pyrolysis, jet fuels, catalytic conversion gasification, biofuel cell, carbon capture storage, and nano-based precision forming technologies can all benefit from nanotechnology (Nizami et al. 2018) bacterial biohydrogen generation is boosted in anaerobic environments when Nanoparticles (NPs) are used. By enhancing the electron transfer rate in the microbial cell, nanoparticles improve biohydrogen production in the microbial system. The major biological components for the synthesis of Nanoparticles are given in Fig. 4.5.

The advantage of Nanomaterials are cost-effective recovery/recyclability; therefore, features like as strong electro-conductivity, bigger and high surface area-to-volume ratio have been researched in the silver, gold, palladium, iron, nickel, copper, and other metals (Mughal et al. 2021). Several NPs have been reported to improve dark fermentation for hydrogen production, including gold NPs, which have a stimulatory effect on substrate utilization by 56% and increase yield by 46%. This increased surface area to volume ratio provides better accessibility to the binding site with bacteria and enzymes (Zhang and Shen 2007; Sekoai et al. 2019). Figure 4.6 shows the microbial enzymes in bioreduction of metal, metalloid, and nonmetal ions to nanoparticles (Dasgupta et al. 2017).

To overcome the obstacle of increased nanoparticle dosage for enhancing biofuel production, various mesoporous NPs have been created. When NPs (zero-valent metals, metal oxide, and carbon-based NPs) are combined with anaerobic fermenting bacteria, the hydrolysis process of organic material is accelerated by increased substrate consumption, and lipase catalyzes biodiesel generation through

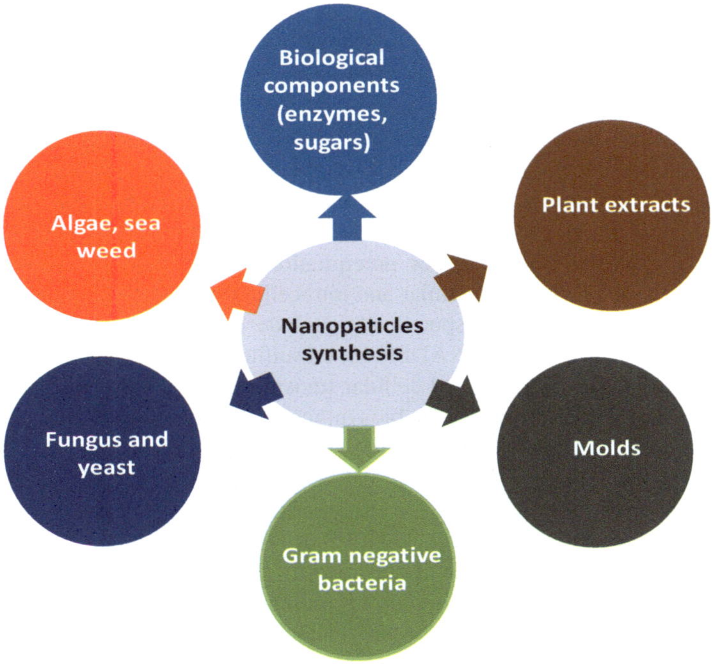

Fig. 4.5 The major biological components for the synthesis of Nanoparticles

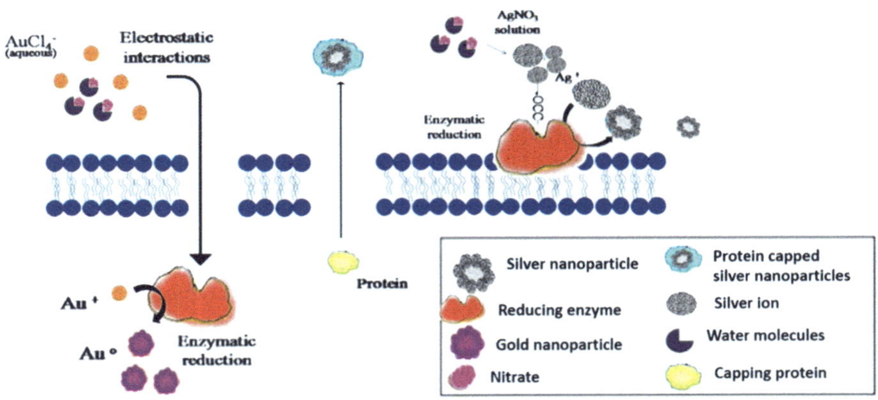

Fig. 4.6 Microbial enzymes in bioreduction of metal, metalloid, and on metal ions to nanoparticles

transection. Various mesoporous NPs have been developed to overcome the barrier of increased NP dosage for improving biofuel production. When nanoparticles (NPs) (zero-valent metals, metal oxide, and carbon-based NPs) are combined with anaerobic fermenting bacteria, the hydrolysis of organic material is accelerated by increased substrate consumption, and lipase catalyzes biodiesel generation via transection.

NPs modified enzymes, on the other hand, can withstand harsh environmental conditions and support increased product synthesis with reusability. Because nanotechnological implementation is still in the laboratory, it necessitates a higher level of knowledge and technology.

4.3.1 *Effect of Nanoparticles on Biohydrogen Production*

The applications of NPs include protein immobilization, biosensors, and biofuels production (Patel et al. 2016, 2018; Mohanraj et al. 2016; Kim et al. 2016; Otari et al. 2016). Biosensors are used for enhancing electron transfer to acceptors. NPs can also significantly influence the microbial metabolic activity for H_2 production through similar phenomenon under aerobic conditions by efficient transfer of electrons (Beckers et al. 2013). The positive effect of various NPs, including silver (Ag), gold (Au), Copper (Cu), Fe, Ni, Palladium (Pd), Silica (SiO_2), Titanium (Ti), activated carbon, carbon nanotubes (CNTs) and composite were observed on BHP (Mohanraj et al. 2014; Patel et al. 2018; Mohanraj et al. 2016; Zhao et al. 2013a, b). Briefly, these NPs might be stimulating BHP by their surface and quantum size effect (Zhang et al. 2007). As surface effect, the smaller the size of NPs larger the specific surface area, which thus enables a strong ability to adsorb electrons. The extent of the quantum size is directly co-related with the rate of electron transfer between NPs and enzyme molecules, such as hydrogenase, which catalyzes the conversion of H_2 to proton and vice versa, either to act as electron sinks or deliver reducing power from H_2 oxidation, as follow: $H_2 \leftrightarrow 2H^+ + 2e^-$.

4.3.2 *The Influence of Nanoparticles on Dark Fermentative for Hydrogen Generation*

The influence of nanoparticles on dark fermentative for hydrogen generation was presented in two ways.

(1) Chemical-assisted synthesis of nanoparticles and (2) Green synthesis of Nanoparticles.

4.3.2.1 Chemical-Assisted Synthesis of Nanoparticles Dark Fermentative H2 Production

Nickel Nanoparticles

In order to increase the hydrogen (H_2) production by dark fermentation, the excess amount of nickel ferrite nanoparticles ($NiFe_2O_4$ NPs) above 400 mg/L decreased the H_2 productivity, whereas moderate quantities (50–200 mg/L) boosted up H_2 production. The 100 mg/L (37 °C) and 200 mg/L NiFe2O4 NPs (55 °C) groups produced the greatest H2 yields of 222 and 130 mL/g glucose, respectively, which were 38.6 and 28.3% greater than the control groups (37 °C and 55 °C).

Soluble metabolites revealed that the $NiFe_2O_4$ NPs boosted up the butyrate pathway, which corresponded to an increase in *Clostridium butyricum* abundance in mesophilic fermentation. Using residual algal biomass from the Cyanobacterium *Lyngbya limnetica* as a substrate, nickel ferrite nanoparticles ($NiFe_2O_4$ NPs) increased crude cellulase enzyme production. After 72 h, the residual algal substrate and crude cellulase mediated by $NiFe_2O_4$ NPs exhibit roughly 2.5 times increased filter paper activity, as well as improved pH and thermal stability. Using the *Bacillus subtilis* PF 1 bacterial strain, a total of 1820 mL/L of hydrogen was produced in the dark fermentation. During the dark fermentation, the addition of charcoal (BC) and metal cofactor nanoparticle Ni0 production of hydrogen is improved. To optimize hydrogen production, a new hybrid strategy combining artificial neural networks and response surface methods was used. The impact of operational parameters, such as BC, metal cofactor Ni0, pH, and microbe dose, on hydrogen production as well as concentrations of other metabolites, such acetic acid, propionic acid, butyric acid, and ethanol, were studied extensively.

Cobalt Nanoparticles

The impact of cobalt ferrate nanoparticles on hydrogen production from dark fermentation was investigated by sol- gel method. These nanoparticles were added to the fermentation process in small amounts (0.1–0.4 g/L). The highest hydrogen yield of 205.24 mL/g glucose was obtained at a dosing level of 0.4 g/L, representing a 31.8% improvement over the control, while excess (0.5 g/L) nanoparticles reduced hydrogen production by 9.7%. The microbial community demonstrated that the moderate amount of cobalt ferrate nanoparticles boosted up the abundance of *Clostridium* sensu stricto 1, which was dominating and supported the long-term conversion of glucose wastewater into clean hydrogen, from 14.49 to 18.84%. The addition of NiO and CoO nanoparticles (NPs) to dark fermentation of rice mill wastewater employing *Clostridium beijerinckii* DSM 791 enhanced biohydrogen generation. Intrinsically generated NiO (26 nm) and CoO (50 nm) NPs with polyhedral morphology and high purity were made using a simple hydrothermal approach. The characteristics of biohydrogen generation were discovered in dose-dependent experiments for 1.5 mg/L concentrations of both NPs. Optimal NiO and CoO dosages increased biohydrogen output by 2.09 and 1.9 times, respectively, as compared to a control run without NPs.

Iron Nanoparticles

In terms of H_2 yield, productivity rate, and metabolites distribution, FONPs effects on the hydrogen fermentation from glucose and pretreated cassava starch employing *E. aerogenes* and FONPs were used to aid dark hydrogen fermentation by increasing hydrogenase activity of *E. aerogenes* and electron transfer with 200 mg/L FONPs, pretreated starch and glucose improved hydrogen yields by 17.0% and 63.1%, respectively (Lin et al. 2016). When the concentration of FONPs was increased from 0 to 200 mg/L, the hydrogen output of glucose increased from 164.5 $\pm$ 2.29 to 192.41 $\pm$ 0.14 mL/g. When the concentration of FONPs was increased to 400 mg/L, the hydrogen yield of glucose fell to 147.2 2.54 mL/g. FONPs improved the acetate pathway but weakened the ethanol pathway of hydrogen production, according to soluble metabolic products. More NAD (nicotinamide adenine dinucleotide) was available for reducing proton to hydrogen as a result of the metabolic pathway shift (Lin et al. 2016). *Enterobacter aerogenes* conducted dark fermentation utilizing glucose as a carbon source at 37 °C in a mesophilic environment. The morphological organization of nano-sized zero-valent iron on chitosan was validated by TEM investigation. 243 mL H_2/glucose was the greatest biohydrogen yield, which was 30% greater than the control. The acetic route was used to make biohydrogen, resulting in a favorable association between hydrogen yield and acetic acid content, according to the distribution of volatile fatty acids (VFA). Dark fermentation using Enterobacter aerogenes has the potential to produce biohydrogen. The addition of iron oxide nanoparticles (Fe_3O_4 NPs) and date seed- activated carbon nanocomposites (Fe_3O_4/DSAC) to the fermentation media boosted hydrogen yield and productivity. For both additions, the optimal dosage and fermentation duration were found to be 150 mg/L and 24 h, respectively. Fe3O4/DSAC nanocomposites outperformed Fe_3O_4 NPs in terms of hydrogen generation (Nurul Sakinah Engliman et al. 2017). The 150 mg/L nanocomposites produced a maximum hydrogen yield of 238.7 mL/g, which was 65.7% greater than the standalone Fe_3O_4 NPs and three times higher than the yield of the control run without any NPs (78.4 mL/g). The Fe_3O4/DSAC nanocomposites showed great promise for producing biohydrogen from date fruit wastes (Rambabu et al. 2021). A study using mixed cultures grown under a thermophilic setting in which iron oxide NPs were added at a very low concentration (50 mg/L) was successful in improving biohydrogen generation. When compared to the control test without the addition of iron NPs, the production rates increased by up to 34%. Both iron (II) oxide and nickel oxide were examined and found to increase hydrogen output by 34.38% and 5.47%, respectively, above the control test. The studies on the influence of starting pH were conducted without the addition of nanoparticles in order to establish the optimum pH for maximum hydrogen production, which was found to be around 1.78 mol H_2/mol glucose at pH

5.5. These findings indicate that adding iron (II) oxide nanoparticles to the system is the most important component in increasing hydrogen generation (Engliman et al. 2017). The feasibility of increasing the rate of H_2 gas production from glucose through anaerobic fermentation by mixed bacteria was investigated using ferric oxide/carbon nanoparticles (FOCNPs). The results showed that an acceptable dose of FOCNPs enhanced H_2 production, whereas an excessive concentration of FOCNPs reduced it. The greatest yield of 218.63 mL H_2/g glucose was obtained with the addition of 200 mg/L FOCNPs, which was 33.7% greater than the 163 mL H_2/g glucose obtained in the control test without the addition of FOCNPs. However, 400 mg/L FOCNPs reduced H_2 evolution to 154 mL/g glucose, which was 5.5% less than the control. The acetic routes were discovered to be used in the development of H_2. FOCNPs could help anaerobes proliferate by providing more attachment sites. FOCNPs may also boost hydrogenase activity and electron transfer efficiency, which is good for bio-H_2 evolution. Excessive addition of FOCNPs, on the other hand, could be harmful to microorganisms, further decreasing H_2 generation. Using heat-shock pretreatment anaerobic sludge in batch reactors, the effects of Fe0 and Ni0 nanoparticles on mesophilic dark hydrogen fermentation from starch were examined. In the presence of 0–50 mg/L of each Fe0 and Ni0 nanoparticle, a starch concentration of 0–20 g/L was utilized, with the initial pH of all experiments set to 7. At starch, Fe0, and Ni0 concentrations of 5.37 g/L, 37.5 mg/L, and 37.5 mg/L, respectively, the maximal yield of hydrogen generation was 149.8 mL/g-VS. This was over 200% greater than the results from the control group (Taherdanak et al. 2015). The study on the effects of grass hydrogen fermentation by using biochar and zero-valent iron nanoparticles (Fe0 NPs) results showed that combining biochar (600 mg/L) with Fe0 NPs (400 mg/L) increased and enriched the microbial activity of Clostridium species, resulting in excellent hydrogen yielding efficient metabolic pathways. When biochar + Fe0 NPs added, hydrogen productivity reached 50.6 mL/g -dry grass, which was 89.8%, 53.3%, and 15.9% highest than the control, individual biochar, and individual Fe0 NPs groups, respectively. Fe0 NPs supplementation increased hydrogen production and hastened the grass fermentation process. Hydrogen yield and maximal hydrogen yield are best achieved at a dosage of 400 mg/L. The production rates were 64.7 mL/g dry grass and 12.1 mL/h, respectively, which were 73.1% and 128.3% greater than the control group. Fe0 NPs enhanced microbial activity and altered the dominating bacteria in the system from Enterobacter sp. to Clostridium sp., resulting in a more efficient pathway for enhancing hydrogen production. Most critical factors affecting the hydrogen production from starch via mesophilic dark fermentation were found to be starch and Fe0 NPs concentrations. There was some interface between the concentrations of Fe0 NPs and Ni0 NPs. At a starch concentration of 5 g/L, Fe0 NPs and Ni0 NPs concentrations of 37.5 mg/L found the greatest experimental hydrogen yield of 147.3 mL/g VS was obtained (Taherdanak et al. 2015). Fe0 NPs supplementation increased microbial activity and shifted the main microbial community from *Enterobacter* to *Clostridium* sp., resulting in an extra efficient metabolic pathway that produced more. At a Fe0 NPs dosage of 400 mg/L, the maximum hydrogen yield and hydrogen generation rates were 64.7 mL/g -dry grass and 12.1 mL/h, respectively, which were 73.1% and 128.3% greater than the control group.

Gold Nanoparticles

Gold nanoparticles improve biohydrogen production by attracting microbial cells to active regions. Because of their tiny size and huge surface area AuNPs (5 nm) improve the substrate utilization efficiency by 56% and biohydrogen generation yield by 46% (Zhang and Shen 2007). The AuNPs increase the activity of catalytic hydrogenase, which plays an important role in H_2 production.

Ag Nanoparticles

Silver nanoparticles were added to anaerobic batch reactors to boost acidogenesis while also producing fermentative hydrogen. Using glucose-fed mixed bacteria dominated *by Clostridium butyricum*, the effects of silver nanoparticle concentration and inorganic nitrogen concentration on cell growth and hydrogen production were investigated. The silver nanoparticle tests produced significantly highest H_2 yields than the control, with the maximum hydrogen yield (2.48 mol/mol glucose) obtained at a silver concentration of 20 nmol L^{-1}. The presence of silver nanoparticles reduced ethanol yield while increasing acetic acid yield. The higher the concentration of silver nanoparticles, the higher the rate of cell biomass production (Zhao et al. 2013a, b).

4.3.2.2 Green Synthesis of Nanoparticles for Dark Fermentative H_2 Production

Phytogenic iron nanoparticles (FeNPs) on dark fermentative hydrogen (H_2) production by *Enterobacter cloacae*, a mesophilic soil bacterium FeNPs were made from $FeSO_4$ utilizing aqueous leaf extract of *Syzygium cumini* in a quick green with 100 mg L^1 FeNPs supplementation, the greatest H_2 output of 1.9 mol mol^1 glucose was found, with twofold improvement in the conversion efficiency process of glucose. On batch fermentative H_2 generation from glucose, the effects of $FeSO_4$ and FeNPs were compared. Fermentation tests revealed that FeNPs supplementation considerably boosted the proportion and yield of H_2 compared to the control (no supplementation) and $FeSO_4$-containing media. FeNPs which was biogenerated by *Murraya koenigii* leaf extract as reducing and stabilizing. The influence of FeNPs was tested using the *Clostridium acetobutylicum* NCIM 2337 for fermentative hydrogen production from glucose. The maximum hydrogen yield in the FeNPs supplemented experiment was 2.33 ± 0.09 mol H_2/mol glucose at 175 mg/L of FeNPs, whereas the highest yield of hydrogen in the control experiment was 1.74 0.08 mol H_2/mol Glucose. The comparative investigation revealed that iron oxide nanoparticles have a greater enhancing effect on fermentative hydrogen production than $FeSO_4$. With 125 and 200 mg/L iron oxide nanoparticles, the maximal yields of hydrogen 2.07 ± 0.07 mol H_2/mol with glucose and 5.44 ± 0.27 mol H_2/mol with sucrose were reached, respectively. All the findings suggest that the improvement of hydrogen production from iron oxide nanoparticle supplementation is significantly greater than that from ferrous iron supplementation (Lin et al. 2016). The synthesis and characterization of magnetite nanoparticles (NPs) made from *Water hyacinth* (WH) extract, as well as their regulating influence on fermentative hydrogen generation by *Klebsiella* sps. from lignocellulosic hydrolysate. With WH-magnetite-NP at 20 mg/L, the greatest cumulative hydrogen generation was achieved with a 23.49% increase and an optimal Y(H2/S) of 83.20 2.19 mL/g substrate.

Paladium Nanoparticles

PdNPs generated using *Coriandrum sativum* leaf enhanced the mixed culture hydrogen generation from glucose. Adding PdNPs to *E. cloacae* resulted in a negligible hydrogen output. But due to the metabolite shift, $PdCl_2$ had an inhibitory influence on hydrogen generation. The hydrogen yields of *E. cloacae* and mixed culture with 5.0 mg/L of $PdCl_2$ supplemented trials were 1.39 ± 0.07 and 2.11 ± 0.11 mol H_2/mol glucose, respectively. At 5.0 mg/L PdNPs supplementation, the greatest yield of hydrogen with *E. cloacae* and mixed culture were recorded as1.48 ± 0.04 and 2.48 ± 0.09 mol H_2/mol glucose (Mohanraj et al. 2014).

Cu-NPs

Cu-NPs on biohydrogen generation from glucose and the results were compared to $CuSO_4$. The reducing agent used in the production of copper nanoparticles (Cu-NPs) is *Murraya koenigii* leaf extract which was used as a reducing agent in the production of copper nanoparticles (Cu-NPs). *Enterobacter cloacae* and *Clostridium acetobutylicum* were used to investigate the effects of phytogenic Cu-NPs on biohydrogen generation from glucose, and the results were compared to $CuSO_4$. The effects of Cu-NPs and $CuSO_4$ (2.5, 5.0, 7.5, 10.0, and 12.5 mg/L) on biohydrogen production were investigated using *E. cloacae* at pH 7.0 and *C. acetobutylicum* at pH 6.0. The findings imply that adding Cu-NPs to fermentation at a concentration of less than 2.5 mg/L could boost hydrogen production.

NiO-NPs

The green synthesis of nickel oxide nanoparticles (NiO-NP) from Eichhornia crassipes (Ec) extract was carried out for the first time, and the regulatory influence of this NP on fermentative hydrogen production was assessed(Ec-NiO-NP). In the presence of 20 mg/L Ec-NiO-NP, the highest cumulative hydrogen generation and hydrogen yield $Y(H_2/S)$ were 4842.19 23.43 mL/L and 101.45 3.32 mL/g substrate, respectively, which were 47.29% and 37.78% greater than the control without NPs addition. The results show that supplementing with green-synthesized Ec-NiO-NP enhances fermentative hydrogen production while also regulating important node metabolites and functional gene expression (Zhang et al. 2021).

Fe_3O_4 NPs

The biohydrogen generation capability of date-palm fruit waste was investigated utilizing *Enterobacter aerogenes* and dark fermentation. In terms of hydrogen yield, production rate, and metabolites distribution, the effects of FONPs on hydrogen fermentation from glucose and pre-treated cassava starch employing *E. aerogenes* were investigated. Fe_3O_4/DSAC nanocomposites outperformed Fe_3O_4 NPs in terms of hydrogen generation. The 150 mg/L nanocomposites produced a maximum hydrogen yield of 238.7 mL/g, which was 65.7% greater than the standalone Fe_3O_4 NPs and three times higher than the yield of the control run without any NPs (78.4 mL/g). In addition to the stimulatory actions of Fe_3O_4 NPs, the carbon in the nanocomposites is used as an adsorbent buffer, favoring the medium pH. The Fe_3O_4/DSAC nanocomposites showed great promise for producing biohydrogen from date fruit wastes (Rambabu et al. 2021).

Green FeNP was synthesized by using lignin from empty oil palm fruit. Using lignin-mediated FeNPs, biohydrogen is generated from wastewaters. In the realm of bioenergy, soft drink wastewater and corn steep liquor are used. In presence of lignin-mediated FeNPs, after 48 h of fermentation at 37 °C, $BioH_2$ production was 17.67 0.54 mL under ideal circumstances. $BioH_2$ generation was raised by 91.0% and 74.3%, respectively, when LMNP and LNMNP were added. Furthermore, adding 200 mg/L LMNP and LNMNP to the fermentation medium increased $BioH_2$ yields (mL H_2/g COD removed) by 2.8 and 2.3 times, respectively (do Nascimento Junior et al. 2021).

Ag NPs

Henna (*Lawsonia inermis*) and DUC2 *Actinomycete* were used to make green AgNPs in a batch utilizing varied concentrations of AgNP (0, 40, 80, and 120 g/mL). All results indicate that the henna-mediated synthesis of AgNP enhanced the rate of hydrogen yield 1.71 mol H_2/mol. Glucose source additionally supplied after 72 h of AgNPs incubation period along with culture *C beijerinckii*. The results showed that henna-mediated AgNPs increased the bacterial cell population, whereas DUC2 AgNPs reduced the bacterial cell population. As a result, the current research provides a clear picture of how metallic NPs can be used as H_2 synthesis catalysts to boost bioH_2 production (Khan et al. 2020). Overall, the influence of individual and mixtures of NPs on the BHP yield by different organisms has been presented in Table 4.1.

4.4 The Large-Scale Production, Merits, and Demerits

Organic wastes from sewage or agriculture can be fed into large anaerobic bioreactors for the production of H_2 through dark fermentative process. This also helps in achieving the goal of waste treatment. (Antonopoulou et al. 2011). Several studies reported that the different temperatures required for dark fermentative H_2 production are ambient (15–27 °C), mesophilic (30–45 °C), thermophilic (50–60 ° C), and extremely thermophilic (>60 °C) (Chandrasekhar et al. 2015).

In the acidification stage, hydrogen is produced as an intermediate metabolic product to be later used as an electron donor by the methanogens for methane production (methanogenesis). Although the transitional hydrogen metabolite is quite unstable, it is possible to recover this hydrogen if the hydrogen-consuming methanogenesis can be tactfully repressed by lowering the pH or inhibiting the methanogens by heat or chemicals. Figure 4.7 shows different applications of hydrogen outlined (Source—Google).

Hydrogen is taken into account as carbon-free fuel because the only by-product after its combustion is water. Supply of low-carbon energy for heat, balancing of electricity at national grid, and application in combustive engine preferred hydrogen over other hydrocarbon-based gaseous biofuels. From the statistical point of view, the fertilizers and petroleum companies are considered to be the largest users of H_2 which almost account for 37% and 50%, respectively (Nath et al. 2015). The H_2 fuel demand is increased due to the development of H_2 engine (Dicks et al. 2004). It is important to know that hydrogen energy can be used along with many other energies and hence promote the general efficiency of distributing clean fuel for a variety of applications. H_2 utilization in the longer term depends on the specific needs of the community.

There are wide applications of current hydrogen utilization described as followed:

- Hydrogen can be used for food hydration, alkali hydration for the production of fertilizers, and hydration of substantial oils for fuel production.
- H2 can be used for diminishing nitrate, perchlorate, selenite, and a set of other oxidized water contaminations, due to its electron donor ability (Das and Veziroglu 2001).

4.4.1 Merits

The following are the merits of the production of biohydrogen by the dark fermentation route (Antonopoulou et al. 2011; Chandrasekhar et al. 2015):

1. The dark fermentation process is a viable process—most promising as compared to all biological hydrogen production processes.
2. Effective method, its rate being higher than the processes of photofermentation and photolysis.

Table 4.1 Biohydrogen production in the presence of different types of inorganic and organic nanoparticles

Nanoparticles				Process parameters						
Composition	Conc. (mg/L)[a]	Organisms	Feed	Mode	WV (L)	pH	Temp. (°C)	Yield[c]	H_2 yield efficiency (%)	References
Inorganic										
Ag	20.0[b]	Mixed culture	Glucose	Batch	0.08	8.5	35	2.48	67.6	Zhao et al. (2013a, b)
Au	1.0 mm	Anaerobic sludge	Acetate	EAB	0.2	7.2	35	1.76	na[i]	Khan et al. (2013)
	5.0[b]	Anaerobic culture	Wastewater	Batch	0.08	7.2	35	2.24	50	Zhang et al. (2007)
Cu	2.5	*Clostridium acetobutylicum* NCIM 2337	Glucose	Batch	0.2	7	37	1.39	3.5[j]	Mohanraj et al. (2016)
		Enterobacter cloacae 811101	Glucose	Batch	0.2	7	37	1.69	2.9	
Fe	5	Anaerobic sludge	Glucose	Batch	0.08	5.5	37	338[d]	37	Taherdanak et al. (2015)
	100	*Enterobacter cloacae* DH-89	Glucose	Batch	0.1	7	37	1.9	100	Nath et al. (2015)
	400	Mixed bacterial consortium	Glucose	Batch	0.25	7	30	1.23	38	Zhang et al. (2015)
	250	Mixed culture and *Clostridium butyricum* TISTR	Water hyacinth	Batch	0.08	7	35	57.0[d]	55	Mahmood et al. (2013)
	312	*Rhodobacter sphaeroides* NMBL-02 + *Escherichia coli* NMBL-04	Malate	Batch	0.06	5.6	32	3.10[e]	19.4	Dolly et al. (2015)
Fe_2O_3	175	*C. acetobutylicum* NCIM2337	Glucose	Batch	0.2	6	37	2.33	33.9	Mohanraj et al. (2016)
	800	*Clostridium pasteurianum* CH5	Glucose	Batch	0.1	7	35	2.2	10	Hsieh et al. (2016)
	200	*Enterobacter aerogenes* ATCC13408	Glucose	Batch	0.2	6	37	1.55	17	Taherdanak et al. (2015)

(continued)

Table 4.1 (continued)

Nanoparticles				Process parameters						
Composition	Conc. (mg/L)[a]	Organisms	Feed	Mode	WV (L)	pH	Temp. (°C)	Yield[c]	H_2 yield efficiency (%)	References
	125	*E. cloacae* 811101	Glucose	Batch	0.2	7	37	2.07	21.8	Mohanraj et al. (2014)
	50	Anaerobic sludge	Glucose	Batch	0.2	5.5	60	1.92	53.6	Engliman et al. (2017)
	200	*E. cloacae* 811101	Sucrose	Batch	0.2	7	37	2.72	4.8	Han et al. (2011)
	200	Mixed culture	Sucrose	Batch	0.8	6	35	1.78	33	Lin et al. (2016)
	200	*E. aerogenes* ATCC13408	Cassava starch	Batch	0.2	6	37	124[d]	63.1	Gadhe et al. (2015a, b)
	50	Anaerobic sludge	Dairy wastewater	Batch	0.1	5.5	37	16.75[f]	24	Gadhe et al. (2015a, b)
	200	Anaerobic sludge	Molasses wastewater	Batch	0.1	5.5	37	7.85[f]	44	Nasr et al. (2015)
	25.0 mg/g VSS	Anaerobic sludge	Starch wastewater	ABR	30	6.7	30	0.9	57.8	Zhao et al. (2011)
Fe_3O_4	400	Anaerobic sludge	Glucose	Batch	0.08	7	35	1.53	26.4	Malik et al. (2014)
	50	Mixed culture	Wastewater	Batch	0.1	6	37	44.3[d]	83.3	Reddy et al. (2017)
	200	Anaerobic sludge	Sugarcane bagasse	Batch	0.1	5	30	1.21	69.6	Taherdanak et al. (2015)
Ni	2.5	Anaerobic sludge	Glucose	Batch	0.08	5.5	37	250[d]	0.9	Mullai et al. (2013)
	5.7	Anaerobic sludge	Glucose	Batch	0.7	5.6	33	2.54	22.7	Elreedy et al. (2017)
	60	Anaerobic sludge	Wastewater	Batch	0.25	7	55	24.7[d]	23	Engliman et al. (2017)

NiO	200	Anaerobic sludge	Glucose	Batch	0.2	5.5	60	1.3	4.8	Gadhe et al. (2015a, b)
	10	Anaerobic sludge	Dairy wastewater	Batch	0.1	5.5	37	15.7[f]	16	Gadhe et al. (2015a, b)
	5	Anaerobic sludge	Molasses wastewater	Batch	0.1	5.5	37	6.73[f]	23.5	Mohanraj et al. (2014)
Pd	5	*E. cloacae* 811101	Glucose	Batch	0.2	7	37	1.48	0.6	Giannelli and Torzillo (2012)
	5	Mixed culture	Glucose	Batch	0.2	7	37	2.48	6.4	
SiO_2	40	*Chlamydomonas reinhardtii* CC124	Air: CO_2 (97:3)	PBR	110	4.2	28	0.61[g]	45.2	Beckers et al. (2013)
	5.1	*C. butyricum* CWBI1009	Glucose	Batch	0.2	7.6	30	0.96	4.3	Venkata Mohan et al. (2008)
	120	Acidogenic mixed culture	Wastewater	Cont.	0.16	5.5	28	7.02[h]	666	Hsieh et al. (2016)
TiO_2	50	*C. pasteurianum* CH5	Glucose	Batch	0.1	7	35	2.1	5	Zhao et al. (2011)
	100	*Rhodopseudomonas palustris*	Waste sludge	Batch	0.3	8	30	1.01[f]	46.1	Pandey et al. (2015)
	60	*R. sphaeroides* NMBL-02	Malate	Batch	0.1	8	32	1.75[g]	69.9	Taherdanak et al. (2015)
Inorganic mixtures										
Fe + Ni	37.5 + 37.5	Anaerobic sludge		Batch	0.8	7	37	150[d]	200	Gadhe et al. (2015a, b)
Fe_2O_3 + NiO	200 + 5.0	Anaerobic sludge		Batch	0.1	5.5	37	8.83[f]	62	Gadhe et al. (2015a, b)
	50.0 + 10.0	Anaerobic sludge		Batch	0.1	5.5	37	17.2[f]	27	Liu et al. (2012)

(continued)

Table 4.1 (continued)

Nanoparticles				Process parameters						
Composition	Conc. (mg/L)[a]	Organisms	Feed	Mode	WV (L)	pH	Temp. (°C)	Yield[c]	H_2 yield efficiency (%)	References
Organic										
Granular activated carbon	100	Anaerobic sludge	Glucose	UASB	5	6.5	25	1.42	Na	Venkata Mohan et al. (2008)
	10,000	Acidogenic mixed culture	Starch wastewater	Cont.	0.16	5.5	28	2.12[h]	94.5	Wimonsong and Nitisoravut (2015)
Powdered activated carbon	33	Anaerobic sludge	Sucrose	Batch	0.06	5.5	37	1.3	62.5	Wimonsong and Nitisoravut (2014)
	33.3	Anaerobic sludge	Sucrose	UASB	0.06	5.5	37	1.3	73	Venkata Mohan et al. (2008)
	5000	Acidogenic mixed culture	Starch wastewater	Cont.	0.16	5.5	28	1.57[h]	44	Liu et al. (2012)
Carbon nanotubes	100	Anaerobic sludge	Glucose	UASB	5	6.5	25	2.45	Na	Liu et al. (2012)

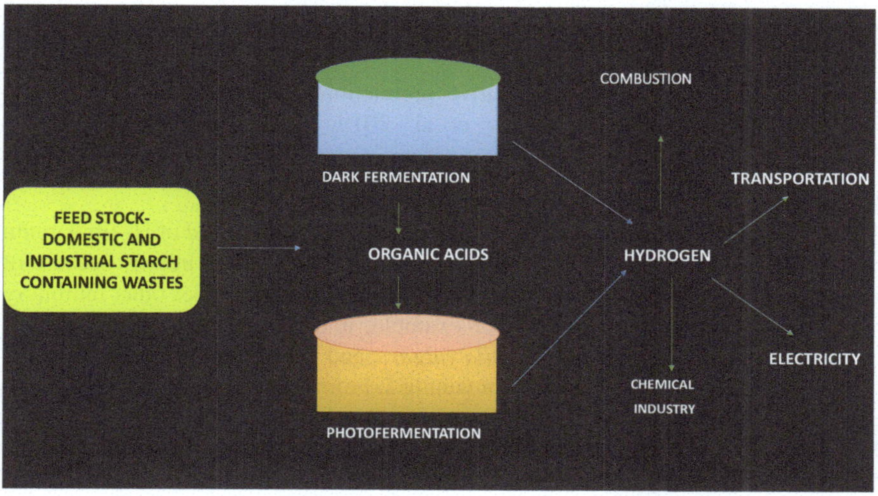

Fig. 4.7 Application of hydrogen synthesized from renewable sources

3. A wide range of feedstock can be applied as substrates for H2 production by this method.
4. Simple in operation—does not require expensive and complicated photobioreactors, as needed for direct biophotolysis and photofermentation reactions. Oxygen removal is also not difficult when the anaerobic conditions are chosen.
5. Reactor design is not affected by light supply; thus, dark bioreactors exploit volume more efficiently.
6. It is carried out at ambient temperatures and pressures, without photoenergy, reducing the cost of hydrogen production 340 times with regard to the photosynthetic processes. So, it is a low-cost production process.
7. Requires low energy input.
8. Exhibits good sustainability.
9. There is a high organic content of the dark fermentation spent. So, dark fermentation can be linked to bioelectrochemical systems such as microbial fuel cells (MFCs) and microbial electrolysis cells (MECs) as the spent makes an ideal substrate for them. Hence, the overall energy recovery can be majorly improved by the conjugation of dark fermentation with bioelectrochemical systems. For example, in an integrated system using dark fermentation and a single-chambered MEC with 0.6 V voltage, the general hydrogen recovery was 96% (Singh and Das 2019).

4.4.2 Demerits

The disadvantages of dark fermentative production of biohydrogen are enlisted below (Koutra et al. 2020; Show et al. 2019; Antonopoulou et al. 2011; Chandrasekhar et al. 2015):

1. Compromised reliability.
2. Low substrate conversion efficiency and a consequent low yield of H_2 (maximum 4 H_2/glucose; 50% less than the theoretical hydrogen yield), primarily because of the formation of various by-products, in particular, gases, besides other inhibitory factors like product inhibition. For example, it has been reported that the H_2 produced from green algae *Chlorella vulgaris* and *Dunaliella tertiolecta* biomass by anaerobic enriched cultures containing 2-bromoethanesulfonic acid (BESA) was subsequently consumed by non-methanogenic microorganisms.
3. The process suffers from thermodynamic limitations.
4. Difficulty in the separation of H_2 from its mixture with 50–60% CO_2. Other gases present in the mixture may be H_2S, CO, CH_4, H_2O vapor (moisture), and NH_3. Hence, purification of the H_2 is essential before it can be utilized.
5. Production of residual COD which is to be removed before discharging that treated wastewater into the water reservoirs.
6. Production of residual COD which is to be removed before discharging that treated wastewater into the water reservoirs.
7. Changes in system redox conditions and buffering capacities (there can be unwanted pH fluctuations).

4.5 Safety Issues and Concerns

Everyone in the lab is responsible for their own safety and the safety of others. One should be cognizant of potential hazards by conducting all experiments and demonstrations prior to their implementation. The introductory laboratory should engage each student in significant experiences with experimental processes, including some experience designing investigations. We should develop a wide array of basic skills and tools for experiment and data analysis and help the students to master basic concepts (Shrivastava 2014). We should understand the role of direct observation and distinguish between inferences based on theory and the outcomes of experiments and help students to develop collaborative learning skills that are vital to success in many lifelong endeavors. The laboratory should develop a broad array of basic skills, tools of experiment, data analysis, and help the students to master basic concepts (Shrivastava 2014). Laboratory accidents are mainly caused by humans which includes physical and mental characteristics, knowledge, skill, and attitudes of the individuals (American Association of Physics Teachers 1998).

4.6 Conclusion

In this chapter, hydrogen synthesis from renewable biological sources proved to be an essential way for hydrogen generation. Among several biological hydrogen production approaches, the Dark Fermentative has been proven to be a practical approach in biological hydrogen production, especially when it is strengthened by the addition of chemicals. The involvement of nanoparticles (Ni, Fe) was found to be more efficient than other approaches as it helps in the transportation of electrons between the ferredoxin and the hydrogenase. The order of effectiveness in promoting the activities of hydrogenase is based on substrate conversion efficiency and hydrogen evolution rate which follows the order of metal NPs > metal/metal ions/ metal oxides > other metals (other than Ni, Fe). In order to build the biological hydrogen production process more feasible and economical for industrial applications, future endeavors should be centered on the optimized combination of several hydrogen manufacturing processes with the energy cascade utilization, material recycling, and recovery. By the combination of wide approaches, it is potentially possible to produce the maximum hydrogen yield. These novel ways of intensification and integration of several hydrogen generation processes, such as photocatalysis, photofermentation, and dark fermentation processes, will facilitate the large-scale synthesis of hydrogen.

References

American Association of Physics Teachers (1998) Goals of the introductory physics laboratory. Am J Phys 66(6):483–485

Antonopoulou G, Ntaikou I, Stamatelatou K, Lyberatos G (2011) Biological and fermentative production of hydrogen. In: Handbook of biofuels production. Woodhead Publishing, Sawston, pp 305–346

Ayodele FO, Mustapa SI, Ayodele BV, Mohammad N (2020) An overview of economic analysis and environmental impacts of natural gas conversion. Sustainability 12:10148. https://doi.org/10.3390/su122310148

Beckers L, Hiligsmann S, Lambert SD, Heinrichs B, Thonart P (2013) Improving effect of metal and oxide nanoparticles encapsulated in porous silica on fermentative biohydrogen production by *Clostridium butyricum*. Bioresour Technol 133:109–117

Chandrasekhar K, Lee YJ, Lee DW (2015) Biohydrogen production: strategies to improve process efficiency through microbial routes. Int J Mol Sci 16(4):8266–8293

Das D, Veziroglu TN (2001) Hydrogen production by biological processes: a survey of literature. Int J Hydrog Energy 26:13–28

Dasgupta A, Sarkar J, Ghosh M, Bhattacharya A, Mukherjee A, Chattopadhyay D (2017) Green conversion of graphene oxide to graphene nanosheets and its biosafety study. PLoS One 12: e0171607. https://doi.org/10.1371/journal.pone.0171607

Dicks AL, da Costa JD, Simpson A, McLellan B (2004) Fuel cells, hydrogen and energy supply in Australia. J Power Sources 131(1–2):1–12

Dolly S, Pandey A, Pandey BK, Gopal R (2015) Process parameter optimization and enhancement of photo-biohydrogen production by mixed culture of Rhodobacter sphaeroides NMBL-02 and Escherichia coli NMBL-04 using Fe-nanoparticle. Int J Hydrog Energy 40:16010–16020 do Nascimento Junior JR, Torres LAZ, Medeiros ABP, Woiciechowski AL, Martinez-Burgos WJ, Soccol CR (2021) Enhancement of biohydrogen production in industrial wastewaters with vinasse pond consortium using lignin-mediated iron nanoparticles. Int J Hydrog Energy 46:27431–27443. https://doi.org/10.1016/j.ijhydene.2021.06.009

Elreedy A, Ibrahim E, Hassan N, El-Dissouky A, Fujji M, Yoshimura C, Tawfik A (2017) Nickel-graphene nanocomposite as a novel supplement for enhancement of biohydrogen production from industrial wastewater containing mono-ethylene glycol. Energy Convers Manag 140:133–144

Engliman NS, Abdul PM, Wu S-Y, Jahim JM (2010) Influence of iron (II) oxide nanoparticle on biohydrogen production in thermophilic mixed fermentation. Int J Hydrog Energy

Engliman NS, Abdul PM, Wu S-Y, Jahim JM (2017) Influence of iron (II) oxide nanoparticle on biohydrogen production in thermophilic mixed fermentation. Int J Hydrog Energy. https://doi.org/10.1016/j.ijhydene.2017.05.224

Gadhe A, Sonawane SS, Varma MN (2015a) Enhancement effect of hematite and nickel nanoparticles on biohydrogen production from dairy wastewater. Int J Hydrog Energy 40:4502–4511

Gadhe A, Sonawane SS, Varma MN (2015b) Influence of nickel and hematite nanoparticle powder on the production of biohydrogen from complex distillery wastewater in batch fermentation. Int J Hydrog Energy 40:10734–10743

Giannelli L, Torzillo G (2012) Hydrogen production with the microalga Chlamydomonas reinhardtii grown in a compact tubular photobioreactor immersed in a scattering light nanoparticle suspension. Int J Hydrog Energy 37:16951–16961

Goyal Y, Kumar M, Gayen K (2013) Metabolic engineering for enhanced hydrogen production: a review. Can J Microbiol 59:59–78

Grasso G, Zane D, Dragone R (2019) Microbial nanotechnology: challenges and prospects for green biocatalytic synthesis of nanoscale materials for sensoristic and biomedical applications. Nanomaterials 10:11. https://doi.org/10.3390/nano10010011

Hallenbeck PC, Lazaro CZ, Sagir E (2018) Photosynthesis and hydrogen from photosynthetic microorganisms

Han H, Cui M, Wei L, Yang H, Shen J (2011) Enhancement effect of hematite nanoparticles on fermentative hydrogen production. Bioresour Technol 102:7903–7909

Hsieh PH, Lai YC, Chen KY, Hung CH (2016) Explore the possible effect of TiO2 and magnetic hematite nanoparticle addition on biohydrogen production by *Clostridium pasteurianum* based on gene expression measurements. Int J Hydrog Energy 41:21685–21691

Kadier A, Kalil MS, Abdeshahian P, Chandrasekhar K, Mohamed A, Azman NF, Hamid AA (2016) Recent advances and emerging challenges in microbial electrolysis cells (MECs) for microbial production of hydrogen and value-added chemicals. Renew Sustain Energy Rev 61:501–525

Khabirul Islam AKM, Dunlop PSM, Hewitt NJ, Lenihan R, Brandoni C (2021) Bio-hydrogen production from wastewater: a comparative study of low energy intensive production processes. Clean Technol 3(1):156–182

Khan MM, Lee J, Cho MH (2013) Electrochemically active biofilm mediated bio-hydrogen production catalyzed by positively charged gold nanoparticles. Int J Hydrog Energy 38(13):5243–5250

Khan I, Anburajan P, Kumar G, Yoon J-J, Bahuguna A, de Moura AGL, Pugazhendhi A, Kim S-H (2020) Comparative effect of silver nanoparticles (AgNPs) derived from actinomycetes and henna on biohydrogen production by *Clostridium beijerinckii* (KTCC1737). https://doi.org/10.1002/er.6076

Kim TS, Patel SKS, Selvaraj C, Jung W-S, Pan C-H, Kang YC, Lee J-K (2016) A highly efficient sorbitol dehydrogenase from *Gluconobacter oxydans* G624 and improvement of its stability through immobilization. Sci Rep 6:34–38

Koutra E, Tsafrakidou P, Sakarika M, Kornaros M (2020) Microalgal biorefinery. In: Microalgae cultivation for biofuels production. Academic, pp 163–185

Kumar G, Mudhoo A, Sivagurunathan P, Nagarajan D, Ghimire A, Lay CH, Chang JS (2016) Recent insights into the cell immobilization technology applied for dark fermentative hydrogen production. Bioresour Technol 219:725–737

Lahiri D, Nag M, Sheikh HI, Sarkar T, Edinur HA, Pati S, Ray RR (2021) Microbiologically-synthesized nanoparticles and their role in silencing the biofilm signaling cascade. Front Microbiol 12:636588

Li X, Xu H, Chen ZS, Chen G (2011) Biosynthesis of nanoparticles by microorganisms and applications. J Nanomater. https://doi.org/10.1155/2011/270974

Lin R, Cheng J, Ding L, Song W, Liu M, Zhou J, Cen K (2016) Enhanced dark hydrogen fermentation by addition of ferric oxide nanoparticles using Enterobacter aerogenes. Bioresour Technol 207:213–219

Liu Z, Lv F, Zheng H, Zhang C, Wei F, Xing X-H (2012) Enhanced hydrogen production in a UASB reactor by retaining microbial consortium onto carbon nanotubes (CNTs). Int J Hydrog Energy 37:10619–10626. https://doi.org/10.1016/j.ijhydene.2012.04.057

Mahmood T, Zada B, Malik SA (2013) Effect of Iron Nanoparticles on Hyacinth Fermentation. Int J Sci 10:106–121

Malik SN, Pugalenthi V, Vaidya AN, Ghosh PC, Mudliar SN (2014) Kinetics of nano-catalysed dark fermentative hydrogen production from distillery wastewater. Energy Procedia 54:417–430

Marone A, Massini G, Patriarca C, Signorini A, Varrone C, Izzo G (2012) Hydrogen production from vegetable waste by bioaugmentation of indigenous fermentative communities. Int J Hydrog Energy 37(7):5612–5622

Mohanraj S, Anbalagan K, Kodhaiyolii S, Pugalenthi V (2014) Comparative evaluation of fermentative hydrogen production using Enterobacter cloacae and mixed culture: effect of Pd (II) ion and phytogenic palladium nanoparticles. J Biotechnol 192:87–95

Mohanraj S, Anbalagan K, Rajaguru P, Pugalenthi V (2016) Effects of phytogenic copper nanoparticles on fermentative hydrogen production by Enterobacter cloacae and Clostridium acetobutylicum. Int J Hydrog Energy 41:10639–10645. https://doi.org/10.1016/j.ijhydene.2016.04.197

Mughal B, Zaidi SZJ, Zhang X, Hassan SU (2021) Biogenic nanoparticles: synthesis, characterisation and applications. Appl Sci 11(6):2598

Mullai P, Yogeswari MK, Sridevi K (2013) Optimisation and enhancement of biohydrogen production using nickel nanoparticles—a novel approach. Bioresour Technol 141:212–219

Nasr M, Tawfik A, Ookawara S, Suzuki M, Kumari S, Bux F (2015) Continuous biohydrogen production from starch wastewater via sequential dark-photo fermentation with emphasize on maghemite nanoparticles. J Ind Eng Chem 21:500–506

Nath D, Manhar AK, Gupta K, Saikia D, Das SK, Mandal M (2015) Phytosynthesized iron nanoparticles: effects on fermentative hydrogen production by Enterobacter cloacae DH-89. Bull Mater Sci 38:1533–1538

Nizami AS, Rehman M, Anwar MN, Fayyaz A, Sohail NF, Khokhar MF, Baqar M, Khan WD, Rasool K (2018) CO2 capture and storage: a way forward for sustainable environment. J Environ Manag 226:131–144. https://doi.org/10.1016/j.jenvman.2018.08.009. Epub 2018 Aug 14

Otari SV, Patel SKS, Jeong JH, Lee JH, Lee J-K (2016) A green chemistry approach for synthesizing thermostable antimicrobial peptide-coated gold nanoparticles immobilized in an alginate biohydrogel. RSC Adv 6:86808–86816

Pandey A, Gupta K, Pandey A (2015) Effect of nanosized TiO2 on photofermentation by Rhodobacter sphaeroides NMBL-02. Biomass Bioenergy 72:273–279

Patel SKS, Choi SH, Kang Y-C, Lee JK (2016) Large-scale aerosol-assisted synthesis of biofriendly Fe_2O_3 yolk-shell particles: a promising support for enzyme immobilization. Nanoscale 8:6728–6738

Patel SKS, Lee JK, Kalia VC (2018) Nanoparticles in biological hydrogen production: an overview. Indian J Microbiol 58(1):8–18

Preethi, Usman TMM, Rajesh Banu J, Gunasekaran M, Kumar G (2019) Biohydrogen production from industrial wastewater: an overview. Bioresour Technol Rep 7:100287

Rambabu K, Bharath G, Banat F, Hai A, Show PL, Nguyen THP (2021) Ferric oxide/date seed activated carbon nanocomposites mediated dark fermentation of date fruit wastes for enriched biohydrogen production. Int J Hydrog Energy 46(31):16631–16643. https://doi.org/10.1016/j.ijhydene.2020.06.108

Reddy K, Nasr M, Kumari S, Kumar S, Gupta SK, Enitan AM, Bux F (2017) Biohydrogen production from sugarcane bagasse hydrolysate: effects of pH, S/X, Fe2+, and magnetite nanoparticles. Environ Sci Pollut Res Int 24:8790–8804

Show KY, Yan YG, Lee DJ (2019) Biohydrogen production from algae: perspectives, challenges, and prospects. In: Biofuels from algae. Elsevier, pp 325–343

Shrivastava SK (2014) Safety measures in physics laboratory procedures. Anusandhan 14(40): 11–23

Siddiqi KS, Husen A, Rao RAK (2018) A review on biosynthesis of silver nanoparticles and their biocidal properties. J Nanobiotechnol 16:14. https://doi.org/10.1186/s12951-018-0334-5

Singh V, Das D (2019) Potential of hydrogen production from biomass. In: Science and engineering of hydrogen-based energy technologies, pp 123–164

Sinha P, Roy S, Das D (2015) Role of formate hydrogen lyase complex in hydrogen production in facultative anaerobes. Int J Hydrog Energy 40(29):8806–8815

Srirangan K, Pyne ME, Chou CP (2011) Biochemical and genetic engineering strategies to enhance hydrogen production in photosynthetic algae and cyanobacteria. Bioresour Technol 102(18): 8589–8604

Sun C, Xia A, Qian F, Huang Y, Lin R, Murphy JD (2019a) Energy Convers Manag 185:431–441

Sun Y, He J, Yang G, Sun G, Sage V (2019b) A review of the enhancement of bio-hydrogen generation by chemicals addition. Catalysts 9(4):353

Sundaresan Mohanraj, Krishnasamy Anbalagan, Palanisamy Rajaguru, Velan Pugalenthi, (2016) Effects of phytogenic copper nanoparticles on fermentative hydrogen production by Enterobacter cloacae and Clostridium acetobutylicum. International Journal of Hydrogen Energy,41:10639-10645. https://doi.org/10.1016/j.ijhydene.2016.04.197

Taherdanak M, Zilouei H, Karimi K (2015) Investigating the effects of iron and nickel nanoparticles on dark hydrogen fermentation from starch using central composite design. Int J Hydrog Energy 40:12956–12963

Venkata Mohan S, Mohanakrishna G, Reddy SS, Raju BD, Rao RKS, Sarma PN (2008) Self-immobilization of acidogenic mixed consortia on mesoporous material (SBA-15) and activated carbon to enhance fermentative hydrogen production. Int J Hydrog Energy 33:6133–6142

Wimonsong P, Nitisoravut R (2014) Biohydrogen enhancement using highly porous activated carbon. Energy Fuel 28:4554–4559

Wimonsong P, Nitisoravut R (2015) Comparison of different catalyst for fermentative hydrogen production. J Clean Energy Technol 3:128–131

Zada B, Mahmood T, Malik SA (2013) Effect of iron nanoparticles on hyacinths fermentation. Int J Sci 2:106–121

Zhang Y, Shen J (2007) Enhancement effect of gold nanoparticles on biohydrogen production from artificial wastewater. Int J Hydrog Energy 32:17–23

Zhang P, Zhang R, McGarvey JA, Benemann JR (2007) Biohydrogen production from cheese processing wastewater by anaerobic fermentation using mixed microbial communities. Int J Hydrog Energy 32(18):4761–4771

Zhang L, Zhang L, Li D (2015) Enhanced dark fermentative hydrogen production by zero-valent iron activated carbon microelectrolysis. Int J Hydrog Energy 40:12201–12208

Zhang Q, Xu S, Li Y, Ding P, Zhang Y, Zhao P (2021) Green-synthesized nickel oxide nanoparticles enhances biohydrogen production of Klebsiella sp. WL1316 using lignocellulosic hydrolysate and its regulatory mechanism. Fuel 305:121585. https://doi.org/10.1016/j.fuel.2021.121585

Zhao W, Zhao J, Chen G, Feng R, Yang J, Zhao Y, Wei Q, Du B, Zhang Y (2011) Anaerobic biohydrogen production by the mixed culture with mesoporous Fe3O4 nanoparticles activation. Adv Mater Res 306–307:1528–1531

Zhao W, Zhang Y, Du B, Wei D, Wei Q (2013a) Enhancement effect of silver nanoparticles on fermentative biohydrogen production using mixed bacteria. Bioresour Technol 142:240–245

Zhao W, Zhang Y, Dua B, Wei D, Wei Q, Zhao Y (2013b) Enhancement effect of silver nanoparticles on fermentative biohydrogen production using mixed bacteria. Bioresour Technol 142:240–245

Chapter 5
Green Route Synthesized Iron Nanoparticles for Biohydrogen Production

Dharm Pal, Dhruti Sundar Pattanayak, Jyoti Mishra, and Naresh Kumar Sahoo

Abstract In today's contemporary culture, detecting and perceiving a sustainable source of energy is a big issue. Since the previous decade, the majority of academics have suggested hydrogen as one of the cleanest fuels, and its demand has been increasing ever since. Hydrogen has the highest energy density and hence outperforms all other fuels. As a consequence of using fossil fuels, hydrogen emits carbon dioxide, which has a detrimental impact on the environment. However, biohydrogen is credited as the productive and cleanest energy source due to no emission of harmful greenhouse gases and different feedstocks and biomass can be used to create biohydrogen. Although dark fermentation and photobiological processes have low yield and rise as a key barrier to industrial production of biohydrogen. Because of its precise structural and physiochemical characteristics, nanotechnology stands out as having a strong promise for improving biohydrogen generation. The usage of iron-based nanoparticles (NPs) enhances the biohydrogen formation from biomass and may lower the costs of production due to their use as a catalyst in extremely small amounts. Furthermore, iron-based NPs synthesized in a green way is promoted for renewable energy generation from biomass with a low costs via thermochemical and biochemical pathways, where the primary expense is likely to be related to catalyst synthesis. Despite the fact that the green synthesis method of iron-based NPs is sustainable and nontoxic, a lack of precise knowledge about green synthesis and its process remains an issue. In this study, several known methods to synthesize iron-based NPs, like utilizing microorganisms and green plants, have been thoroughly explored, as well as the potential processes involved. Finally, we show and discuss the uses of catalytic nanomaterials (NMs) based on iron in biochemical as well as thermochemical energy generation (e.g., biohydrogen).

D. Pal · D. S. Pattanayak (✉)
Department of Chemical Engineering, National Institute of Technology Raipur, Raipur, Chhattisgarh, India

J. Mishra · N. K. Sahoo
Department of Chemistry (Environmental Science and Technology Program), ITER, Siksha'O'Anusandhan (Deemed to be) University, Bhubaneswar, Odisha, India

© The Author(s), under exclusive license to Springer Nature Singapore Pte Ltd. 2023

M. Srivastava, P. K. Mishra (eds.), *NanoBioenergy: Application and Sustainability Assessment*, Clean Energy Production Technologies,
https://doi.org/10.1007/978-981-19-6234-9_5

This thorough study focused on the green route synthesis of iron-based NPs and its uses to propel biohydrogen generation methods that exists previously toward sustainable commercialization through wastes to value-added technologies. Besides, the possible mechanism of NPs synthesis via green route and future direction in the production of sustainable bioenergy has also been critically reviewed.

Keywords Iron nanoparticles · Green synthesis · Biohydrogen · Bioenergy · Sustainable biofuels

5.1 Introduction

Owing to the world's fast-rising population and industrialization, there has been an increase in global energy consumption. With growing worries about the global environment, rising oil costs, and the health risks connected with airborne contaminations, investigators are working effortful to create durable refineries to get hygienic sources of energy to substitute existing conventional fossil-based fuels (Shanmugam et al. 2020). Renewable energy has the ability to solve many of the problems associated with fossil fuels. Biofuels are regarded as promising and sustainable renewable energy solutions in this context since they are green and nontoxic, easy to manufacture, eco-friendly, highly efficient, and cost-effective (Sekoai et al. 2019). Liquid and gaseous biofuels are bioethanol, biobutanol, biodiesel, bio-oil, biogas, biomethane, bioethane, biobutane, and biohydrogen (Ladole et al. 2017). Biohydrogen has attracted the most attention among the available biofuels due to many benefits. These include the ability to produce nonpollution by-products in the form of vaporized water, an energy of 120 kJ/g, the capability to use a broad range of bacteria and feedstock found in various natural environments, and at normal ambient pressure and temperature great capability for large-scale production (Zilouei and Taherdanak 2015). Biohydrogen may be created by several biological operations such as bio-photolysis (direct and indirect), microbial electrolysis, and fermentation techniques like dark fermentation and photo-fermentation (Srivastava et al. 2019; Sekoai et al. 2019). The biological routes to generate biohydrogen employ a variety of organic substrates and microorganisms. Lignocellulosic biomass is regarded as the most flexible substrate in the group of organic species due to its high cellulose content, renewable nature, and wide availability and that may be utilized for large-scale biological production of biohydrogen (Wang et al. 2017a). The technology to produce biohydrogen through biomass is one of justifiable the most sustainable idea of biofuel production from biomass; there are certain drawbacks that must be addressed before its industrial-scale implementation (Nikolaidis and Poullikkas 2017).There are some effective pressing obstacles to the production of sustainable cellulosic biohydrogen, the technology requires incomplete or partial conversion of cellulose to sugar, the presence of lignin in biomass of cellulose, high cost of cellulolytic enzymes, and low conversion of substrate to product (Dincer and Acar 2014). The process to convert biomass into biohydrogen is significantly influenced by the participation of cellulolytic enzymes. Nonetheless, there are numerous problems included, such as high costs for enzyme, less enzyme production and working efficiency, resulting in low sugar yield and therefore a significant influence on biohydrogen generation yield (Kogo et al. 2017). As a result, boosting the availability of cellulose for bioconversion reactions and increasing in enzyme productivity and efficiency are the primary bottlenecks that must be addressed in order to make cellulosic biohydrogen generation method more viable and optimal on an actual level. A variety of techniques were developed to enhance the production of enzyme and its efficiency, and pace of biohydrogen generation; however, they are still far from sustainable and realistic application in terms of economics (Cipolatti et al. 2016). The function of nanotechnology in many biological areas has been thoroughly covered in this reference. Because of their unique characteristics, nano-materials (NMs) involvement could be critically significant in improving the entire biomass to biohydrogen conversion process (Kumar et al. 2019; Pugazhendhi et al. 2019). Each phase of cellulosic biohydrogen production technology can be catalyzed by properties such as strong electro-conductivity, a high surface-to-volume ratio, and greater surface area (Taherdanak et al. 2015). Some nanoparticles(NPs), including nickel, iron, copper, silver, gold, and titanium, have been shown to improve biohydrogen generation by diverse biological routes such as dark and photo-fermentation and bio-photolysis (Taherdanak et al. 2015; Lin et al. 2016). NMs, particularly iron, play an important character in improving biohydrogen generation yield by serving as cofactors on the active sites of nitrogenase and hydrogenase enzymes (Srivastava et al. 2020). As a result, the efficiency of a process may be dramatically altered on the nanoscale by employing iron-based nano-materials (Kandel et al. 2014). The iron content of several categories of biofuels preparation from biomass methods might impact the metabolic and enzymatic activity of microorganisms by modifying the rate of electron transfer. For example, the cellulolytic enzyme's stability and synthesis for the formation of sugars capable to get fermented have been extensively described (Srivastava et al. 2015), where iron impacts the hydrogenase and nitrogenase enzyme activity of the biological hydrogen production process. Depending on the reaction circumstances, iron nano-materials having a wide range of sizes and morphologies may be produced by a variety of pathways including chemical, physical, biological, and hybrid techniques (Huang et al. 2014). Physical and chemical pathways have become the most widely used techniques for synthesizing iron-based nano-materials for utilizing widely, and are thus referred to as standard methods. These physical techniques of nano-material preparation are both costly and time-consuming, rendering them unsuitable for large-scale manufacturing (Stefaniuk et al. 2016). Every stage of the chemical techniques of NMs production involves the usage of hazardous substances. Solvents include reducing agents like sodium borohydride ($NaBH_4$), stabilizing agents like Thiolates, polyethylene glycol (PEG), and other organic-based toxic chemicals were used frequently (Ahmmad et al. 2013). Furthermore, a chemical approach to NMs preparation has a number of disadvantages, including the production of toxic intermediates as well as by-products are highly harmful to the ecosystem and fatal to living creatures (Stefaniuk et al. 2016). Furthermore, this technique has slow rate of formation of NMs, and small size of NMs with deformed structures are often produced with enhanced reactivity and toxicity (Ahmmad et al. 2013). Furthermore, it is susceptible to agglomeration because of attractive factors like van der waals pressures. Agglomeration may be addressed by the use of particular capping agents, which raises the manufacturing cost and makes it less viable for industrial scale usage (Narayanan et al. 2012). The disadvantages of traditional techniques, such as high costs and toxicity concerns, may be solved by a green and sustainable synthesis pathway, which has gained a lot of interest from people all over the world (Ali et al. 2016). The green method to NMs production is extremely efficient, quick, less priced, and toxic free, all while being environmentally benign. Under ambient physicochemical circumstances, NMs may be produced utilizing a variety of living sources such as bacteria, animal tissues, fungus, plant extracts, biomass, and algae (Saratale et al. 2018; Srivastava et al. 2021). Given these facts, the aim of this investigation is to explore the many obtainable green and sustainable methods for synthesizing iron-based NMs. Different techniques to green NMs production employing microbes and plants have been thoroughly addressed. Finally, the utilization of iron-based NMs in biohydrogen production, as well as the influence of its synthesis process, have been investigated, as have current limits and possible solutions.

5.2 Green Routes Synthesis of Iron Nanoparticles

Metallic NPs are prepared involving a variety of techniques, including physicochemical and biological approaches, as shown in Fig. 5.1. The bottom-up and top-down approaches are including physicochemical procedures and biological methods (Pattanayak et al. 2020). The term "green route" refers to a safe, minimal hazardous, and environmentally friendly method of synthesizing NPs that primarily use renewable energy sources. Therefore, in this synthesis method, microorganisms and plants are employed as operating bio-components, resulting in minimal energy usage and eco-friendly solvents, mostly water (Pattanayak et al. 2021). Synthesis aided by plants is more practicable than microorganism-aided synthesis owing to several intrinsic benefits. The NPs are fabricated by combining bioactive ingredients that serve as both capping and reducing agent. Thus, in green production of iron NPs, bioactive chemicals play an important role in lowering iron ions (Fe^{+2} and Fe^{+3}) (Sadhasivam et al. 2020). Green-produced NPs are often maintained in a liquid solution at a particular pH to ensure consistent size and stability. Certain biocompatible agents and surface coatings are utilized to improve stability, converting it into inert material when it comes into touch with natural ambient fluids. To minimize moisture contact the maintaining of vacuum for nanopowders is required, which might induce a cluster of iron NPs produced (utilizing plant extracts). By employing a green synthesis methodology for NPs production, the hazardous by-product was minimized (Ali et al. 2020). Greenly synthesized iron NPs can also be used in biological, catalytic, environmental, and energy applications (Mondal et al. 2020). Ionic solvent (a more hazardous organic solvent) can be substituted in certain

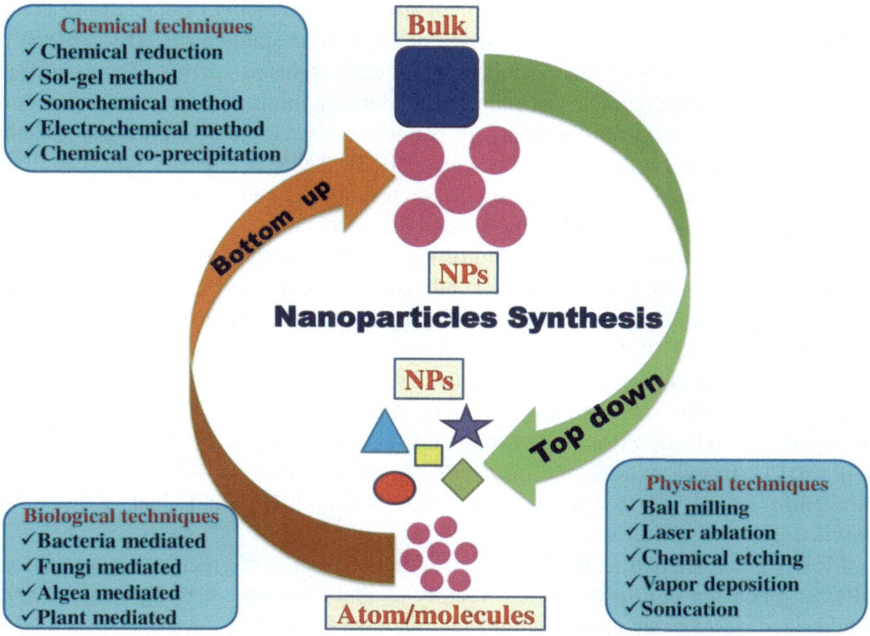

Fig. 5.1 Bottom-up and top-down approaches for nanoparticle synthesis

biosynthetic processes utilizing living entities (Bolade et al. 2020). This chapter examines several green approaches for producing iron-based NPs. The use of green produced iron NPs to accelerate biohydrogen production is also thoroughly explored.

5.2.1 Synthesis of Iron NPs from Plant Biomaterials

NPs synthesized by microorganisms are with restricted dispersion and a slow rate evolution in comparison to plant-aided preparation (Dhillon et al. 2012). Phyto-synthesized metallic NPs are cost-effective, simpler, and has good reproducibility and repeatability. It is widely known that plant extracts are best suited for generating extremely steady metallic NPs at a high rate and in huge amount (Iravani 2011). Production of NMs aided with plants is favored to environment as strong biomolecular species causing reduction can be derived by using different plants (Mukunthan and Balaji 2012). In plants and some variety of herbs, antioxidants are the active phytochemical components in leaves, stems, barks, seeds, and flowers. Plant metabolites like protein, enzymes, amino acids, cyclic peptides, polyphenols, flavonoids and other compounds, and caffeine have the potential to function as green bioreductants and stabilizers in the synthesis of iron NPs. Summarized as iron ions attach to these metabolites and formation of atomic iron by reduction, resulting in the creation of an iron nucleus. Iron nuclei develop into stable iron NPs. At room temperature, this reaction generally completes with vigorous stirring. The synthesis of NMs using phytochemicals of plants considerably minimizes the environmental pollution, making a current standard in economically viable and ultra-sustainable clean and green technology (Zambre et al. 2013). Table 5.1 shows how different plant parts have been used as sources of green reducing and capping agent used in the synthesis of iron-based NMs by diverse researchers.

The synthesis of iron NPs by using different plant extracts has been reported by many researchers. The iron NPs were synthesized by reducing $Fe^{+2/+3}$ by taking leaf extract of *Green, oolong, and black tea* (Huang et al. 2014); *Green tea* (Asghar et al. 2018; Xiao et al. 2020). Plant sources caused reduction of metallic precursors for the preparation of Zero-valent Iron nano-materials (nZVI), e.g., the *Camellia sinensis* (Hoag et al. 2009), *Dodonaea viscosa* (Kiruba Daniel et al. 2013), *Eucalyptus* (Wang et al. 2014a), extract of shoots and leaves of blueberry (Manquián-Cerda et al. 2017), flower extract of *Calotropisgigantean (CG) and Hibiscus sabdariffa* (Sravanthi et al. 2018; Khan and Al-Thabaiti 2018) and leaf extract of *Emblicaofficinalis* (Kumar et al. 2015).

Mostly green tea extract has been utilized in the process of green synthesis of iron NPs, as green tea is a low-cost and readily available resource. Hoag et al. (2009) synthesized nZVI from a polyphenol-rich green tea (*Camellia sinensis*) extract. NPs with good stability were produced at ambient temperature without any polymer or surfactant addition. Plant polyphenols function as reducing agents as well as a capping agent, and we have green nZVI of nanoscale of particular properties with good stability (Hoag et al. 2009). Machado et al. (2013) investigated the possibility of producing nZVI from various tree leaves. Furthermore, the antioxidant capability of leaf extracts was calculated. According to the findings, extracts generated by dried leaves with greater antioxidizing capabilities than extracts from non-dry leaves (Machado et al. 2013). In second research, Machado et al. (2014) reduced ferric ions using fruit waste component-based extracts like albedo, peel, pulp of lemon, orange, mandarin, and lime. Their findings revealed that the fruit fractions-based extracts generated nZVI of varying shapes and sizes as well as reactivity (Machado et al. 2014). Wang et al. (2014a) used Eucalyptus leaves in similar work to manufacture iron oxide NPs using a relatively simple and efficient approach. During the experiments, the leaf extract was mixed with 0.10 Molar $FeSO_4$ solution in a 2:1 volume ratio. The transformation of the hue of solution from yellow color to black color indicated the production of iron oxides (NPs). Circular NPs having a diameter of 20–80 nm were discovered. Furthermore, the prepared iron NPs were used to remediate swine wastewater, and get 71.7% removal of whole nitrogen (N) and chemical oxygen demand was removed up to 84.5% (Wang et al. 2014a).

As a result of its renewable nature, near-zero cost availability, high concentration of bioactive compounds, and ease of operation in ambient conditions, the green path for synthesizing the NMs based on iron are economical as well as eco-friendly, that may cause the lowering of the cost of production as well as the reduction in operational cost in which these NMs are used. Furthermore, in the comparison

Table 5.1 Green synthesis of iron nanoparticles using plant sources

S. no.	Plant name	Plant part	Biomolecule	Nanoparticle	Precursor	Morphology	Size (nm)	References
1.	*Camellia sinensis*	Leaves	Polyphenols, caffeines	nZVI	$Fe(NO_3)_3$	Spherical	5–15	Hoag et al. (2009)
2.	*Dodonaea viscosa*	Leaves	Carbonyl group, polyhydroxy group in santin	nZVI	$FeCl_3$	Spherical	27	Kiruba Daniel et al. (2013)
3.	*Grape marc (GM), Black tea (BT), Vine leaves (VL)*	Leaves	Polyphenols	nZVI	Fe(III) solution	NA	15–45	Machado et al. (2013)
4.	*Green, oolong and black tea*	Leaves	Caffeine and polyphenols	Fe-NPs	$FeSO_4 \cdot 7H_2O$	Hexagonal	40–50	Huang et al. (2014)
5.	*Citrine juice waste*	Pulp and peel	Polyphenols and antioxidants	nZVI	$FeCl_3 \cdot 6H_2O$	Spherical, cylindrical, or irregular	3–300	Machado et al. (2014)
6.	*Eucalyptus*	Leaves	Polyphenols, epicatechin and glucuronide	nZVI	$FeSO_4 \cdot 7H_2O$	NA	20–80	Wang et al. (2014a)
7.	*Eucalyptus and green tea*	Leaves	Polyphenols and antioxidants	nZVI	$FeSO_4$	Quasi-spherical	20–80	Wang et al. (2014b)
8.	*Emblica officinalis*	Leaves	Polyphenols and ascorbic acid	nZVI	$FeCl_3.6H_2O$	Spherical	22.6	Kumar et al. (2015)
9.	*Rosa damascene, Thymus vulgaris, and Urticadioica*	Leaves	Polyphenols, terpenoids, proteins, and organic acids	nZVI	$FeCl_2.4H_2O$	Nonuniform shape	100	Fazlzadeh et al. (2016)
10.	*Blueberry*	Leaves and shoots	Polyphenols, chlorogenic acid, ellagic acid	nZVI	$FeCl_3.6H_2O$	Irregular shaped	80.2	Manquián-Cerda et al. (2017)
11.	*Nettle and Thyme*	Leaves	NA	nZVI	$FeCl_2.4H_2O$	Irregular shaped	20–80	Leili et al. (2017)
12.	*Tea*	Leaves	Polyphenol	nZVI	$FeSO_4.7H_2O$	Irregular shaped	20–80	Wang et al. (2017b)

(continued)

Table 5.1 (continued)

S. no.	Plant name	Plant part	Biomolecule	Nanoparticle	Precursor	Morphology	Size (nm)	References
13.	*Moringaoleifera*	Seed	–	Fe-NPs	$FeCl_3.6H_2O$	Spherical	2.6–6.2	Katata-Seru et al. (2017)
14.	*Green tea and black tea*	Leaves	Polyphenols, caffeine, and terpenoids	Fe-NPs	$FeCl_3$	Spherical	42–60	Asghar et al. (2018)
15.	*Tea*	Waste tea	–	Fe-NPs	$FeSO_4.7H_2O$	Round shape	98.786	Gautam et al. (2018)
16.	*Calotropisgigantea (CG)*	Flowers	Polyphenol	nZVI	$FeNO_3.9H_2O$	Spherical	50–90	Sravanthi et al. (2018)
17.	*Hibiscus sabdariffa*	Flowers	Polyphenol	nZVI	$FeCl_3.6H_2O$	NA	NA	Khan and Al-Thabaiti (2018)
18.	*Nephrolepisauriculata*	NA	Flavonoids, polyphenols, organic acids	Fe-NPs	$FeCl_3.6H_2O$	Spheroidal shaped	40–70	Yi et al. (2019)
19.	*Grape*	Seed	Polyphenol	Fe-NPs	$FeSO_4.7H_2O$	Quasi-spherical	50–100	Guo et al. (2020)
20.	*Green tea*	Leaves	Polyphenols and caffeine	Fe-NPs	$FeSO_4.7H_2O$	Regular Spherical or ellipsoidal shape	75–100	Xiao et al. (2020)
21.	*Korla fragrant pear (Pyrussinkiangensis Yu)*	Peel	Polyphenols, flavonoids	nZVI	$FeSO_4.7H_2O$	Irregularly shaped	20–90	Rong et al. (2020)

Fe-NPs Iron nanoparticles, *nZVI* Nano zero-valent iron nanoparticles, *NA* Not available with microbial-based preparation technique, based on plant approaches are significantly low-cost, free of maintenance, practically nontoxic, and feasible as natural capping agents, and so might be selected over the microbial preparation techniques.

5.2.2 Synthesis by Microorganisms

The manufacturing of microbe-assisted NPs has acquired in crowd owing to its advantages over the conventional chemical-based processes. The advantages involve the preparation at normal room temperature, efficient in energy conversion, easily found in nature and sustainable precursors, simplicity in the similitude and robustness (Park et al. 2016). Manufacturing of NPs through extracellular or intracellular mechanisms could be done by the microorganisms such as yeast, fungus, and bacteria. These techniques involve enzymes reducing metallic ions and generating well-dispersed NPs having reduced distribution of mean particle diameter. Peptides, natural proteins, capping agents, and tannins get integrated over the surface of NPs. By minimizing agglomeration, dispersion and stability of NPs are enhanced by such surface coating (Singh et al. 2016). The internal process involves metallic ions diffusion inside the cell, where ions to create NPs are decreased by enzymes. On the other hand, the extracellular operation involves the attraction of metallic ions to the cell membrane by electrostatic forces and the reduction of metallic ions by enzymes. Table 5.2 shows the microorganisms used by different research groups, as well as the synthesis method used to synthesize iron and iron-based NMs as reducing and capping agents.

An investigation on the extracellular production of maghemite (Fe_2O_3) with a mean size of particles 5–7 nm was reported by the action of sulfate and iron reductase enzymes consisted in Actinobacter sp. after a 48 h of incubation at ambient temperature (Bharde et al. 2008). The fungus was used to synthesize iron NMs with particle sizes ranging from 10 to 24.6 nm in the exposure of 10.3 molar Aspergillus oryzae TFR9 fungal culture to $FeCl_3$ precursor salt for 12 h at a temperature of 28 °C in a rotatory shaker having rotational speed of 150 rpm (Tarafdar and Raliya 2013). Subramaniyam et al. (2015) described the production of iron NPs having a spherical shape with size ranges from 20 to 50 nm using, Chlorococcum sp., soil microalgae, and chlorides of iron precursor in a separate research. The study shows that the carbonyl and amine functional groups present on glycoproteins and polysaccharides consisting of the algal cells were important in the formation of NPs, which was revealed by FTIR analysis (Subramaniyam et al. 2015). In another work, Das et al. (2018) stated that by utilizing the native hypersaline sulfate-reducing bacterial strain (LS4), maghemite NPs are produced via anaerobic method having a mean diameter of 18 nm. The bacterial strain is obtained and grown using saltpan sediments from Goa, India (Das et al. 2018). After a 96-h incubation period, iron NMs were synthesized using bacteria Bacillus pasteurii (PTCC1645), Bacillus subtilis (PTCC1254), and Bacillus licheniformis (PTCC1350) in concentrations of 1.34 g, 1.62 g, and 1.26 g, respectively, the production of nano-materials based on magnetite

Table 5.2 Green synthesis of iron nanoparticles using microorganism sources

S. no.	Microorganism	Precursor	Mode of synthesis	Nanomterial	Shape	Size (nm)	Reducing agent	References
1.	*Actinobacter sp.*	$FeCl_3$	Extracellular	Fe_2O_3	NA	5–7	Iron reductase	Bharde et al. (2008)
2.	*Aspergillus oryzae TFR9*	$FeCl_3$	Extracellular	Fe-NPs	Spherical	10–24.6	Extracellular proteins	Tarafdar and Raliya (2013)
3.	*Chlorococcum sp. MM11*	$FeCl_4$	Intra- and extracellular	Fe-NPs	Spherical	20–50	NA	Subramaniyam et al. (2015)
4.	*Escherichia coli*	$FeSO_4$	Extracellular	Fe-NPs	Spherical	23 ± 1	NA	Crespo et al. (2017)
5.	*Desulfovibrio sp.*	$FeCl_3$+ $FeSO_4$	Extracellular	Fe_2O_3	Rounded	18	Ferritin protein, reductase enzyme	Das et al. (2018)
6.	*Bacillus licheniformis (BL), B. subtilis (BS), B. pasteurii(BP)*	$FeCl_2$+ $FeCl_3$	Intracellular	Fe_3O_4	NA	37–97	Urease enzyme	Daneshvar and Hosseini (2018)
7.	*Bacillus cereus HMH1*	$FeCl_3{\cdot}6H_2O$	Extracellular	Fe_3O_4	Spherical	29.3	Glucose, fructose and sucrose	Fatemi et al. (2018)
8.	*Lactobacillus Fermentum*	$FeSO_4$	Extracellular	Fe_3O_4	Spherical	10–15	NADH reductase, nitrate reductase, cysteine, histidine, aldehydes, and ketones	Fani et al. (2018)

NA Not available

of size range from 37 nm to 97 nm via the reducing action on the mixture of $FeCl_2$ and $FeCl_3$ precursor salts with the ratio of 1:2.5 and the reduction caused by the reducing agents ammonia and urease enzyme (Daneshvar and Hosseini 2018).

NMs using microorganisms are synthesized by the following bottom-up approach. This technique allows the iron-based NMs synthesis at ambient settings, decreases energy requirements, less production or no hazardous wastes is justifiable, and may thus be used to synthesize the NMs in an environmentally benign way. Although abovementioned techniques entail a complex and time-consuming operation in the perspective of microorganism isolation, preservation, maintenance, and purification of their cell cultures, necessitate highly germfree circumstances for microbial culture, exhibit a low rate of formation of NMs, and may pose bio-hazardous conditions. Aside from that, microorganisms such as bacteria have a limited life cycle with a brief phase of immobilization thus they generate comparably fewer secondary metabolites for a small period of time and are thus unsuitable for the production in terms of long duration. Furthermore, owing to the microorganism's brief life cycle, its metabolites are depleted as it completes its life cycle. All of the aforementioned disadvantages make this technique to synthesize NMs extremely difficult for wide-ranging production.

5.3 Mechanism of Green Route Synthesized Iron Nanoparticles

Though the exact process of NPs formation by a living organism is unknown, studies suggest that enzymes generated by bacteria and fungus, as well as biomolecules, especially phenolic chemicals found in plants, promote the formation of NPs based on metallic iron (Saif et al. 2016). The green approach to NPs manufacturing is a simple and highly efficient technology. To begin, the extracted solution the biomolecules are extracted from the biomaterials like animal tissues, plant materials, or biomass by boiling at a high temperature in the deionized water, and after that by filtering (Ali et al. 2016). The solution extracted by biomaterials is extremely high in secondary metabolites, the concentration of which varies depending on the kind of biomaterial employed. These secondary metabolites, which include nucleoproteins, proteins, polyphenols, amino acids, flavonoids, and others are excellent capping and reducing species and cause reduction of the precursor salts of metals which are mixed to the extracted solution in a specific ratio (volume/volume) at ambient temperature for the formation of the NPs based on corresponding metals (Wang et al. 2017b). The production of iron NPs may be simplified into three steps: activation, growth, followed by termination. The initial phase in the production of NMs is the activation phase, in which the precursor of metallic salt is ionized and gets dissociated to free metallic ions with a positive charge. These metallic ions bind to the bioactive components having negative charge present in the extracted solution like alkaloids, amino acids, polyphenols, enzymes, flavonoids, vitamins, proteins, quinines, and so on and form an intermediate complex (Malik et al. 2014). These bioactive chemicals cause chelation and decrease the oxidation state of metallic ions, which nucleates the reduced metallic atoms and the creation of the NMs. Furthermore, the second stage is the growth, in which growth of random sizes, formation of metallic NMs with diverse morphologies by the combining of the nucleated metallic atoms of the initial stage, the shape may be spherical, rods, pentagonal, hexagonal, cubical, and so on, it depends upon the different parameters such as pH, metallic precursors nature, temperature, incubation time, types and sizes, and so on. However, because of the prolonged nucleation, particle agglomeration occurs, resulting in huge clusters with deformed shape (Narayanan et al. 2012). Thus, in order to generate particles with the required shape, the thermodynamic energy of the particles must be reduced in order to prevent agglomeration and stabilize the NMs formed. The termination phase of NMs production follows the growth phase. The termination phase characterizes the thermodynamically stable produces NMs (Shamaila et al. 2016). Random development of NMs is slowed by covering the nucleated particles with a capping agent or surfactant. Energy consisted in the particle's surface is lowered by the capping agents, preventing nucleated ligands from developing further and therefore preventing aggregation (Stefaniuk et al. 2016). In this approach, capping agents enable us to produce NMs with the required shape, which is dependent on the reaction conditions. Furthermore, plant-based synthesis is more suitable in the comparison of synthesis enabled by the microbes due to the formation of a greater amount and varieties of different metabolites that can aid in the functions of the capping and reducing during NPs formation (Srivastava et al. 2021). Figure 5.2 depicts a possible process to produce the biogenic iron-based NPs.

5.4 Applications of Green Iron Nanoparticles for Biohydrogen Production

In the present anthropocene, energy generation and consumption are unsustainable in terms of environmental, economic, and social effects. Through climatic change, loss of ozone layer, and biosphere and devastation of geosphere, fossil fuels, and other nonrenewable energy source increases environmental harm while also affecting human health. Almost all the required energy (>95%) in the transportation sector (which contributes 14% to global warming) is satisfied by fossil fuels. Especially, energy production accounts for more than 80% of worldwide CO_2 emissions. Since burning fossil fuels meet more than 90% of world energy demand, CO_2 concentrations in the atmosphere have risen from 280 to 370 ppm during the last 150 years. It is predicted to reach a dangerously high level of 550 ppm by the late twenty-first century. As a result, research has attempted to adapt the nano-revolution to the renewable energy industry in order to alleviate such environmental devastation. Bioenergy may be regarded as an eco-technological breakthrough in order to achieve

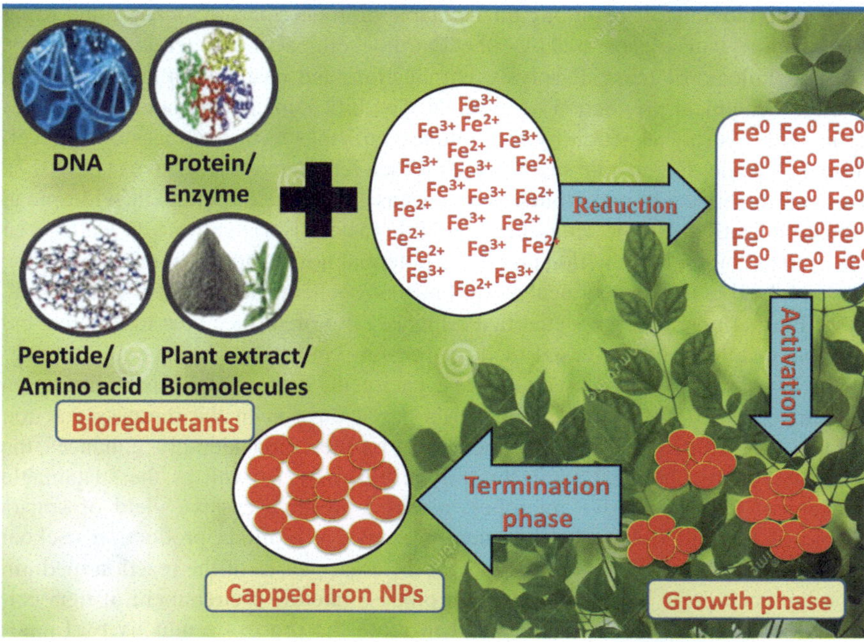

Fig. 5.2 Mechanism of green route synthesized iron nanoparticles

energy sustainability. The sunlight is an eventual source of energy and it may be used indirectly as a source of plant-based energy (bioenergy or biofuels). In the framework of green renewable energy, some nations modified legal elements in response to the energy crisis caused by the over-exploitation of nonrenewable energy sources and the consequent environment degradation, giving encouragement to sustainable energy production techniques. In December 2015, 147 member nations of the United Nations Framework Convention on Climate Change (UNFCCC), Paris (COP21), and agreed to keep the limitation of global warming to 1.5 °C. Global researchers have been on the hunt for green sustainable bioenergy generation to decarbonize the environment in order to fulfill the COP21 objective. Biomass presently accounts for a sizable percentage (9–14 wt%) of the world's primary energy. The US Department of Energy Biomass Program has proposed a scenario in which biofuels may provide 30% of motor fuel demand by 2030 (Rai et al. 2018). Because rising population and industrialization need a large demand for energy, energy is a key prerequisite for a nation's development goals globally. Due to the loss of fossil fuel sources and their negative impact on the ecosystem and living things, there is an ongoing quest for a renewable, environmentally benign, and economically viable source of energy (Liu et al. 2019). Biofuel is a possible replacement for fossil fuels among existing renewable energy alternatives since it is eco-friendly, clean, emits no pollutants, and is a renewable source of energy sustained by nature itself. Biohydrogen, biobutenol, biogas, biomethane, biodiesel, bio-oil, and bioethanol are among the available biofuels (Ibarra-Gonzalez and Rong 2019). Biofuel can be generated by either thermochemical or biochemical hydrolysis of lignocellulosic biomass. Pyrolysis, combustion, and gasification are all steps in the thermochemical production of biofuels. The biochemical production of biofuel begins with hydrolysis action by the cellulase enzymes to the biomass of lignocellulose or organics to generate sugar, which is then fermented (Srivastava et al. 2017). However, the huge cost and limited output act as a key impediment to its widespread commercialization. By incorporating iron NMs (as catalyst) synthesized via a green route into the biofuel production process, a potential novel approach for lowering the costs involved in the biochemical as well as thermochemical methods (Sekoai et al. 2019). Furthermore, it is expected that a higher ratio of surface area to volume and super-paramagnetic nature-based facile segregation influenced by an external magnetic field will result in increased loading capacity and decreased diffusion limitation (Vaghari et al. 2016). Implementing NMs based on iron in the biomass reaction medium for the production of fermentative biofuels considerably enhances the efficiency, thermochemical stable hydrolytic enzymes, as well as the sustainable engagement of fermentative microbes, resulting in an excessive yield of sugars capable of fermentation and subsequently excessive biofuels production (Sekoai et al. 2019). NMs based on iron may affect different phases of the reaction medium in the lignocellulosic biofuels production process like the pretreatment of lignocellulosic biomass to biofuels synthesis, to increase the process productivity. Lignocellulosic biomass pretreatment process employed by the iron-based NMs increases sugar production, which may assist to minimize the quantity of chemical reagent needed, lowering the total costs involved in the pretreatment operation (Arora et al. 2020; Srivastava et al. 2021).

There are several environmentally friendly techniques for producing iron-based NMs that employ various bio-reducing and biochemical agents. Details about energy generation, namely biohydrogen production utilizing greener iron NPs, are briefly given here.

5.4.1 Biohydrogen

Anaerobic bacteria are used to provide molecular hydrogen via several metabolic routes. Time of retention, pH, substrate concentration, and temperature influence the kinetics behind the process of metabolism. In the procedure to prepare NPs in the anaerobic biohydrogen, the capability to quick reaction with electron donors and electron switch is improved (Rittmann and Herwig 2012). Therefore, addition of NPs stimulates the microorganism's activities in dark or photo-fermentation biohydrogen processes. In the current decades, we have three well-advanced methods which are applied for biohydrogen yielding named as (1) dark fermentation, (2) photo-fermentation, and (3) photocatalysis (Chandrasekhar et al. 2015; Kumar et al. 2021). Generally, there are numerous approaches for producing biohydrogen, such as direct and indirect photolysis, microbial electro-hydrogenesis cells, dark fermentation, photo-fermentation, and hybrid systems. One of the benefits of producing biohydrogen using the methods described above is the high process effectiveness, and one of the familiar sources is biomass (Osman et al. 2020). Lower eukaryotes (Protista and Chlorophyceae algae), facultative aerobic and anaerobic bacteria, archaea, and cyanobacteria have all been utilized in various biological methods for biohydrogen generation (Chandrasekhar et al. 2015). Using trash from diverse activities as a substrate not only simplifies waste management but also results in the cheap generation of hydrogen as a clean and green fuel. Every technology has its own working conditions and forms, as well as benefits and downsides, which are mentioned below (Nabgan et al. 2021).

5.4.2 Mechanism for Biohydrogen Production Using Iron Nanoparticles

Because of their high selectivity, greater catalytic performances, and wide surface area, NPs improve the performance of dark fermentation process involved in the biohydrogen production. NPs increase the productivity of hydrogen and production in anaerobic bacteria aiming directly at the hydrogenase enzymes and improving electron transfer efficiency (Lin et al. 2016). A prior study described a method for iron NPs to produce hydrogen in the dark fermentation utilizing microbes (Bordetella, Bacillus, Proteus, Pseudomonas sp., and Enterobacter) (Porwal et al. 2008). Initially, NPs enter into the cell of bacteria via the cell wall at biohydrogen generation, and this process and characteristic affect the rate of electron transfer. As a result of increased surface area and quantum size, ferredoxin oxidoreductase performance is enhanced. The activity of ferredoxin oxidoreductase increases hydrogen generation during dark fermentation (Mohanraj et al. 2014a). Lin et al. also proposed that the conductivity of FeO NPs which is efficient, may be responsible for the improvement in the rate of electron transfer, preparing them as beneficial in the generation of hydrogen (Lin et al. 2016). However there are two strategies for hydrogen generation metabolically by fermentation of glucose, first strategy is the formation of formate hydrogen and the second is the decreased nicotinamide adenine dinucleotide (NADH) dependent hydrogen manufacturing pathway (Cai et al. 2011). In the formate hydrogen strategy, formate is decomposed by the enzyme formate hydrogenlyase. Reoxidation of NAD happens in the NADH-established hydrogen generating strategy with the help of the enzyme hydrogenase. Lin et al. found that adding FeO NPs to dark fermentation enhanced acetate synthesis while decreasing ethanol concentration (Lin et al. 2016). The basic mechanism of biohydrogen production by using green synthesized iron NPs is shown in Fig. 5.3 and discussed briefly.

The conversion of 1 mole of glucose into ethanol requires 4 moles of NADH, but acetate production does not. A change in the metabolic strategy speeds the H^+ to hydrogen conversion through NADH, ensuring a higher yield of biohydrogen.

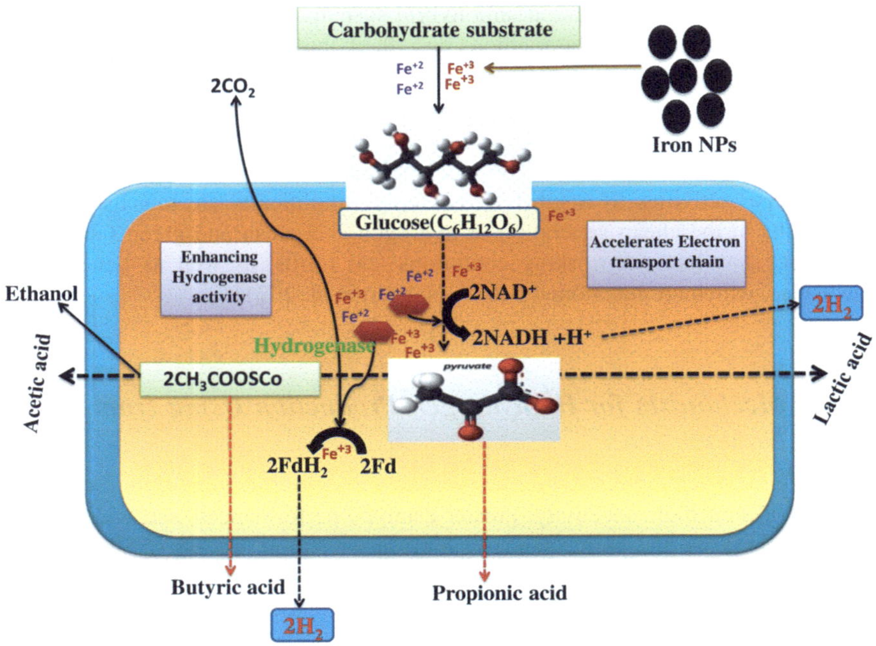

Fig. 5.3 General mechanism of biohydrogen production using iron nanoparticles (NPs)

According to Jiang et al. NPs help in the effective transmission of electrons among microbial cells. Previous research found that the ideal dosage of NPs employed was toxic less or inhibiting to microorganisms and was efficient for creating greater quantities of hydrogen through dark fermentation (Jiang et al. 2014). Nonetheless, it is important to remember that greater NP concentrations reduce hydrogen output and cause cell death. Extracellular appendages may serve as electron transfer routes between bacteria. Electrical impulses, were transported by bacterial nanowires for example, in Shewanella oneidensis (El-Naggar et al. 2010). Effectiveness of methane formation has been increased up to 33% in syntrophic bacteria by NPs like nanosized magnetic NPs, apart from the hydrogen generation (Kumar et al. 2019).

5.4.3 Green Iron Nanoparticles for Biohydrogen Production

With better-designed catalytic transformations, to produce biohydrogen, continuous progress has been observed in the development of processes that decompose naturally depending on the numerous forms of biomass. The many approaches to synthesize catalysts and its applications all have been aimed at increasing the particular and general characteristics of the catalysts therefore boosting the overall efficiency of the system of biomass conversion aided by catalysts. Such improvements in catalysis activity may be obtained very easily by choosing and customizing optimal metallic combinations, as well as any other conditions or green components such as NPs. This section of the review is reviewed by choosing research papers and focused on significant findings from studies that address the topic of improving the circumstances of catalytic synthesis of biohydrogen utilizing green-produced iron NPs (Mudhoo et al. 2018) (Table 5.3).

Mohanraj et al. claimed that pure cultures of E. cloacae 811101 fed with glucose demonstrated a 21.8% increase in H_2 generation yield with Fe_2O_3 NPs concentration of 200 mg/L. It has been proposed that the kind of sugar used as feed has a substantial impact on H_2 generation by E. cloacae 811101 with the existence of Fe_2O_3 NPs (200 mg/L). Glucose was discovered to be a superior diet, yielding 21.8% more H_2 than sucrose (4.8%) (Mohanraj et al. 2014a). This work demonstrated that Fe(0) NPs may efficiently accelerate hydrogen generation and the fermentation route of biomass materials. Fe_2O_3 NPs of concentration 175 mg/L, C. acetobutylicum NCIM2337 improved its H_2 generation yield by 33.9% when fed glucose. C. acetobutylicum NCIM2337 produced a high yield of approximately 2.33 mol of H_2 per mol of glucose (Mohanraj et al. 2014b). On the opposite side, Nath et al. (2015) demonstrated the effect of green synthesized Fe-NPs utilizing extracts of Syzygiumcumini leaf and bark and at a higher concentration of Fe^{+2} ions up to 200 mg/L on H_2 generation by E. cloacae DH-89. In the presence of Fe-NPs rather than Fe^{+2} ions, a comparable beneficial effect on H_2 yield was found. Fe-NPs (100 mg/L) outperformed the control (0.95 moles of H_2 per mol of glucose) by 100% (1.9 mol H_2/mol hexose). However, with the Fe^{+2} ions concentration of 25 mg/L, the H_2 is obtained with maximum output and we have 1.45 moles of H_2 per mole of glucose. Interestingly, Fe-NPs increased cell proliferation in E. cloacae DH-89. As a result of these findings, Fe-NPs appear to improve the metabolic pathway of E. cloacae DH-89 for H_2 production (Nath et al. 2015). On the other hand, Sinharoy and Pakshirajan (2020) proposed that the iron NPs produced from green tea be utilized in the generation of biohydrogen. The existence of iron(III) oxide hydroxide (FeO(OH)) and iron oxide (Fe_3O_4) in the green tea-produced iron NPs was verified by X-ray diffraction analysis. The change of biomass in the microbial composition is because of the presence of iron NPs. Gamma-proteobacter, Bacteroidia, and Clostridia classes grew more with iron NPs than in the original inoculum. Clostridia sp. class was discovered to dominate among other classes. More than 56.32% increase in aqueous solubility of CO was reached in the existence of iron NPs at 1000 mg/L, coupled with a maximum H_2 generation of 1.58 $\pm$ 0.13 mmol/L by anaerobic biomass, and this generation is 44% higher than the production of H_2 in the absence of any iron NPs. Continuous research on bioreactors employing a gas lift reactor confirmed iron NPs in increasing biohydrogen generation from CO, and due to the addition of iron NPs in the system maximum production of H_2 quantitatively 30.7 mmol/L (Sinharoy and Pakshirajan 2020). Another study shows that the improvement of biohydrogen generation in the course of dark fermentation of industrial effluents (waste) due to lignin-mediated NPs addition to the system. Green NPs based on iron were also effectively produced utilizing lignin taken from empty fruit bunches of oil palms, a waste from the industries of palm oil.

Table 5.3 Applications of green iron nanoparticles for biohydrogen production

S. no.	Inoculum source	Substrate	Nanoparticle	Concentration of NPs (mg/L)	Hydrogen yield	Efficiency (%)	References
1.	*E. cloacae 811101*	Sucrose	Fe_2O_3	200	2.07 mol mol^{-1}	21.8	Mohanraj et al. (2014a)
2.	*C. acetobutylicum NCIM2337*	Glucose	Fe_2O_3	175	2.33 mol mol^{-1}	33.9	Mohanraj et al. (2014b)
3.	*Enterobacter cloacae DH-89*	Glucose	Fe-NPs	100	1.9 mol mol^{-1}	100	Nath et al. (2015)
4.	*Clostridia sp.*	CO	Fe-NPs	1000	1.58 $\pm$ 0.13 mmol/L	85.9	Sinharoy and Pakshirajan (2020)
5.	*C6 consortium*	Soft drink wastewater	Fe-NPs	200	22.6 mmol/g COD	91	do Nascimento Junior et al. (2021)

There is more production of biohydrogen by using lignin magnetic NPs than using lignin nonmagnetic NPs. Similarly, a greater yield of biohydrogen (506 and 410 mL of H_2 per gram of eliminated COD) was obtained if Fe-NPs were used while in the control test (180 mL of H_2 per gram of eliminated COD) by following the gas generation procedure (do Nascimento Junior et al. 2021).

Thus, iron-based NMs have a significant impact on the production of biofuel by lowering the required temperature to degrade biomass, decreasing the tar as by-product, and increasing biofuel output. However, none of the investigations published so far have used iron-based NPs produced through green methods to catalyze the reaction, implying that the NPs synthesis is both expensive and environmentally unfriendly. Because iron is a key catalyst in the technology of biofuel manufacturing from biomass, to phyto-synthesize by utilizing biomass and plants may also assist in the reduction of total costs involved in the existing technology, making them more sustainable, environmentally manageable, and practical.

5.5 Conclusions

Emerging approaches include the use of microbes and plant extracts in iron-based NMs synthesis. It has been discovered that distinct bioactive species present in the extraction solution serve as capping agent as well as reducing agent simultaneously, decreasing the use of hazardous compounds necessary in chemical synthesis. Each approach however has advantages and downsides. Synthesis of iron-based NMs via microorganisms confronts problems like slow production rate, maintenance and preparation of culture leading to high costs, and the possibility of infection; hence, we can have the plant-mediated technique as a superior option. Furthermore, being environmentally benign, nontoxic, having simple and free access to plant materials, and continuously producing bioactive chemicals in the photosynthesis process can lower the total costs involved in the synthesis of NMs, making it more useful and affordable. Organic wastes are created by various industries, on the other hand, it can serve as raw materials for the production of NMs based on iron. Nonetheless, in comparison with chemical approaches, shape-controlled synthesis, and finally impacts the characteristics of iron-based NMs catalysts, is one of the primary problems generally confronted by biological routes, and therefore further study in this regard should be carried out. This chapter concentrated on the application of iron-based NPs in biohydrogen generation. Although in late years, NPs have been increasingly employed in the manufacture of biofuels, until now, iron-based NMs have been used to create liquid fuels like biodiesel, biobutanol, bioethanol, etc. Given the importance of iron-based NPs in the manufacture of liquid fuels, experiments with NPs to generate biohydrogen on a laboratory scale have been done by researchers. Despite worries about NPs toxicity at higher concentrations, optimum NPs concentrations might significantly increase hydrogen generation after breakdown of the substrates for conversion of bacterial hydrogen. According to the available data, a critical role can be played by the NPs in improving the sustainability of this process in order to obtain a large biohydrogen yields at a realistic scale. Because this subject is still in its initial phases, additional investigations are required to bridge the existing gap in its practical applications and economic viability.

5.6 Future Perspectives

Despite substantial research on the green route synthesized iron-based NPs have been increasingly employed for the production of biohydrogen, their adaptation and evolution in biofuel production from waste biomass are still in their early phases, with several challenges ahead. Further research into the production of biohydrogen using NPs is required.

The following are some of the key problems and future prospects:

- In the literature, green synthesis of NMs based on iron from different biosynthetic and green routes like microorganisms, extract of plants have been described. Plants, and microorganisms as well as their specific kinds, differ substantially in the content of bioactive chemicals; these features have tremendous potential during NMs production and should thus be investigated.
- Furthermore, the chemical composition of the plant with same species found in various geographical locations differs, yielding diverse findings under comparable experimental settings that should be thoroughly studied.
- A significant work is needed to reduce the excess costs involved with the maintenance and growth of microbe cultures for the production of iron-based NMs. One more disadvantage is the creation of poly-dispersed NMs and the parameter optimization associated with the synthesis process.
- Aside from that, compared to traditional physical and chemical approaches for synthesis, the shape (e.g., nanorods, nanoflower, nanostar, nanosphere)-controlled NMs based on iron through green paths remains with great difficulties.
- Several investigations have been reported on the laboratory-scale production of iron-based NMs utilizing plant extracts and waste biomass. Nonetheless, there is no pilot plant synthesis of NMs based on iron that has been outlined to the best of our knowledge. The potential of iron NPs and vast variety of uses necessitates their extensive manufacturing, which lasts a significant problem.
- Iron NPs have been discovered as a vital role player in improving the quality as well as output of production of biofuels, and they have tremendous possibilities to solve the energy crisis situation, but they have large manufacturing costs.
- Nonetheless, the use of iron NPs synthesized using a green pathway in biofuel production is a novel and creative technique that has the potential to lower total synthesis costs but need more research in this area.

Declarations **Conflict of interest:** The authors declare no competing interests.

Funding: This research did not receive any specific grant from any funding agencies.

Authors' contributions: Manuscript was conceptualized and edited by Dr. Dharm Pal. Literature survey and manuscript drafting were performed by Dhruti Sundar Pattanayak and Jyoti Mishra. The draft correction and critical revision were performed by Dr. Naresh Kumar Sahoo. All authors read and approved the final manuscript.

Acknowledgments: The work greatly acknowledges the department of chemical engineering of the National Institute of Technology Raipur for providing research opportunities.

References

Ahmmad B, Leonard K, Shariful Islam M, Kurawaki J, Muruganandham M, Ohkubo T, Kuroda Y (2013) Green synthesis of mesoporous hematite (α-Fe2O 3) nanoparticles and their photocatalytic activity. Adv Powder Technol 24:160–167. https://doi.org/10.1016/j.apt.2012.04.005

Ali I, Al-Othman ZA, Alwarthan A (2016) Green synthesis of functionalized iron nano particles and molecular liquid phase adsorption of Ametryn from water. J Mol Liq 221:1168–1174. https://doi.org/10.1016/j.molliq.2016.06.089

Ali M, Haroon U, Khizar M, Chaudhary HJ, Munis MFH (2020) Facile single step preparations of phyto-nanoparticles of iron in Calotropis procera leaf extract to evaluate their antifungal potential against Alternaria alternata. Curr Plant Biol 23:100157. https://doi.org/10.1016/j.cpb.2020.100157

Arora A, Nandal P, Singh J, Verma ML (2020) Nanobiotechnological advancements in lignocellulosic biomass pretreatment. Mater Sci Energy Technol 3:308–318. https://doi.org/10.1016/j.mset.2019.12.003

Asghar MA, Zahir E, Shahid SM, Khan MN, Asghar MA, Iqbal J, Walker G (2018) Iron, copper and silver nanoparticles: green synthesis using green and black tea leaves extracts and evaluation of antibacterial, antifungal and aflatoxin B1 adsorption activity. LWT Food Sci Technol 90: 98–107. https://doi.org/10.1016/j.lwt.2017.12.009

Bharde AA, Parikh RY, Baidakova M, Jouen S, Hannoyer B, Enoki T, Prasad BLV, Shouche YS, Ogale S, Sastry M (2008) Bacteria-mediated precursor-dependent biosynthesis of superparamagnetic iron oxide and iron sulfide nanoparticles. Langmuir 24:5787–5794. https://doi.org/10.1021/la704019p

Bolade OP, Williams AB, Benson NU (2020) Green synthesis of iron-based nanomaterials for environmental remediation: a review. Environ Nanotechnol Monit Manag 13:100279. https://doi.org/10.1016/j.enmm.2019.100279

Cai G, Jin B, Monis P, Saint C (2011) Metabolic flux network and analysis of fermentative hydrogen production. Biotechnol Adv 29:375–387. https://doi.org/10.1016/j.biotechadv.2011.02.001

Chandrasekhar K, Lee YJ, Lee DW (2015) Biohydrogen production: strategies to improve process efficiency through microbial routes. Int J Mol Sci 16:8266–8293. https://doi.org/10.3390/ijms16048266

Cipolatti EP, Valério A, Henriques RO, Moritz DE, Ninow JL, Freire DMG, Manoel EA, Fernandez-Lafuente R, De Oliveira D (2016) Nanomaterials for biocatalyst immobilization-state of the art and future trends. RSC Adv 6:104675–104692. https://doi.org/10.1039/c6ra22047a

Crespo KA, Baronetti JL, Quinteros MA, Páez PL, Paraje MG (2017) Intra- and extracellular biosynthesis and characterization of iron nanoparticles from prokaryotic microorganisms with anticoagulant activity. Pharm Res 34:591–598

Daneshvar M, Hosseini MR (2018) From the iron boring scraps to superparamagnetic nanoparticles through an aerobic biological route. J Hazard Mater 357:393–400. https://doi.org/10.1016/j.jhazmat.2018.06.024

Das KR, Kowshik M, Praveen Kumar MK, Kerkar S, Shyama SK, Mishra S (2018) Native hypersaline sulphate reducing bacteria contributes to iron nanoparticle formation in saltpan sediment: a concern for aquaculture. J Environ Manag 206:556–564. https://doi.org/10.1016/j.jenvman.2017.10.078

Dhillon GS, Brar SK, Kaur S, Verma M (2012) Green approach for nanoparticle biosynthesis by fungi: current trends and applications. Crit Rev Biotechnol 32:49–73. https://doi.org/10.3109/07388551.2010.550568

Dincer I, Acar C (2014) Review and evaluation of hydrogen production methods for better sustainability. Int J Hydrog Energy 40:11094–11111. https://doi.org/10.1016/j.ijhydene.2014.12.035

do Nascimento Junior JR, Zevallos Torres LA, Medeiros ABP, Woiciechowski AL, Martinez-Burgos WJ, Soccol CR (2021) Enhancement of biohydrogen production in industrial wastewaters with vinasse pond consortium using lignin-mediated iron nanoparticles. Int J Hydrog Energy 46:27431–27443. https://doi.org/10.1016/j.ijhydene.2021.06.009

El-Naggar MY, Wanger G, Leung KM, Yuzvinsky TD, Southam G, Yang J, Lau WM, Nealson KH, Gorby YA (2010) Electrical transport along bacterial nanowires from Shewanella oneidensis MR-1. Proc Natl Acad Sci U S A 107:18127–18131. https://doi.org/10.1073/pnas.1004880107

Fani M, Ghandehari F, Rezaee M (2018) Biosynthesis of iron oxide nanoparticles by cytoplasmic extract of bacteria lactobacillus fermentum. J Med Chem Sci 1:28–30

Fatemi M, Mollania N, Momeni-Moghaddam M, Sadeghifar F (2018) Extracellular biosynthesis of magnetic iron oxide nanoparticles by Bacillus cereus strain HMH1: Characterization and in vitro cytotoxicity analysis on MCF-7 and 3T3 cell lines. J Biotechnol 270:1–11. https://doi.org/10.1016/j.jbiotec.2018.01.021

Fazlzadeh M, Rahmani K, Zarei A, Abdoallahzadeh H, Nasiri F, Khosravi R (2016) A novel green synthesis of zero valent iron nanoparticles (NZVI) using three plant extracts and their efficient application for removal of Cr(VI) from aqueous solutions. Adv Powder Technol 28:122–130. https://doi.org/10.1016/j.apt.2016.09.003

Gautam A, Rawat S, Verma L, Singh J, Sikarwar S, Yadav BC, Kalamdhad AS (2018) Green synthesis of iron nanoparticle from extract of waste tea: An application for phenol red removal from aqueous solution. Environ. Nanotechnology. Monit Manag 10:377–387. https://doi.org/10.1016/j.enmm.2018.08.003

Guo B, Xu T, Zhang L, Li S (2020) A heterogeneous fenton-like system with green iron nanoparticles for the removal of bisphenol A: Performance, kinetics and transformation mechanism. J Environ Manag 272:111047. https://doi.org/10.1016/j.jenvman.2020.111047

Hoag GE, Collins JB, Holcomb JL, Hoag JR, Nadagouda MN, Varma RS (2009) Degradation of bromothymol blue by "greener" nano-scale zero-valent iron synthesized using tea polyphenols. J Mater Chem 19:8671–8677. https://doi.org/10.1039/b909148c

Huang L, Weng X, Chen Z, Megharaj M, Naidu R (2014) Green synthesis of iron nanoparticles by various tea extracts: comparative study of the reactivity. Spectrochim Acta Part A Mol Biomol Spectrosc 130:295–301. https://doi.org/10.1016/j.saa.2014.04.037

Ibarra-Gonzalez P, Rong BG (2019) A review of the current state of biofuels production from lignocellulosic biomass using thermochemical conversion routes. Chinese J Chem Eng 27:1523–1535. https://doi.org/10.1016/j.cjche.2018.09.018

Iravani S (2011) Green synthesis of metal nanoparticles using plants. Green Chem 13:2638–2650. https://doi.org/10.1039/c1gc15386b

Jiang X, Hu J, Lieber AM, Jackan CS, Biffinger JC, Fitzgerald LA, Ringeisen BR, Lieber CM (2014) Nanoparticle facilitated extracellular electron transfer in microbial fuel cells. Nano Lett 14:6737–6742. https://doi.org/10.1021/nl503668q

Katata-Seru L, Moremedi T, Aremu OS, Bahadur I (2017) Green synthesis of iron nanoparticles using Moringa oleifera extracts and their applications: removal of nitrate from water and antibacterial activity against Escherichia coli. J Mol Liq 256:296–304. https://doi.org/10.1016/j.molliq.2017.11.093

Kandel K, Anderegg JW, Nelson NC, Chaudhary U, Slowing II (2014) Supported iron nanoparticles for the hydrodeoxygenation of microalgal oil to green diesel. J Catal 314:142–148. https://doi.org/10.1016/j.jcat.2014.04.009

Khan Z, Al-Thabaiti SA (2018) Green synthesis of zero-valent Fe-nanoparticles: catalytic degradation of rhodamine B, interactions with bovine serum albumin and their enhanced antimicrobial activities. J Photochem Photobiol B Biol 180:259–267. https://doi.org/10.1016/j.jphotobiol.2018.02.017

Kiruba Daniel SCG, Vinothini G, Subramanian N, Nehru K, Sivakumar M (2013) Biosynthesis of Cu, ZVI, and Ag nanoparticles using Dodonaea viscosa extract for antibacterial activity against human pathogens. J Nanopart Res 15:1319. https://doi.org/10.1007/s11051-012-1319-1

Kogo T, Yoshida Y, Koganei K, Matsumoto H, Watanabe T, Ogihara J, Kasumi T (2017) Production of rice straw hydrolysis enzymes by the fungi Trichoderma reesei and Humicola insolens using rice straw as a carbon source. Bioresour Technol 233:67–73. https://doi.org/10.1016/j.biortech.2017.01.075

Kumar R, Singh N, Pandey SN (2015) Potential of green synthesized zero-valent iron nanoparticles for remediation of lead-contaminated water. Int J Environ Sci Technol 12:3943–3950. https://doi.org/10.1007/s13762-015-0751-z

Kumar G, Mathimani T, Rene ER, Pugazhendhi A (2019) Application of nanotechnology in dark fermentation for enhanced biohydrogen production using inorganic nanoparticles. Int J Hydrog Energy 44:13106–13113. https://doi.org/10.1016/j.ijhydene.2019.03.131

Kumar Y, Yogeshwar P, Bajpai S, Jaiswal P, Yadav S, Pathak DP, Sonker M, Tiwary SK (2021) Nanomaterials: stimulants for biofuels and renewables, yield and energy optimization. Mater Adv 2:5318–5343. https://doi.org/10.1039/d1ma00538c

Ladole MR, Mevada JS, Pandit AB (2017) Ultrasonic hyperactivation of cellulase immobilized on magnetic nanoparticles. Bioresour Technol 239:117–126. https://doi.org/10.1016/j.biortech.2017.04.096

Leili M, Fazlzadeh M, Bhatnagar A (2017) Green synthesis of nano-zero-valent iron from Nettle and Thyme leaf extracts and their application for the removal of cephalexin antibiotic from aqueous solutions. Environ Technol (United Kingdom) 39:1158–1172. https://doi.org/10.1080/09593330.2017.1323956

Lin R, Cheng J, Ding L, Song W, Liu M, Zhou J, Cen K (2016) Enhanced dark hydrogen fermentation by addition of ferric oxide nanoparticles using Enterobacter aerogenes. Bioresour Technol 207:213–219. https://doi.org/10.1016/j.biortech.2016.02.009

Liu CG, Xiao Y, Xia XX, Zhao XQ, Peng L, Srinophakun P, Bai FW (2019) Cellulosic ethanol production: progress, challenges and strategies for solutions. Biotechnol Adv 37:491–504. https://doi.org/10.1016/j.biotechadv.2019.03.002

Machado S, Stawiński W, Slonina P, Pinto AR, Grosso JP, Nouws HPA, Albergaria JT, Delerue-Matos C (2013) Application of green zero-valent iron nanoparticles to the remediation of soils contaminated with ibuprofen. Sci Total Environ 461–462:323–329. https://doi.org/10.1016/j.scitotenv.2013.05.016

Machado S, Grosso JP, Nouws HPA, Albergaria JT, Delerue-Matos C (2014) Utilization of food industry wastes for the production of zero-valent iron nanoparticles. Sci Total Environ 496:233–240. https://doi.org/10.1016/j.scitotenv.2014.07.058

Malik P, Shankar R, Malik V, Sharma N, Mukherjee TK (2014) Green chemistry based benign routes for nanoparticle synthesis. J Nanoparticles 2014:1–14. https://doi.org/10.1155/2014/302429

Manquián-Cerda K, Cruces E, Angélica Rubio M, Reyes C, Arancibia-Miranda N (2017) Preparation of nanoscale iron (oxide, oxyhydroxides and zero-valent) particles derived from blueberries: reactivity, characterization and removal mechanism of arsenate. Ecotoxicol Environ Saf 145:69–77. https://doi.org/10.1016/j.ecoenv.2017.07.004

Mohanraj S, Kodhaiyolii S, Rengasamy M, Pugalenthi V (2014a) Phytosynthesized iron oxide nanoparticles and ferrous iron on fermentative hydrogen production using Enterobacter cloacae:

evaluation and comparison of the effects. Int J Hydrog Energy 39:11920–11929. https://doi.org/10.1016/j.ijhydene.2014.06.027

Mohanraj S, Kodhaiyolii S, Rengasamy M, Pugalenthi V (2014b) Green synthesized iron oxide nanoparticles effect on fermentative hydrogen production by Clostridium acetobutylicum. Appl Biochem Biotechnol 173:318–331. https://doi.org/10.1007/s12010-014-0843-0

Mondal P, Anweshan A, Purkait MK (2020) Green synthesis and environmental application of iron-based nanomaterials and nanocomposite: a review. Chemosphere 259:127509. https://doi.org/10.1016/j.chemosphere.2020.127509

Mudhoo A, Torres-Mayanga PC, Forster-Carneiro T, Sivagurunathan P, Kumar G, Komilis D, Sánchez A (2018) A review of research trends in the enhancement of biomass-to-hydrogen conversion. Waste Manag 79:580–594. https://doi.org/10.1016/j.wasman.2018.08.028

Mukunthan KS, Balaji S (2012) Silver nanoparticles shoot up from the root of Daucus carrota (L.). Int J Green Nanotechnol Biomed 4:54–61. https://doi.org/10.1080/19430892.2012.654745

Nabgan W, Tuan Abdullah TA, Nabgan B, Jalil AA, Nordin AH, Ul-Hamid A, Hassan NS, Hussain I, Coelho A, Amin A, Ikram M (2021) Catalytic biohydrogen production from organic waste materials: a literature review and bibliometric analysis. Int J Hydrog Energy 46:30903–30925. https://doi.org/10.1016/j.ijhydene.2021.04.100

Narayanan S, Sathy BN, Mony U, Koyakutty M, Nair SV, Menon D (2012) Biocompatible magnetite/gold nanohybrid contrast agents via green chemistry for MRI and CT bioimaging. ACS Appl Mater Interfaces 4:251–260. https://doi.org/10.1021/am201311c

Nath D, Manhar AK, Gupta K, Saikia D, Das SK, Mandal M (2015) Phytosynthesized iron nanoparticles: effects on fermentative hydrogen production by Enterobacter cloacae DH-89. Bull Mater Sci 38:1533–1538. https://doi.org/10.1007/s12034-015-0974-0

Nikolaidis P, Poullikkas A (2017) A comparative overview of hydrogen production processes. Renew Sustain Energy Rev 67:597–611. https://doi.org/10.1016/j.rser.2016.09.044

Osman AI, Deka TJ, Baruah DC, Rooney DW (2020) Critical challenges in biohydrogen production processes from the organic feedstocks. Biomass Convers Biorefinery. https://doi.org/10.1007/s13399-020-00965-x

Park TJ, Lee KG, Lee SY (2016) Advances in microbial biosynthesis of metal nanoparticles. Appl Microbiol Biotechnol 100:521–534. https://doi.org/10.1007/s00253-015-6904-7

Pattanayak DS, Mallick N, Thakur C, Pal D (2020) Plant mediated green synthesis of silver nanoparticles for antimicrobial application: present status. J Indian Chem Soc 97:1108–1114

Pattanayak DS, Pal D, Thakur C, Kumar S, Devnani GL (2021) Bio-synthesis of iron nanoparticles for environmental remediation: status till date. Mater Today Proc 44:3150–3155. https://doi.org/10.1016/j.matpr.2021.02.821

Porwal S, Kumar T, Lal S, Rani A, Kumar S, Cheema S, Purohit HJ, Sharma R, Singh Patel SK, Kalia VC (2008) Hydrogen and polyhydroxybutyrate producing abilities of microbes from diverse habitats by dark fermentative process. Bioresour Technol 99:5444–5451. https://doi.org/10.1016/j.biortech.2007.11.011

Pugazhendhi A, Shobana S, Nguyen DD, Banu JR, Sivagurunathan P, Chang SW, Ponnusamy VK, Kumar G (2019) Application of nanotechnology (nanoparticles) in dark fermentative hydrogen production. Int J Hydrog Energy 44:1431–1440. https://doi.org/10.1016/j.ijhydene.2018.11.114

Rai PK, Kumar V, Lee SS, Raza N, Kim KH, Ok YS, Tsang DCW (2018) Nanoparticle-plant interaction: implications in energy, environment, and agriculture. Environ Int 119:1–19. https://doi.org/10.1016/j.envint.2018.06.012

Rittmann S, Herwig C (2012) A comprehensive and quantitative review of dark fermentative biohydrogen production. Microb Cell Factories 11:1. https://doi.org/10.1186/1475-2859-11-115

Rong K, Wang J, Zhang Z, Zhang J (2020) Green synthesis of iron nanoparticles using Korla fragrant pear peel extracts for the removal of aqueous Cr(VI). Ecol Eng 149:105793. https://doi.org/10.1016/j.ecoleng.2020.105793

Sadhasivam S, Vinayagam V, Balasubramaniyan M (2020) Recent advancement in biogenic synthesis of iron nanoparticles. J Mol Struct 1217:128372. https://doi.org/10.1016/j.molstruc.2020.128372

Saif S, Tahir A, Chen Y (2016) Green synthesis of iron nanoparticles and their environmental applications and implications. Nanomaterials 6:1–26. https://doi.org/10.3390/nano6110209

Saratale RG, Karuppusamy I, Saratale GD, Pugazhendhi A, Kumar G, Park Y, Ghodake GS, Bharagava RN, Banu JR, Shin HS (2018) A comprehensive review on green nanomaterials using biological systems: recent perception and their future applications. Colloids Surfaces B: Biointerfaces 170:20–35. https://doi.org/10.1016/j.colsurfb.2018.05.045

Sekoai PT, Ouma CNM, du Preez SP, Modisha P, Engelbrecht N, Bessarabov DG, Ghimire A (2019) Application of nanoparticles in biofuels: an overview. Fuel 237:380–397. https://doi.org/10.1016/j.fuel.2018.10.030

Shamaila S, Sajjad AKL, Ryma N-u-A, Farooqi SA, Jabeen N, Majeed S, Farooq I (2016) Advancements in nanoparticle fabrication by hazard free eco-friendly green routes. Appl Mater Today 5:150–199. https://doi.org/10.1016/j.apmt.2016.09.009

Shanmugam S, Hari A, Pandey A, Mathimani T, Felix LO, Pugazhendhi A (2020) Comprehensive review on the application of inorganic and organic nanoparticles for enhancing biohydrogen production. Fuel 270:117453. https://doi.org/10.1016/j.fuel.2020.117453

Singh P, Kim YJ, Zhang D, Yang DC (2016) Biological synthesis of nanoparticles from plants and microorganisms. Trends Biotechnol 34:588–599. https://doi.org/10.1016/j.tibtech.2016.02.006

Sinharoy A, Pakshirajan K (2020) A novel application of biologically synthesized nanoparticles for enhanced biohydrogen production and carbon monoxide bioconversion. Renew Energy 147: 864–873. https://doi.org/10.1016/j.renene.2019.09.027

Sravanthi K, Ayodhya D, Yadgiri Swamy P (2018) Green synthesis, characterization of biomaterial-supported zero-valent iron nanoparticles for contaminated water treatment. J Anal Sci Technol 9:3. https://doi.org/10.1186/s40543-017-0134-9

Srivastava N, Singh J, Ramteke PW, Mishra PK, Srivastava M (2015) Improved production of reducing sugars from rice straw using crude cellulase activated with Fe3O4/alginate nanocomposite. Bioresour Technol 183:262–266. https://doi.org/10.1016/j.biortech.2015.02.059

Srivastava N, Srivastava M, Kushwaha D, Gupta VK, Manikanta A, Ramteke PW, Mishra PK (2017) Efficient dark fermentative hydrogen production from enzyme hydrolyzed rice straw by Clostridium pasteurianum (MTCC116). Bioresour Technol 238:552–558. https://doi.org/10.1016/j.biortech.2017.04.077

Srivastava N, Srivastava M, Malhotra BD, Gupta VK, Ramteke PW, Silva RN, Shukla P, Dubey KK, Mishra PK (2019) Nanoengineered cellulosic biohydrogen production via dark fermentation: a novel approach. Biotechnol Adv 37:107384. https://doi.org/10.1016/j.biotechadv.2019.04.006

Srivastava N, Srivastava M, Mishra PK, Kausar MA, Saeed M, Gupta VK, Singh R, Ramteke PW (2020) Advances in nanomaterials induced biohydrogen production using waste biomass. Bioresour Technol 307:123094. https://doi.org/10.1016/j.biortech.2020.123094

Srivastava M, Srivastava N, Saeed M, Mishra PK, Saeed A, Gupta VK, Malhotra BD (2021) Bioinspired synthesis of iron-based nanomaterials for application in biofuels production: a new in-sight. Renew Sustain Energy Rev 147:111206. https://doi.org/10.1016/j.rser.2021.111206

Stefaniuk M, Oleszczuk P, Ok YS (2016) Review on nano zerovalent iron (nZVI): from synthesis to environmental applications. Chem Eng J 287:618–632. https://doi.org/10.1016/j.cej.2015.11.046

Subramaniyam V, Subashchandrabose SR, Thavamani P, Megharaj M, Chen Z, Naidu R (2015) Chlorococcum sp. MM11—a novel phyco-nanofactory for the synthesis of iron nanoparticles. J Appl Phycol 27:1861–1869. https://doi.org/10.1007/s10811-014-0492-2

Taherdanak M, Zilouei H, Karimi K (2015) Investigating the effects of iron and nickel nanoparticles on dark hydrogen fermentation from starch using central composite design. Int J Hydrog Energy 40:12956–12963. https://doi.org/10.1016/j.ijhydene.2015.08.004

Tarafdar JC, Raliya R (2013) Rapid, low-cost, and ecofriendly approach for iron nanoparticle synthesis using Aspergillus oryzae TFR9. J Nanoparticles 2013:141274. https://doi.org/10.1155/2013/141274

Vaghari H, Jafarizadeh-Malmiri H, Mohammadlou M, Berenjian A, Anarjan N, Jafari N, Nasiri S (2016) Application of magnetic nanoparticles in smart enzyme immobilization. Biotechnol Lett 38:223–233. https://doi.org/10.1007/s10529-015-1977-z

Wang T, Jin X, Chen Z, Megharaj M, Naidu R (2014a) Green synthesis of Fe nanoparticles using eucalyptus leaf extracts for treatment of eutrophic wastewater. Sci Total Environ 466–467:210–213. https://doi.org/10.1016/j.scitotenv.2013.07.022

Wang T, Lin J, Chen Z, Megharaj M, Naidu R (2014b) Green synthesized iron nanoparticles by green tea and eucalyptus leaves extracts used for removal of nitrate in aqueous solution. J Clean Prod 83:413–419. https://doi.org/10.1016/j.jclepro.2014.07.006

Wang S, Dai G, Yang H, Luo Z (2017a) Lignocellulosic biomass pyrolysis mechanism: a state-of-the-art review. Prog Energy Combust Sci 62:33–86. https://doi.org/10.1016/j.pecs.2017.05.004

Wang X, Wang A, Ma J, Fu M (2017b) Facile green synthesis of functional nanoscale zero-valent iron and studies of its activity toward ultrasound-enhanced decolorization of cationic dyes. Chemosphere 166:80–88. https://doi.org/10.1016/j.chemosphere.2016.09.056

Xiao C, Li H, Zhao Y, Zhang X, Wang X (2020) Green synthesis of iron nanoparticle by tea extract (polyphenols) and its selective removal of cationic dyes. J Environ Manag 275:111262. https://doi.org/10.1016/j.jenvman.2020.111262

Yi Y, Tu G, Tsang PE, Xiao S, Fang Z (2019) Green synthesis of iron-based nanoparticles from extracts of Nephrolepis auriculata and applications for Cr(VI) removal. Mater Lett 234:388–391. https://doi.org/10.1016/j.matlet.2018.09.137

Zambre A, Upendran A, Shukla R, Chanda N, Katti KK, Cutler C, Kannan R, Katti KV (2013) Green nanotechnology—a sustainable approach in the nanorevolution. RSC Green Chem 144–156

Zilouei H, Taherdanak M (2015) Biohydrogen from lignocellulosic wastes. https://doi.org/10.1007/978-3-319-14033-9_7

Chapter 6
Synthesis and Application of Nanoengineered Cellulosic Biomass in Biohydrogen Production

Diksha Verma, Tanvi Sahni, Sachin Kumar, and Loveleen Kaur Sarao

Abstract Hydrogen is a ray of hope under drastically increasing energy needs that are affecting the environmental conditions on earth. Hydrogen is clean, fuel efficient, and has a higher energy density and here biohydrogen is more environment friendly, and therefore seems more lucrative yet there are certain limitations in its development on large scale. Nano-engineered partials (NPs) can overcome these limitations to contribute a significant part in biohydrogen production. Production of biohydrogen entirely depends upon the success rate of pretreatment, enzymatic hydrolysis, and fermentation and here nano-engineered particles help in increasing the efficiency of the whole process. In the pretreatment method, the use of NPs decreases the processing cost by eliminating the utilization of acid/base. Enzymatic hydrolysis of pretreated material has a slow reaction rate, nonreproducibility of enzymes, and incomplete conversion of substrates therefore enzyme immobilization of enzymes using magnetic NPs makes the enzyme reusable with improved temperature stability. Fermentation is the last step and here also the biohydrogen production can also be increased via the incorporation of NPs. This chapter provides significant details for the synthesis of NPs and their implementation in H_2 production using different cellulosic biomass materials to enhance the efficiency of biohydrogen production at each step along with the synthesis of some important nano particles derived from Iron, Silver, and Palladium.

6.1 Introduction

An increase in energy demands is a continuous process at present time that is responsible for accelerating fossil-based emissions throughout the world. Prices of oil and natural gas have increased to a certain level (Gong et al. 2020) and here various countries are working towards alternative energy sources. Biofuels are considered as the best alternative fuels and here biohydrogen has given us very

D. Verma · T. Sahni · S. Kumar · L. K. Sarao (✉)
Punjab Agricultural University, Ludhiana, Punjab, India

© The Author(s), under exclusive license to Springer Nature Singapore Pte Ltd. 2023
M. Srivastava, P. K. Mishra (eds.), *NanoBioenergy: Application and Sustainability Assessment*, Clean Energy Production Technologies,
https://doi.org/10.1007/978-981-19-6234-9_6

promising results (Shanmugam et al. 2020). Hydrogen has been a topic of discussion due to its higher energy density and complete clean combustion (Navlani-Garcia et al. 2018). Biochemical and microbial processes to generate hydrogen provide an environment-friendly, resource renewable efficient alternative (Das and Veziroğlu 2001). Commercial hydrogen production is predominantly done by thermal technologies such as partial oxidation and auto thermal reforming. Steam reforming of CH_4 (methane) is a widely utilized and economical approach that accounts for more than 85% of the global generation of hydrogen (Sun et al. 2019). Though more environmental friendly methods based on bioactivity of the microorganisms for production of hydrogen are available in the literature, they need to be polished for an effective and stable supply of hydrogen over a long period (Otto et al. 2008). Lignocellulosic materials such as plant-based materials, organic wastewater, and industry waste could be potential resources for biohydrogen production (Kapdan and Kargi 2006) and nano-engineered particles can bridge the gap between the effectiveness and economic viability of cellulosic biomass in biohydrogen production. During the production of biohydrogen from cellulosic biomass, several steps are involved including pretreatment followed by enzyme-aided hydrolysis and fermentation process for production of biohydrogen. The amount of biohydrogen produced is greatly determined by the success rate of these steps during which a sufficient amount of substrate will be produced for each successive step. In general, an important step in biohydrogen production is the hydrolysis of cellulose that occurs by the acid or enzymatic catalyzes of lignocellulosic biomass. Enzymatic hydrolysis is proven to be more effective than acid hydrolysis due to (1) more sugar yield and (2) being eco-friendly as nonuse of hazardous chemicals (Taherzadeh and Karimi 2007), but several factors put a constraint on the success of hydrolysis step, viz., crystallinity of lignocellulosic fibers, presence of hemicelluloses and lignin components in biomass in addition to cellulosic fibers that reduced the accessibility of cellulose to substate during the fermentation process (Orts and McMahan 2016). Therefore, the pretreatment step is considered as one of the crucial and necessary steps that include the effective removal of hemicellulose and lignin from the cellulosic component of biomass and conversion of the complex matrix of cellulose into simple sugars to make the substrate accessible for enzymatic hydrolysis (Chozhavendhan et al. 2020).

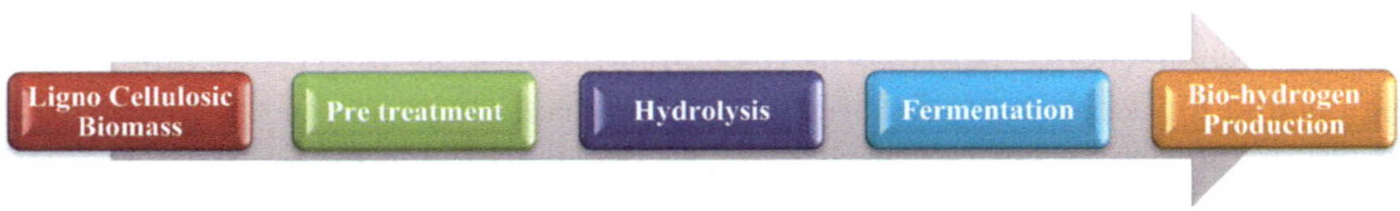

6.2 Pretreatment Process

Since the direct consumption of cellulosic component of the lignocellulosic biomass does not take place for effective biohydrogen production due to its crystalline structure, pretreatment method is employed whose basic aim is to open the structure of the cell wall for making access to cellulose and hemicelluloses to carry out enzymatic hydrolysis to get their monomeric form. Various types of pretreatment processes are employed either individually or in combination that (1) remove hemicelluloses and lignin contents, (2) enhance the porosity of lingo cellulosic material, and (3) bring down the crystallinity of cellulose which is the major constraint in biohydrogen production (Chozhavendhan et al. 2020). The pretreatment steps vigorously impact the impending costs by adjudging the rate of enzymatic hydrolysis and its loading, fermentation toxicity, mixing power, and other variables of the process. All the components of biomass need to be recovered in an effective pretreatment, to further upgrade them to valuable products, loss of a few but important components often occurs that somehow affects the process economically (Rocha et al. 2012).

6.2.1 Types of Pretreatment Process

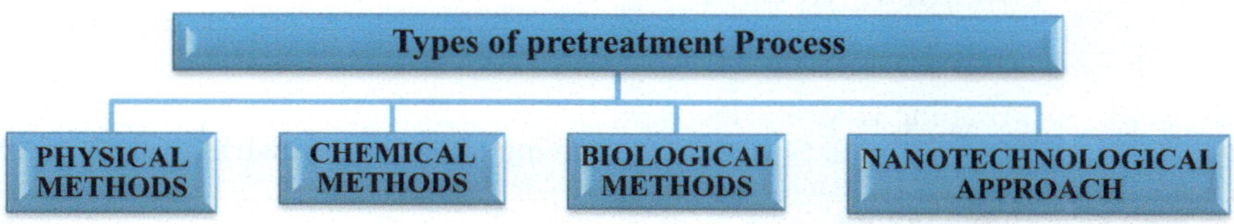

6.2.1.1 Physical Methods

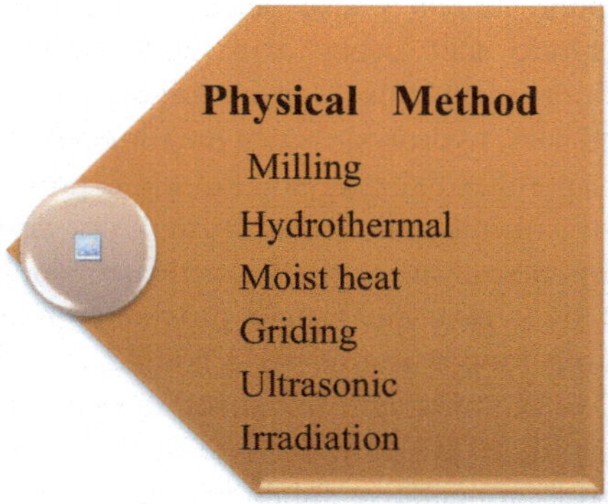

Physical methods include various processes like use of the different type of radiations, mechanical process, and physiochemical process (steam explosion, microwave) that not only surges the surface area along with the pore size of cellulosic material but also decrease its degree of crystallinity that ultimately increase its accessibility towards the better outcome of following steps and hence to the production of biohydrogen. Owing to these benefits, the various physical methods have certain drawbacks, during the mechanical process, different types of mills including centrifugal mills, ball mills, knife and pin mills, vibratory mills, colloid mills, hammer mills, and extruders are used (Cheng and Timilsina 2011) that needs more energy requirement and therefore is not considered good as one of the pretreatment economically. In the steam explosion method, though it is economically beneficial, it leads to the condensation and precipitation of lignin components because of fragmentary lysis of the lignin-carbohydrate matrix, making the biomass least digestible (Amin et al. 2017). In contrast to this, the microwave method is also an economical method but its use on an industrial scale is limited.

6.2.1.2 Chemical Methods

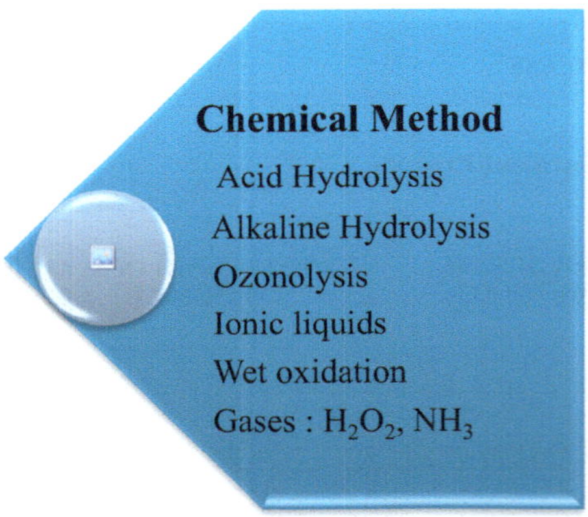

Owing to various drawbacks in physical methods, chemically pretreatment is also being carried out using various acidic (sulfuric acid, hydrochloric acid, acetic acid) and alkaline (sodium hydroxide, potassium hydroxide, lime) chemicals (Gonzalez et al. 2005). An important difference between acidic and basic hydrolysis is that the former breakdown the polymeric structures while latter saponifies the ester bond of lipids present in the cell wall that contribute to cell matrix disintegration and crystallinity destruction of cellulosic biomass. The author reported in his study that activated sludge subjected to acidic treatment (0.5% (w/v) HCl) gave the highest production of hydrogen in approximately 6 h while treating food waste at pH 2 and 12 increased the biohydrogen production by 3- and 2.6-fold times in comparison to (Kim et al. 2014) and control (Assawamongkholsiri et al. 2013). But, because of the involvement of hazardous chemicals, production of fermentative inhibitors during acid treatment (furfural,

5-HMF, phenolic acids, and aldehydes) and involvement of various steps like detoxification, neutralization (to remove inhibitors and lignin), and waste disposal cost, address to environmental issues, chemical treatment is not used in a broader sense (Rocha et al. 2012).

6.2.1.3 Biological Methods

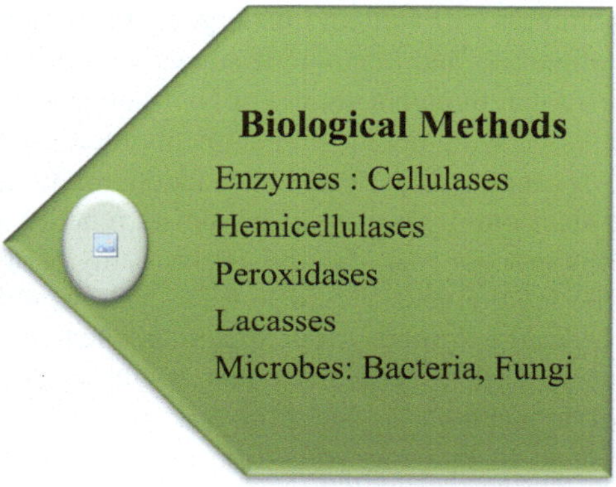

Since the biological treatment does not require the application of harmful chemicals and high-temperature employment, it is most commonly used as an efficient and greener pretreatment method in the whole process of biohydrogen production. In the biological method, various bacteria, fungi, and especially their metabolite, i.e., enzymes, are utilized for the effective hydrolysis of various components of LB to yield an adequate amount of sugars. Depending upon the types of organisms used, different effects were observed, some increased the hydrolysis yield (Taha et al. 2015), some minimized cellulose loss during pretreatment (Cianchetta et al. 2014) while others were effective in removing lignin (Suhara et al. 2012). After the application of various physical and chemical treatments, enzymatic actions are done on the cellulosic matrix for biohydrogen production on large scale with lesser by-products. During biological pretreatment, various enzymes (cellulose, xylanases, and laccases) aid the process of conversion of lignin, cellulose, and hemicellulose into simple sugars after carrying out the acidic treatment, resulting in the biomass swelling, distortion of lignin structure, and ultimately the degradation of its cell wall using cellulolytic enzyme thus releasing free cellulose which can undergo hydrolysis to transform into simpler sugars (glucose, xylose, arabinose, etc.) by the action of cellulase enzyme (Chozhavendhan et al. 2020). It was reported in one study that hydrolysis of rice straw by using cellulose enzyme produces 54.18 g/L sugar and 2.58 LH_2/L of hydrogen production. Owing to various advantages over physical and chemical methods, an extremely slow process that has no industrial scope to date and very little contribution in the production of biohydrogen due to inadequate conversion of biomass into the respective sugars as most of the carbohydrates got consumed by microorganisms is the biological one (Srivastava et al. 2017).

6.2.1.4 Nanotechnological Approach

To overcome the drawback of the biological method, a potential, new and eco-friendly approach has been employed, that has increased the output of preexisting pretreatment methods. Such an approach includes nanotechnology that has been incorporated to raise the content of glucose production by introducing nanoparticle induced enzymes for the successful use of lignocellulosic biomass in the production of biohydrogen. Smaller size and large surface area of NPs, facilitate easy penetration to the lignocellulosic biomass cell wall and undergoes interaction with its forming elements to discharge oligomers and monomeric sugars with better yield (Rai et al. 2017). It has been reported that though a variety of NPs can be used for the hydrolysis of biomass, NPs used for carrying out the pretreatment process are preferable to magnetic as it helps to recover easily from the reaction mixture, repeated use of nanocatalysts in more than one hydrolysis step and its repetitive use with immobilized enzymes, make the process cost-effective (Periyasamy et al. 2018; Verma et al. 2019). Moreover, the requirement of acid in each cycle would also be eliminated which in short will reduce the rigorness of pretreatment due to the use of acid functionalized magnetic nanoparticles (AFMN); this results in the formation of an appreciable amount of fermentable sugars from LB. Because of the high surface-to-volume ratio, low level of enzymes would be required to be encapsulated or coated on NPs which ultimately enhanced the process of hydrolysis along with easy recovery of sugar (Pena et al. 2012). In one of the studies, it was noted that using carbon-based solid catalyst (functionalized with sulfonic acid) as a biocatalyst helps in increasing the efficacy of pretreatment of corncob with the release of 78.1% xylose (Qi et al. 2018).

Production of Carbon-Based Solid Catalyst

In an autoclave reactor, the sulfuric acid was added to microcrystalline cellulose, the mixture was reacted at a temperature ranging from 120 to 200 °C with variation in the volume of cellulose mass/sulfuric acid (1:1–1:20) followed by 24 h of heating process taking place in a muffle furnace. The product so obtained undergoes repeated washing with ethanol and hot water till the time sulfate ions detection gets nullified using $BaCl_2$. After this, the product was placed in a vacuum oven for drying at 105 °C. After this, the resulting carbon-based solid acid catalyst was subjected to grinding to obtain powder form and further sieved using a 120 mesh screen. Then the reaction was carried out by treating carbon-based solid acid catalyst (0.5 g) with corncob (0.25 g) in deionized water (25 mL) and stirred on a magnetic stirrer in an oil bath at 130 °C for complete 4 h. After the completion of the reaction, the supernatant was collected for further analyses (Fig. 6.1).

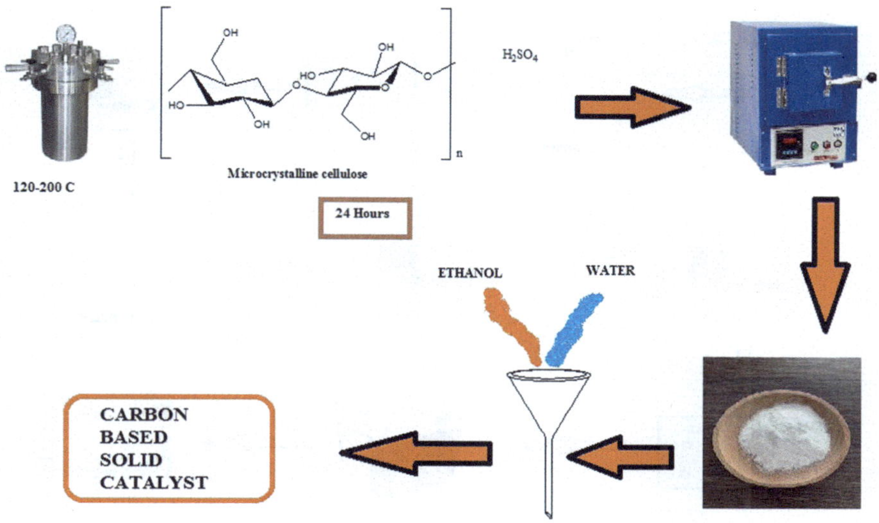

Fig. 6.1 Synthesis of carbon-based solid nanocrystal synthesis

Solid Acid Catalysts

Solid acid catalysts, also known as acid functionalized nanoparticles, have a strong ability to carry out the hydrolysis process with an advantage that it has replaced the use of hazardous mineral acids during chemical pretreatment method as they possess similar catalytic activity as mineral acids. Due to the magnetic nature of these catalysts, they are easily recoverable from the reaction mixture which ultimately leads to cost-effective production of fermentable sugars and biohydrogen production. The author has used alkylsulfonic (AS) acid functionalized magnetic nanoparticles and Perfluoroalkylsulfonic (PFS) for the evaluation of degradation of hemicelluloses into oligosaccharides and it was observed that PFS solubilized higher more amount of hemicelluloses (24.0% ± 1.1%) in contrast to AS NPs (9.1% ± 1.7%) at 80 °C while at 160 °C, PFS and AS gave a maximum yield of oligosaccharides (45.3% ± 0.5% and 46 ± 1.3%, respectively) as compared to control ((36.0% ± 1.9%) (Pena et al. 2012) (Fig. 6.2).

Synthesis of Acid Functionalized Magnetic Nanoparticles

To synthesize acid functionalized NPs, firstly SiMNPs (silica-coated NPs) were produced by the technique reported in studies (Gill et al. 2007; Phan and Jones 2006). In Fig. 6.3, the synthesis model of silica-coated cobalt ferrite ($CoFe_2O_4$) is given using microemulsion method. After the formation of SiMNPs, its PFS acid functionalized NPs were synthesized following the procedure reported in the

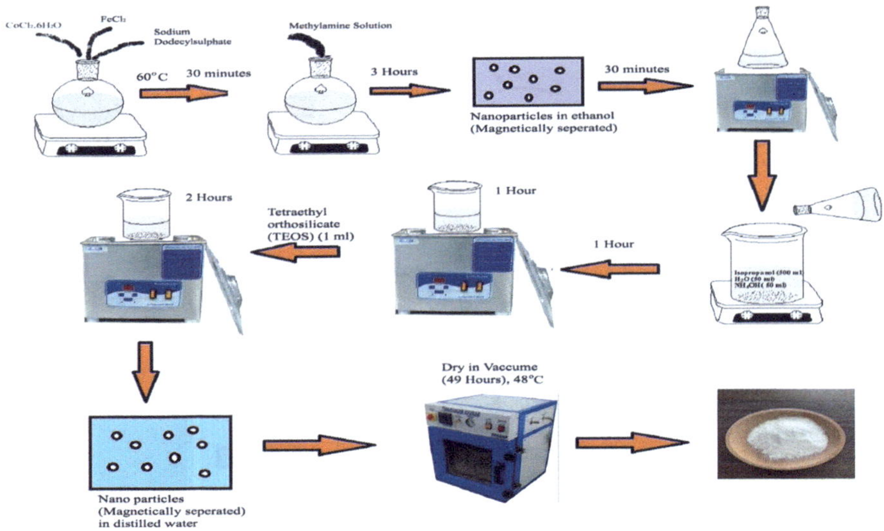

Fig. 6.2 Formation of acid functionalized magnetic NPs (nanoparticles)

literature (Ingle et al. 2017). Complete synthesis is presented in Fig. 6.3 (Ingle et al. 2017).

Synthesis of Alkylsulfonic Acid Functionalized (AS) NPs

Figure 6.4 represented step by step formation of AS NPs similar to the method reported for PFS NPs synthesis (Ingle et al. 2017). The overall procedure followed for synthesizing SiMNPs, AS NPs, and PFS NPs can be depicted in Fig. 6.4 in step 1, 2, and 3, respectively.

In another similar study, silica-coated propyl sulfonic and perfluoropropyl sulfonic acid functionalized NPs were used and results showed that they gave 90% and 58% glucose yield (Dutta and Saha 2019).

Synthesis of Propyl Sulfonic Acid Functionalized NPs

Firstly, silica-mediated magnetic NPs were produced by applying a similar method as given in Fig. 6.1 followed by synthesis of acid functionalized MNPs in which dry SiMNPs (500 mg) were sonicated with (3-mercaptopropyl) trimethoxysilane (MPTMS) (1 mL) using ethanol as reaction medium for 1 h followed by addition of 100 mM (100 mL) acetate buffer solution at pH 4.8 and the mixture was heated at 75 °C after stirring. The synthesized mercaptopropyl functionalized NPs were subjected to the magnetic strength of magnets after 16 h of reaction and washed with ethanol (three times). After this, NPs were placed in a solution (60 mL) of

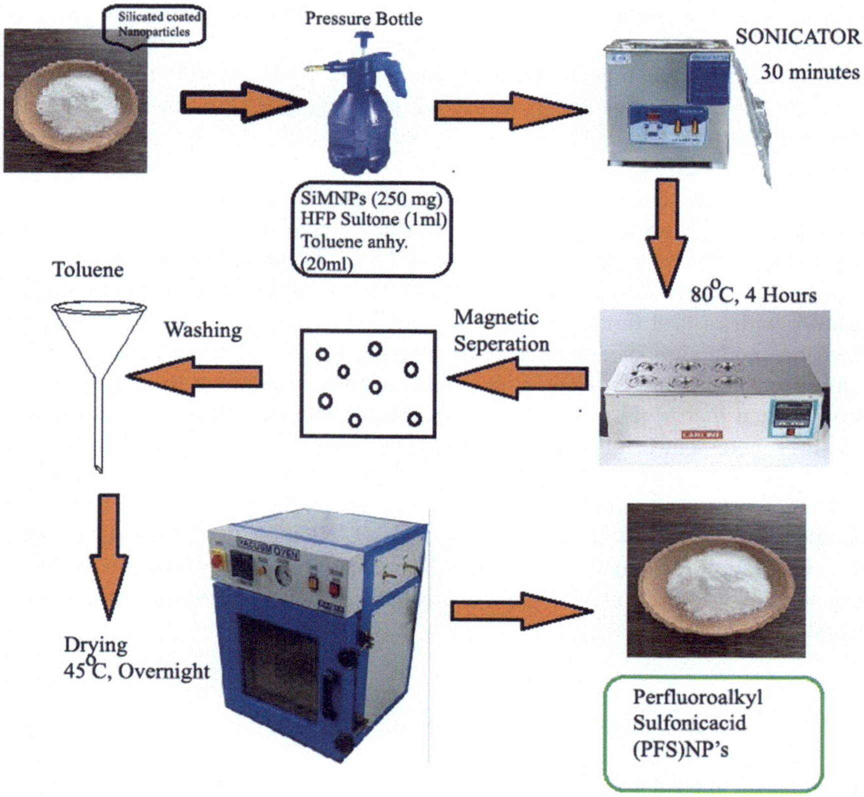

Fig. 6.3 Formation of acid functionalized magnetic NPs (nanoparticles)

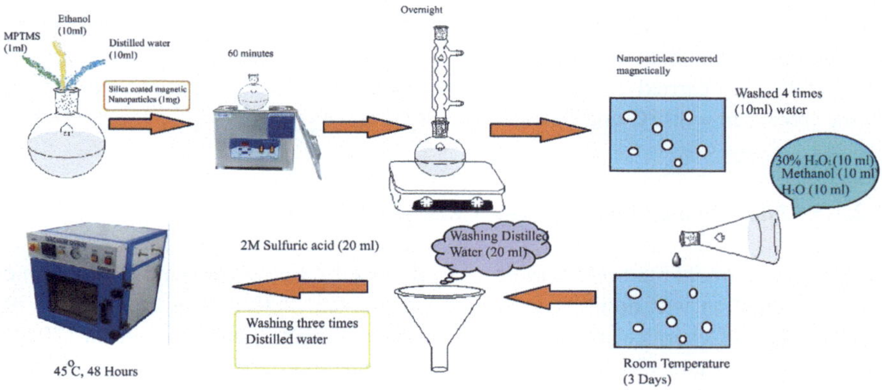

Fig. 6.4 Overall mechanism for the synthesis of SiMNPs, AS NPs, and PFS NPs

hydrogen peroxide, methanol, water, and ethanol in equal proportions to carry out the oxidation of mercapto groups followed by washing out the oxidized product with distilled water (three times) and addition of HCl (0.01 N, 100 mL), and kept overnight for the protonation of respective product. After this, the PS NPs were separated from the acid solution and washed with distilled water which was then dried for 3 h at 120 °C.

Three activated enzymes, viz., protease-activated MgN-Pro (magnesium oxide nanoparticles), MgN-xyl (xylanase-activated magnesium oxide NPs), and (MgN-cel) cellulose-activated magnesium NPs for pretreatment of sugarcane bagasse were used and compared with (MgN-Pro) mediated amino acid production quantity with control (untreated enzyme) and found that at 95 °C, MgN-pro escalated the production of amino acid by roughly 6.18-fold in addition to 18-fold elimination of lignin from the respective feedstock in contrast to control. It was also reported that MgN-xyl pretreated sugarcane bagasse bring out 1.91-fold and 1.82-fold increment in glucose and sugar production respectively at 8 °C as compared to control (Dutta and Saha 2019). The authors examined the synergic effect of alkaline pretreatment and magnetic Fe_3O_4 particles on rice straw and carried out the treatment at four different concentrations of Fe_3O_4 NPs (60, 80, 100, and 120 ppm). The observations recorded were as the pretreatment efficiency and production of biogas were increased at 120 ppm in treated LB as compared to untreated one (Abraham et al. 2014).

Synthesis of Magnetic Oxide (Fe_3O_4) NPs

Synthesis of magnetic oxide NPs was done by carrying out a reaction between aqueous solutions of zinc chloride ($ZnCl_2$), iron (II) chloride tetrahydrate ($FeCl_2.4H_2O$), and iron (III) chloride hexahydrate ($FeCl_3.6H_2O$) in molar ratio (Zn^{2+}:Fe^{2+}:Fe^{3+} = 0.4:0.6:2.0). In order to neutralize the mixture, NaOH (aq) was added to the mixture followed by the formation of precipitates which were given hydrothermal treatment for the duration of 12 h at 150 °C. Then washing of the precipitates was carried out with deionized water followed by the process of drying in the freezer at −80 °C and 0.014 mbar for 24 h.

6.2.1.5 Nano-Scale Shear Hybrid Alkaline Pretreatment

NSHA pretreatment method for biomass of lignocellulose was devised in which one step modification was done by Pedersen and Iverseb (2009), with the addition of another type of chemical reagent acting as pretreatment agents. Lee et al. (2012) devised a rapid and efficient NSHA pretreatment approach in which synergistic effect of chemical and thermal effects with high-velocity shearer was examined in a modified Taylor-Couette reactor that ultimately eliminate the lignin component from LB leaving only cellulose content in the remaining solid. It is a method of lignocellulosic biomass pretreatment in which corn stover was subjected to Taylor-Couette reactor in alkaline medium (NaOH) at room temperature with retention time of 2 min and observed that upto 82% cellulose content was achieved in the left solid along with fourfold increase in enzymatic cellulose conversion and fivefold enhancement in hemicelluloses conversion into respective sugars as compared to untreated corn stover. Due to the synergistic effect and significant removal of both the lignin and hemicelluloses components from LB, this approach gave maximum fermentable sugars that in turn maximize the biohydrogen production by making more access of sugars in the following steps. Apart from the use of nonvolatile compounds like NaOH, KOH, and H_2SO_4 in NSHA pretreatment, different kinds of volatile chemicals, viz., ozone, ammonia, and organic solvents can also be employed whose vaporization and depletion can be prevented by using a sealed reactor in NSHA assembly (Pedersen and Iverseb 2009). Another modification was done in NSHA pretreatment by Wang et al. (2013) in which PDAC (Poly (diallkyldimethylammonium chloride)) cationic polyelectrolyte was used as an additive which has the ability to stabilize lignin and reshape the surface of cellulose. It was observed that PDAC stabilized lignin underwent a change in its cell wall morphology in addition to a reduction in the use of chemicals which are requisite for pretreatment. However, on lignocellulosic biomass pretreatment, little information is available using nanoparticles, and therefore more efforts should be done to explore the nanotechnological approach to make the process more viable at an industrial scale.

6.3 Cellulosic Hydrolysis

After the degradation of hemicellulose and lignin content from lignocellulosic biomass using various pretreatment methods, the next step is subsequent hydrolysis of cellulose left in LB to get fermentable sugars which are then used in biohydrogen production. This step is one of the crucial steps that determine the overall efficiency of the process and to what amount biohydrogen is going to be produced. Chemical and enzymatic hydrolysis are the two important methods used to carry out hydrolytic cleavage of cellulose which is a biopolymer that contains many glucose units linked together by β-1,4-glycosidic bonds that are required to be broken down to get separate glucose molecules or oligosaccharides (Lenihan et al. 2010). During chemical degradation, dilute acids like HCl and H_2SO_4 were used to carry out the hydrolysis process but the use of dilute acids has certain drawbacks like unwanted production of furfural and HMF that step down the efficiency of sugar formation (Mittal et al. 2017), low specificity, poor recyclability, product separation problem, demand of waste effluents treatment and many more. To overcome this issue regarding poor recyclability and difficulty in product separation, different heterogeneous solid acid catalysts like polymer-based catalysts, magnetic solid catalysts, sulfonated carbonaceous-based acids, etc., were used that increased the effectiveness of the hydrolysis process. Takagaki et al. (2008) used layered $HNbMoO_6$ transition-metal oxide in addition to other solid catalysts and observed that hydrolysis over

$HNbMoO_6$ gave 21% more glucose production as compared to amberlyst-15 (3.4%) and the total product yield of glucose and cellobiose was noted to be 8.5%. Much more work was done using heterogeneous catalysts (Zhang et al. 2011; Hu et al. 2015a, b; Onda et al. 2008; Huang and Fu 2013; Lai et al. 2011) yet there is a need to improve the reaction efficiency and selectivity. Therefore, enzymatic hydrolysis came into play which involved the use of enzymes that provide better selectivity; catalytic activity, and use of green approach provide much acceptance to this method as it does not require hazardous and toxic chemicals to put into use. Enzymatic hydrolysis involves the utilization of various enzymes particularly cellulase, hemicellulase, and laccase to carry out the hydrolytic process (Reynaldo et al. 2018) but reusability of enzymes causes another concern in its usage (Chen and Liu 2017) along with high production cost. To deal with this, a new approach has been introduced which involves the use of nanoparticles mediated enzymes/ biocatalysts immobilized using various physical and chemical methods to increase its efficiency. Though less information is available regarding the actual mechanism of immobilization, it is evident that due to large surface area of NPs and the use of magnetic NPs results in much loading of enzymes and easy recovery of NPs, respectively (Ingle et al. 2017).

6.3.1 Nanoparticles in Enzymatic Hydrolysis

To provide better adaptation of enzymes to environmental conditions like temperature and pH and to induce the biocatalysts stability, immobilization of enzymes is carried out on NPs that ultimately access the formation of biofuels in great amount (Taherdanak et al. 2016; Gadhe et al. 2015) along with reusing of expensive enzymes at industrial scale at an economical level. In the process of utilization of NPs as a support for enzyme immobilization, some basic factors need to be considered like (1) efficacious utilization of catalyst, (2) specific activity, (3) minutest enzyme deactivation, (4) cost-effectiveness, and (5) recovered activity (Rizwan et al. 2016). There are mainly four techniques to carry out the immobilization technique—adsorption, cross linking, entrapment, and covalent binding (Cipolatti et al. 2014).

Adsorption: During adsorption method, enzyme is only adsorbed on the surface of magnetic NPs after mixing with a suitable adsorbent under appropriate conditions of temperature and pH. In adsorption principle, weak interatomic forces are responsible for imparting the strength for binding with the disadvantage that adsorbed layers could easily be washed away with mild changes in environmental conditions like change in temperature, substrates, or ionic concentrations (Rizwan et al. 2016).

Covalent Bonding: In covalent bonding, enzyme and substrate (support matrix NPs) are connected to each other via covalent linkage between them. It has been observed that the involvement of different functional groups in enzyme and support matrix increased the covalent linking between the two (Fu et al. 2011; Hartmann and Kostrov 2013).

Entrapment: It is the immobilization of the enzymes over 3D matrix of high molecular compounds. It is a conventional method of physical caging or covalent binding or entrapping of specific amount of enzymes in natural polymers like agarose, agar, or gelatin via thermo reverse polymerization while this encaping in case of alginate and carrageenan is carried by ionotropic gelation (Datta et al. 2013).

Cross Linking: In this method, enzyme/biocatalyst is linked to bi- or tri-functional ligands that results in the formation of insoluble aggregates. The most commonly used bi-functional reagent is glutaraldehyde whose reactive aldehydic groups undergo covalent type interaction with amino groups of enzymes which undergoes a change in the confirmation which results in the activity loss. Among the three immobilization methods, this method is less acceptable as it does not use any kind of solid matrix. The most recent advancement in cross linking process is the use of new biocatalyst, cross linked enzyme aggregates (CLEA's) that are developed via precipitation process of enzyme from aqueous solution by adding nonionic polymers, salts, or water-miscible organic solvents (Sheldon 2007). Due to non-covalent bonding involved between the aggregates, denaturation of enzymes or change in its confirmation is restricted that ultimately preserves its catalytic activity. These kinds of enzymes are very efficacious biocatalysts as they are cost-effective which is a prerequisite in the whole process of enzyme hydrolysis.

While choosing support for enzyme immobilization, several characteristics need to be focused on like its mechanical and chemical resistance, potential interaction with enzyme without remarkably changing its activity, nature of surface (hydrophilic or hydrophobic), defined porous morphology, and cost-effectiveness (Villeneuve et al. 2000). Cellulase and hemicellulase, i.e., immobilized enzymes have shown better catalytic activity with the advantage of being used again and again in several steps with one disadvantage that their recovery from the whole process is difficult to carry out. Different kinds of materials having nanometric dimensions have been used as a support for the immobilization of enzymes like polymeric support which can be biopolymer or inorganic polymer, synthetic organic polymer among which poly (acrylamides), poly (urethane), poly (styrene), poly (methyl-methacrylate) and poly (acrylates) are broadly used for enzyme immobilization (Zhou and Hartmann 2013). Apart from polymeric support, various other NPs are used as support which includes silica NPs, nickel NPs, metal oxide NPs, magnetic NPs, carbon nanotubes, and grapheme-based NPs (Singh et al. 2020). Immobilization of enzymes on NPs increased its affinity towards LB thus increasing the hydrolysis rate which subsequently increased the biohydrogen production. Various NPs were used for immobilization one of which is glutaraldehyde-based (Fe_2O_3) iron oxide immobilized cellulase that revealed a wide range of temperature along with pH stability in comparison with untreated enzymes (Xu et al. 2011).

6.3.2 Synthesis of Iron (II) Oxide NPs

Magnetite or iron (II) oxide NPs were designed using hydrothermal method by carrying out coprecipitation of Fe^{2+} and Fe^{3+} ions with ammonia solution. Both the ferrous and ferric salts of sulfates ($FeSO_4.7H_2O$ and $FeCl_3.6H_2O$) respectively were dissolved in deionized water in stoichiometric ratio (1:2) by keeping the total iron concentration at 0.3 mol/L. After this, the resulting mixture was subjected to sonication for 30 min for deoxidation of stoichiometric mixture followed by heating at 70 °C in the atmosphere of nitrogen. By maintaining the pH of the solution at 9, aqueous ammonia solution (25% w/w) was added dropwise with continuous stirring and the mixture was put in a water bath at a temperature of 85 °C for 1 h to evaporate excess ammonia. Finally, the designed magnetite NPs was washed with anhydrous ethanol and water two and three times, respectively; followed by the process of drying in a vacuum at 40 °C. The product or the dried powder was stored at 4 °C in the freezer for more use.

6.3.3 Synthesis of MnO_2 NPs

Used MnO_2 NPs which involves the use of covalent binding immobilization technique through modification of surface using glutaraldehyde at 70 °C and pH 5.0, providing thermal and pH stability. Using the same immobilization method, another modification in cellulases enzyme using Fe_3O_4/chitosan NPs can provide pH and temperature stability at 5.0 and 60 °C. The respective NPs were formed using the chemical coprecipitation method as reported by (Lineweaver and Burk 1934). A solution of $MnSO_4.H_2O$ (1 mol/L) and NaOH (2 mol/L) was made in deionized water and then stirred for 2 h at 60 °C which resulted in the precipitation of NPs. After this, the precipitates were washed with deionized water 2–3 times, and further they were subjected to drying process in a hot air oven for 12 h at 100 °C. Among various NPs, magnetic NPs are mostly used to (1) overcome the difficulty of recovery of enzymes from the mixture, (2) stability along with specific activity, (3) superparamagnetic nature, and (4) well suspension in the reaction mixture (Zang et al. 2014).

Three functionalized magnetic nanospheres were prepared using three amino silanes 3-(2-aminoethylamino propyl)-trimethoxysilane (AEAPTMES), 3-(2-aminoethylamino propyl)-triethoxysilane (AEAPTES), and 3-aminopropyltriethoxysilane (APTES) and used for immobilization of cellulase enzyme. The results showed that AEAPTES displayed the highest (87%) activity recovery with more temperature stability and wider pH stability than free cellulase (Zhang et al. 2015).

6.3.4 *Synthesis of Magnetite NPs*

Firstly, by using Stober method, nanospheres (silica-coated) were synthesized (Stober and Fink 1968) followed by the synthesis of Fe_3O_4 NPs using the conventional chemical coprecipitation method (Zhang et al. 2013). A mixture of magnetite particles (2 g), alcohol (160 mL), water (40 mL), and $NH_3.H_2O$ (5 mL) was taken and subjected to ultrasonication for 1 h followed by the addition of tetraethyl orthosilicate (TEOS) (5 mL) with continuous stirring for 12 h at room temperature. This resulted in the formation of silica-coated NPs through the process of hydrolysis and condensation of TEOS which were then washed out with distilled water, and underwent drying at 50 °C for 24 h under vacuum conditions. The resulting silica-coated nanospheres (1 g) were added to ethanol (60 mL) and exposed to ultrasonic radiations at room temperature for the duration of 30 min. To the above mixture, $NH_3.H_2O$ (6 mL) and amino silanes— 3-(2-aminoethylamino propyl)-triethoxysilane, 3-(2-aminoethylamino propyl)-trimethoxysilane, and 3-aminopropyltriethoxysilane (4 mL) were put in and the temperature was increased to 50 °C under the atmosphere of nitrogen while stirring the mixture for at least 8 h (Zhang et al. 2014). The resulting NPs were undergone washing with distilled water and drying for 24 h at 50 °C under vacuum conditions. It has been reported that the use of NPs for immobilization of enzymes using different immobilization techniques has improved the different aspects of the scheme and increased biohydrogen production as mentioned in Table 6.1.

For effective bioconversion of LB to biohydrogen production, the former need to be effectively hydrolyzed using enzymes or biocatalysts as they increased the hydrolysis rate (Wang et al. 2015). It was observed that to carry out hydrolysis of various lignocellulosic substances like rice straws, sugarcane bagasse, etc., sulfonated magnetic NPs were used that provided much higher conversion efficiency of respective biomass materials with more percent recovery. *Aspergillus niger* cellulase was immobilized on cyclodextrin-based MNPs to carry out rice straw hydrolysis and it was observed that it yield a much higher concentration of simple sugars in comparison with nonimmobilized enzymes having 85% enzyme recovery for further hydrolysis treatment (Huang et al. 2015).

6.3.5 *Synthesis of Fe_3O_4 NPs*

The respective magnetic NPs were prepared using coprecipitation method in which a mixture of $FeCl_3$ (16 mmol, 2.6 g) and $FeCl_2.4H_2O$ (5 mmol, 1.0 g); after dissolving the mixture in deionized water (400 mL), the reaction proceeded in an inert gas environment for 30 min for the prevention of oxidation of magnetite. Then the resulting solution was heated to 60 °C during which the color of the solution changed from yellow to orange. To this mixture, ammonia solution (1.5 M, 100 mL) was added that resulted in the formation of black colored colloidal solution. Then the

Table 6.1 Type of nanoparticles and their Immobilization technique used for improvement

Synod.	Type of nanoparticles	Immobilization technique	Enzymes/Origin	Improvement	References
1	Alginate-g-PEG Cyclodextrin particle	Encapsulation	Aspergillus terreus lipase	Better pH adaptability and thermal stability compared to fuel CRL	Shang et al. (2015)
2	Attapulgite @ chitosan nanocomposites	Glutaraldehyde activation method (Covalent)	Acremonium Cellulase	High pH and temperature stability, activity	Yang et al. (2016)
3	LiY(Fe_3O_4) Graphene oxide composites	Covalent	Cellulase, Bovine Serum	Better thermal stability and gain of optimal pH with increased half-life of immobilized cellulase as compared to free enzymes	Li et al. (2015)
4	Fe_3O_4 coated on gold electrode surface	Covalent	Superoxide dismutase, Bovine	Rapid estimation of superoxide anions	Thandavan et al. (2013)
5	Fe_3O_4@SiO_2@ P (NIPAM-ω-GMA) double shell micro spheres	Covalent	Endoglucanase, Exoglucanase, and β-Glucosidase	Wider temperature and storage stability as compared to free enzymes with high reproducibility	Han et al. (2018)
6	Gum acacia stabilized MNPs	Adsorption	α-amylase	Enhancement of catalytic activity of α-amylase twofolds, High-temperature stability	Swarnalatha et al. (2013)
7	Magnetic iron oxide NPs	Adsorption	β-Glucosidase from *Aspergillus niger*	Increased thermal stability, 50% regain of enzymatic activity in comparison with free enzymes.	Verma et al. (2013)
8	Amine functionalized cobalt ferrite (AF $CoFe_2O_4$) NPs	Covalent	Cellulase from Trichoderma resei ATCC 26921	High thermal and pH stability	Bohara et al. (2016)
9	Graphine oxide-Fe_3O_4	Physical adsorption	Catalase from bovine serum	Increase in stability	Yang et al. (2015)
10	Alg-G-PEG	Covalent Interaction	Aspergillus terreus lipases	Maintained performances after 20 reuses than free enzymes	Hu et al. (2015a, b)
11	Vinyl functionalized mesoporous silica	Covalent Adsorption	Aspergillus niger	Enhanced cellulase enzyme stability	Harmoko et al. (2016)

12	Fe_3O_4-AP-enzyme	Covalent	Thermomyces lanuginosus lipases	Promising biocatalyst for biodiesel synthesis	Raita et al. (2015)
13	Multi-walled carbon nanotubes via glutaraldehyde	Adsorption	Kluveromyces Lactis-β-galactosidase	More biocatalytic activity	Ansari et al. (2013)

black-colored solution was heated to 80 °C after being reduced by alkali solution followed by the addition of oleic acid with continuous heating until ammonia got evaporated completely. Then, the solution was cooled to room temperature followed by the addition of kerosene (100 mL) to silica-coated suspension, and then the oil layer was collected and water layer was decanted off. This kerosene oil layer was dropped on β-CD's solution to black colloid having 0.02. 0.1, 0.5, and 1.0 (v/v) ratio of β-CD's solution. The resulting magnetite MNPs synthesized was stored in a refrigerator for further use. Su et al. (2015) described the use of cellulase enzyme which was immobilized on sulfonated magnetic carbonaceous acid NPs to carry out the hydrolysis of tropical biomass wastes, i.e., Bagasse, Plukenetia hulls, and Jatropha and the treatment yielded glucose ($C_6H_6O_{12}$) in 58.4%, 35.7%, and 35.9%, respectively.

6.3.6 *Synthesis of Sulfonated Magnetic Carbonaceous Acid NPs*

For a typical procedure, glucose (12–18 g) and nano-Fe_3O_4 (2 g) were mixed in 100 mL water using a three-neck flask under active stirring in an oil bath at a temperature of around 100 °C. After the evaporation of water, the mixture was shifted to a tubular furnace (SGL-1100, Shanghai Daheng Optics and Fine Mechanics Co., Ltd.), heated to 650–750 °C at a rate of 5–7 °C/min, and pyrolyzed for 0.5–1.5 h under N_2 flow (280 mL min^{-1}). The carbonized solid C/Fe_3O_4 was sulfonated by 98% H_2SO_4 with the ratio of 1 g/20 mL in an oil bath at 150 °C for 19–21 h. The sulfonated sample was washed repeatedly with distilled water at 200 ° C for 3 h in a 50 mL high-pressure micro-autoclave (YZPR-50, Shanghai Yanzheng Experimental Instrument Co., Ltd.) until neutral solution was reached (and no SO_4^{2-} was detected using $CaCl_2$). The washed catalyst was dried in an oven (WFO-710, EYELA, Tokyo Rikakikai Co., Ltd.) at 105 °C for 24 h until constant weight, ground and sieved through 200 mesh for hydrolysis. Verma et al. (2013) used β-glucosidase enzyme from *A. niger* immobilized on magnetic Fe_2O_3 NPs through covalent linking technique for cellobiose hydrolysis and it revealed that immobilized enzyme retained 50% enzyme activity up to 16 cycles as compared to free enzymes (Fig. 6.5).

6.3.7 *Synthesis of Magnetic Iron Oxide NPs*

Hydrothermal technique was used to prepare MNPs by carrying out reaction between equimolar ratio of $FeCl_3.6H_2O$ (0.125 M) and $FeCl_2.4H_2O$ (0.125 M) using deionized water (80 mL) as a reaction medium. The aqueous solution so formed was then stirred at 200 rpm followed by the addition of NaOH (30 mL) which resulted in the formation of black precipitates. These precipitates were then

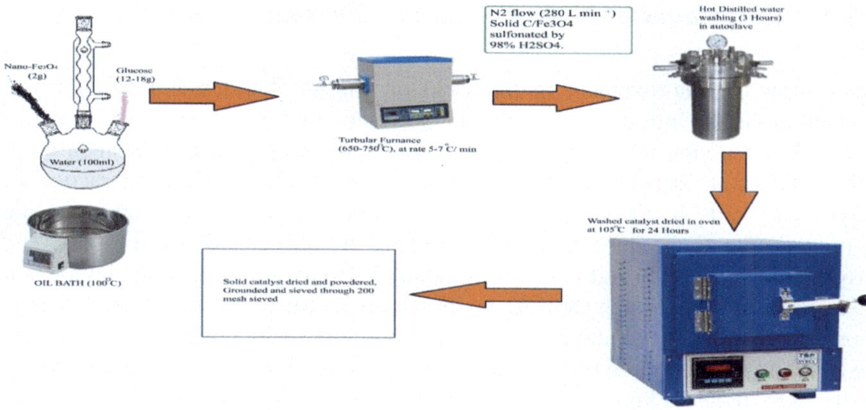

Fig. 6.5 Synthesis of Sulfonated Magnetic Carbonaceous Acid NPs

subjected to autoclaving for 12 h at 150 °C followed by washing and drying (freeze drying at −80 °C) for further use. The same enzyme was also immobilized on magnetic chitosan microspheres for cellulosic corn straw hydrolysis by Zheng et al. (2013) to produce 60.2 g/L reducing sugar with a 78.2% conversion rate that showed the operational stability of the immobilized enzyme.

6.3.8 Synthesis of Magnetic Chitosan Microspheres

Chitosan solution with concentrations (v/v) (1.0%, 2.0%, 3.0%, and 4.0%) was prepared in propionic acid having concentrations (v/v) (1.0%, 1.5%, 2.0%, and 2.5%) (50 mL) to which polyethylene glycol with concentrations (v/v) (0.04%, 0.06%, 0.08%, and 0.1%) (PEG-400) and Fe_3O_4 were added to the solution and stirred at room temperature with subsequent addition of NaOH solution (2 M) to form microspheres. After 10 h of immersion in NaOH solution, the mixture was washed with deionized water to pH 7 and resulted microspheres were collected, filtered, and treated with glutaraldehyde having concentrations (v/v) (0.5%, 1.0%, 1.5%, and 2.0%) at 25 °C for 2 h. The synthesized magnetic chitosan MNPs were recovered after washing with reductant glutaraldehyde. The ratio of chiosan/Fe_3O_4 obtained were 1:1, 1:2, and 1:3. Various scientists used different types of NPs as solid matrix support for the immobilization of various enzymes to carry out NPs mediated enzymatic hydrolysis with better efficiency, reusability, and improvement in biohydrogen production yield.

6.3.9 Advantages of Nanoparticles in Enzyme Properties

Apart from the improvement in the yield of oligosaccharides and biohydrogen production via the introduction of NPs in enzymatic hydrolysis, enzyme immobilization played a vital role in improving temperature and pH stability (Ismail et al. 2011; Zhu et al. 2016) which greatly affects the affinity and the interaction of enzymes towards lignocellulosic biomass substrates that ultimately improve the hydrolysis rate to produce more biohydrogen. A mechanism was proposed by Jariyaboon et al. (2015) and Kirli and Kapdan (2016) to demonstrate the interaction of enzymes and NPs which showed the formation of structure which caused change in protein stability due to alteration in surface properties which is further decided by the type and concentration of NPs utilized for the immobilization of enzyme. Many studies were done for knowing the efficacy of immobilization under variable parameters of temperature and pH and it was observed by Xu et al. (2011) that cellulase enzymes immobilized on glutaraldehyde-based (Fe_3O_4) iron oxide NPs displayed a good range of stability in aspects of both pH and temperature in comparison to free enzymes. Similar studies were done to show the efficiency of immobilized enzymes in the improvement of temperature and pH stability (Bohara et al. 2016; Harmoko et al. 2016; Yang et al. 2016; Lima et al. 2017; Han et al. 2018).

6.4 Microbial Fermentation for Biohydrogen Production

The last step involved in biohydrogen production is the fermentation of microbes of oligosaccharides or simple sugars being obtained in the step of enzymatic hydrolysis. Although a number of researchers carried out the biohydrogen production process by taking into consideration the economic and technical aspects, the outcome is not up to the mark as the quantity of biohydrogen production depends on various factors like the type of feedstock, downstream process configuration, efficiency of pretreatment and enzymatic hydrolysis process, yield of simple sugars obtained in hydrolytic step of cellulose and many more. As there is a continuous increase in energy consumption worldwide and will be increased more in coming future, sustainable biohydrogen production must be carried out. Various types of microorganisms were used for sustainable production of biohydrogen with new metabolic pathways along with nanotechnological approach (Yallappa et al. 2013).

6.4.1 Types of Microorganisms Used in Biohydrogen Production (Dark Fermentation)

Diversity of organisms like facultative aerobic bacteria, anaerobic bacteria, prokaryotes (cyanobacteria and bacteria), and eukaryotes (protists and green algae) are used to produce biohydrogen which may either work as a single entity or work as mixed cultures (Hallenbeck 2012). Fermentation process is a type of heterotrophic metabolism in which carbon-bearing initial substrate is used instead of oxygen as a terminal electron acceptor, the basic requirement to carry out fermentation process. Two types of fermentation namely; photo fermentation (cyanobacteria mediated fermentation) and dark fermentation (bacterial fermentation) were used for biohydrogen production among which bacterial mediated fermentation has gained more attraction as it is the most efficient process in aspect of production yield, less reaction time, choice of using wide range of organic substrates (cellulose-rich biomass), and ambient operational conditions (Azman et al. 2016; Balat and Balat 2009). Apart from the choice of substrate, dark fermentation also depends upon another factor, i.e., type of microorganisms used, as its growth pattern is a key factor for determining the production yield of biohydrogen, micro and macro nutritional sources, operational temperature range, and pH. Two classes of microorganisms, i.e., obligate anaerobic bacterial strain (*Clostridium beijerinckii, Ruminococcus albus,* and many more) and facultative bacterial strain (*Enterobacter aerogenes, E. cloacae*, *Citrobacter intermedius*, *Escherichia coli,* and many more) are used to carry out dark fermentation step (Chandrasekhar and Venkata Mohan 2014). The two main enzymes/biocatalysts used are hydrogenases and nitrogenases (O-Thong et al. 2008). Studies have shown that during the conversion of LB (cellulose-rich agricultural waste) to biohydrogen production, though thermophilic bacteria showed much better yield as compared to mesophilic bacteria as they are more stable at high temperature (Asada et al. 2000), yet not much study has been done on same (Bensah and Mensah 2013). Both the dark and light fermentation differed in the choice of terminal electron acceptor source which is carbon source (organic and inorganic compounds) in case of the former and O_2 in the latter. The most accepted and mostly used fermentative pathway used among acetic acid and butyric acid pathway is acetate mediated pathway whose general reaction is given as:

$$C_6H_{12}O_6 + 2H_2O \rightarrow 2CH_3COOH + CO_2 + 4H_2$$

6.5 Nanotechnology in Biological Fermentation

Due to certain drawbacks of microbial dark fermentation like the low yield of biohydrogen production and its less energy recovery (Ding et al. 2010) less conversion (30–35%) of substrate into biohydrogen in addition to the production of other metabolites (butanol, butyric acid, acetic acid, ethanol) (Logan 2004), certain enzyme modifications were done via supplementing them with metal ions as cofactor like Zn, Cu, Fe, Ni, and many more. This process can also be known as the implementation of nanotechnological approach for improving biohydrogen production efficiency. Due to the introduction of metal ions, bacterial growth ended to be improved as these metal ions provide micro and macronutrients as studied by various researchers (Zheng and Yu 2004; Zhang and Shen 2007; Wang and Wan 2009). Though a little information is available regarding NPs contribution to biological dark fermentation, few studies revealed the efficacy of metal ions in the enhancement of biohydrogen production yield and energy recovery. It has been known the role of hydrogenase enzyme in biohydrogen production as its two active centers (Fe and Ni) played a significant role in the biological fermentation of simple sugars; therefore, its absence or presence revealed the importance of metal cofactors as nutritional supplements (Lin et al. 2012). The quantity of metal cofactors also played an important role in the yield production of biohydrogen as it was mentioned that excess use of metal ions would suppress the activity of enzymes which leads to nullify their positive effect on the yield (Lin et al. 2012). It was studied that introduction of NPs improved the ferrodoxin-reductase activity which in turn improved the electron transfer rate (Bunker and Smith 2005).

To improve the biohydrogen production yield with the help of glucose acting as substrate after enzymatic hydrolysis, Pd, Ag, Cu, and metallic oxide (Fe_xO_y) NPs were encapsulated on porous SiO_2 by *Clostridium butyricum* bacteria and its enhancement effect was studied in fermentative biohydrogen process. It was obtained that yield of H_2 production was 38% that was greater than the control at pH-7.6 and the optimum temperature was 30 °C (Beckers et al. 2013).

6.5.1 Synthesis OF NPs

Four metal ions (Pd, Cu, Fe, and Ag) were used for the synthesis of NPs whose encapsulation was done inside porous matrix using cogelation method to synthesize Cu/SiO_2, Pd/SiO_2, Ag/SiO_2, and Fe/SiO_2 cogel as reported in the literature by (Heinrichs et al. 2008). In cogelation method, one-step doping of inorganic matrix with cations was done at the molecular level. The method included simultaneous condensation and hydrolysis of two alkyl silanes which are SiO_2 network structures that formed a reagent like tetraethoxysilane (TEOS), and alkoxysilane of type $(RO)_3$Si-X-L, in which L is a ligand that formed metal complex (LnM) with metal cation M (M = Ag, Pd, Fe, Cu, etc.) is linked to $(RO)_3$Si alkoxide group via hydrolytically stable and inert spacer X. After the condensation and hydrolysis of cogelates, resulted material possessed metal cations anchored to silica matrix.

Fe/SiO_2 dissol was synthesized using dissolution method as reported by Heinrichs et al. 2008, in which iron salt was mixed in silica gel precursor initially prepared and compared its efficiency with porous SiO_2 alone in biohydrogen production (Lambert et al. 2004). The resulting product was subjected to calcination in air to remove organic moieties (500 °C for Fe/SiO_2 cogel and Fe/SiO_2 dissol, 400 °C for other samples) followed by reduction of Ag/SiO_2 and Cu/SiO_2 using H_2 to obtain metallic NPs. A higher yield of hydrogen production (34.38%) for using α-Fe_2O_3 NPs and 5.47% for NiO NPs than control was reported using glucose as substrate and using glucose-fed thermophilic anaerobic mixed bacteria at 60 ° C (Engliman et al. 2017).

6.5.2 *Synthesis of α-Fe_2O_3 NPs*

In the process, NiO NPs were purchased while α-Fe_2O_3 NPs were synthesized in the laboratory according to the method reported by Matijevic and Scheiner (1978) with few modifications by carrying out hydrolysis of ferric chloride solution. Various scientists used *C. butyricum* bacteria as a bacterial fermentative enzyme and used different types of NPs Cu, Pd, Ag, and Fe_2O_3 to compare the yield of H_2 production. Of these NPs, AgNPs gave maximum 67.5% increment in H_2 yield as compared to the control (Zhao et al. 2013).

6.5.3 *Synthesis of AgNPs*

Silver NPs were synthesized by dissolving $AgNO_3$ (0.0158 g) in double distilled water (40 mL) followed by the addition of sodium citrate (1%, 2 mL) with continuous heating and stirring at 86 ° C. After 30 min of reaction, the color of the solution changed from colorless to olivine after which the mother liquor obtained was diluted to 50 mL and the solution was kept at 0–4 °C for further use (Zheng et al. 2008).

Apart from *C. butyricum* bacteria, various other bacteria (*Enterobacter aerogenes*, *Enterobacter cloacae*, *Enterobacter cloacae* DH-89, *Bacillus anthacis* PUNAJAN 1, *Clostridium pasteurianu*) were used to carry out biological fermentation using different types of NPs (γ Fe_2O_3, Pd (II), Fe_2O_3. Fe_3O_4/carbon, NiO/CoO, α-Fe_2O_3 and TiO_2 NPs) respectively via the use of diverse nature of substrates (mostly glucose), among which *Enterobacter cloacae* DH-89 encapsulated $Fe_2O_3.Fe_3O_4$/carbon NPs using glucose as substrate yielded a much higher percentage of H_2 yield approx. 230% in comparison with control (Lin et al. 2016; Mohanraj et al. 2014; Nath et al. 2015).

6.5.4 *Green Synthesis of PdNPs Using* C. sativum *Leaf Extract*

Firstly, *C. sativum* leaves (20 g) were fine chopped and boiled in double distilled water (100 mL) for 3 min. The resulting extract was filtered and stored at 4 °C. The resulting leaf broth (2 mL) was taken in a flask (100 mL) to which palladium chloride solution (1 mM, 8 mL) was added to yield PdNPs. The contents were incubated in a shaker at 150 ppm at 37 °C for 24 h. To know the effect of *C. sativum* and palladium chloride on the synthesis of palladium NPs, varied concentrations of *C. sativum* broth (1 mL, 2 mL, and 3 mL) using palladium chloride (1 Mm) and different concentrations of palladium chloride (0.25, 0.5, and 1 mM) using *C. sativum* (2 mL). The reaction mixture was centrifuged for 15 min at 11000 rpm for separation of PdNPs.

6.5.5 Synthesis of FeNPs Using S. cumini *Leaf Extract*

Different concentrations of freshly prepared *S. cumini* leaf extract were treated with $FeSO_4$ solution to get optimum concentrations to yield FeNPs. After getting the optimum concentration, plant extract (25 mL) and $FeSO_4$ solution (1 mM, 475 mL) was subjected to stirring at room temperature and after 10 min of reaction, color of the reaction was changed from light yellow to dark black color denoting synthesis of FeNPs. The resulting colloidal solution was centrifuged at 12000 rpm for 10 min at 20 °C. For the further use of synthesized NPs, these are deep freezed after washing with distilled water.

Different yield of H_2 production using different enzymes and NPs merely depends upon the type of their interaction with each other which ultimately enhanced the electron transfer rate in ferrodoxin-reductase chain. It was studied that as small the size of Fe_2O_3 NPs, more will be an increase in reaction rate and hence much improvement in hydrogenase activity via the removal of unnecessary oxygen from fermentation broth. Henceforth, the impact of NPs on the improvement of H_2 production was well established and was incorporated in many studies whose data is compiled in Table 6.2.

6.6 Recent Advancements and Future Perspective

For reducing the dependency on fossil fuels and to provide much cleaner environment, efforts are made on the production of biofuels which is a sustainable approach for meeting the energy requirements of the world. Though biohydrogen production has major advantages, there are certain limitations in its development on large scale. To overcome this issue, nanotechnological approach came into light which is considered as an emerging branch of science. Because of the small size and less surface-to-volume ratio, penetration of nanomaterials into LB becomes very easy which modifies various processes involved in the biohydrogen production starting from pretreatment to dark fermentation. Though this field is emerging day by day, it is required to carry out the process of production of biohydrogen more economically beneficial using a novel nanotechnological approach in every consecutive step. More efforts are made on increasing the efficiency of the two most important steps of biohydrogen production (pretreatment and enzymatic hydrolysis) as the output of these steps determines the actual output of biohydrogen yield. Acid or alkali pretreatment is the most expensive step to be carried out at an industrial scale which ultimately increases the overall cost of biohydrogen production. To deal with this problem, nanoparticles are being used in each step. In the pretreatment method, use of NPs lowers the cost of processing by eliminating the use of acid/base use via the introduction of nano-size particles despite the use of macro-size particles. Though the synthesis of NPs is also carried out chemically which is also an expensive step therefore new synthetic approaches are used for their synthesis

Table 6.2 Types of substrates used for nanoparticles

S. no.	Nanoparticles used	Type of microorganisms	Remarks	Type of substrate	References
1	Nano-titanium dioxide	Anaerobic sludge	Maximum hydrogen production of 101.5 mL/VS (Volatile solids) was obtained at 1 g nano TiO_2/L than control	Sugarcane Baggase	Jafari and Zilouei (2014)
2	Nickel nanoparticles	Anaerobic granular sludge	22.71% increase in hydrogen production	Glucose	Mullai et al. (2013)
3	Zero valent iron nanoparticles	Microbial composition with Fe-400 groups (from *emtero bacter sp*. to *clostrodium sp*.)	Hydrogen yield was increased by 73.1% with respect to control	Grain Biomass (*Lorium perenne L.*)	Yang and Wang (2018)
4	Fe and Ni NPs	Anaerobic sludge	Ni^{+2}, Fe^{0}, Fe^{+2} ions increased hydrogen production by 55%, 37%, and 15%, respectively	Glucose	Taherdanak et al. (2016)
5	Hematite and Ni Nano particles	Anaerobic sludge	59% increase in hydrogen yield, with coaddition of hematite and NiO_2 NPs	Dairy wastewater	Gadhe et al. (2015)
6	Fe^{+2} and magnetite nanoparticles	Enterobacteriaceae, Clostridium sp., and hydrogenase-specific gene	Addition of Fe^{+2} and magnetite increase hydrogen production by 62% and 69.6%, respectively	Sugarcane baggase	Reddy et al. (2017) Yang and Wang (2018)
7	Ferric oxide/ Carbons NPs	Anaerobic fermentation (Mixed bacteria)	33.7% increase in hydrogen production as compared to control	Glucose	Zhang et al. (2018)
8	Mg-Al hydrotalcite (HT), Au/Zn-Mg-Al HT, and nanoporous activated carbon (NAC)	Anaerobic mixed culture	Maximum yield was obtained for Au/Zn-Mg-Al with 2.74 ± 0.14 mol H_2/mol sucrose	Sucrose	Wimonsong and Nitisoravut (2015)
9	Treated activated carbon	Anaerobic sludge	73% increase in hydrogen	Sucrose	Wimonsong and

(continued)

Table 6.2 (continued)

S. no.	Nanoparticles used	Type of microorganisms	Remarks	Type of substrate	References
			production as compared to control for the first cycle followed by drop of 32% after 3 cycles		Nitisoravut (2014)

among which the green methodological approach is the most emerging one which may improve the cost-effectiveness of the whole process (Kumar et al. 2012). Various extracts of plants and microorganisms are used for the economical synthesis of NPs (Yang and Li 2013). Therefore, the introduction of nanomaterials and carrying out its green synthesis eliminates the use of hazardous chemicals, more stable product formation with a reduction in the cost of pretreatment process.

In the second step of H_2 production, enzymatic hydrolysis of pretreated material is done using various biocatalysts which provide another hindrance to the efficient production of H_2 as the whole process requires stability of enzymes under high-temperature conditions, slow reaction rate, nonreproducibility of enzymes, incomplete conversion of substrates to desired products and cost intensiveness (Yang et al. 2011). Therefore, enzyme immobilization of enzymes is done on suitable solid matrix support using magnetic NPs that make the use of enzymes over and over again with improved temperature stability.

Biohydrogen production efficiency is also depending on the last step which is the dark fermentation process whose effectiveness can also be increased via the incorporation of NPs (Wimonsong and Nitisoravut 2014, 2015; Khan et al. 2013). Observations have been made on biologically derived NPs that also reduce the hazardous impact of them on the cells of microbes during biological fermentation. Though a lot of research has been done on studying the effectiveness of NPs in biohydrogen production, little information is available to know the mechanism and type of NPS during the production of cellulosic biohydrogen. It has been observed that mostly powder form of NPs is used in each step of H_2 production, but the introduction of nanosheets and nanotubes could depict better binding efficiency and of course, catalytic activity. Thus, the implementation of the nanotechnological approach can help in achieving efficient, economical, and sustainable production of biohydrogen at a commercial scale with more advanced future perspectives which ultimately will help in the preservation of the most valued natural resources (fossil fuels).

6.7 Conclusion

This chapter provides a complete route for the synthesis of NPs and biohydrogen production and implementation of NPs in each and every step of the whole process of H_2 production using different cellulosic biomass materials. Starting from pretreatment of LB to the dark fermentation step, nanomaterials have shown their major role to make the process more economical, sustainable, and viable. In each step, different types of nanomaterials are used depending on the demand of each step among which mostly magnetic NPs are used as they provide better recovery of biocatalysts with an improved output of each step. As nanotechnology is an emerging field of science, more efforts need to be done to explore it more to meet the greater needs of the world by producing value-added products on large scale with cost-effective approach to meet the sustained requirements.

References

Abraham RE, Verma ML, Barrow CJ, Puri M (2014) Suitability of magnetic nanoparticles immobilized cellulases in enhancing enzymatic saccharification of pretreated hemp biomass. Biotechnol Biofuels 7:90

Amin FR, Khalid H, Zhang H, Rahman S, Zhang R, Liu G, Chen C (2017) Pretreatment methods of lignocellulosic biomass for anaerobic digestion. Appl Microbiol 7:72. https://doi.org/10.1186/s13568-017-0375-4

Ansari SA, Satar R, Chibber S, Khan MJ (2013) Enhanced stability of Kluyveromyced lactis β galactosidase immobilized on glutaraldehyde modified multiwalled carbon nanotubes. J Mol Catal B Enzym 97:258–263

Asada Y, Koike Y, Schnackenberg J, Miyake M, Uemura I, Miyake J (2000) Heterologous expression of clostridial hydrogenase in the cyanobacterium synechococcus PCC7942. Biochim Biophys Acta 1490(3):269–278. https://doi.org/10.1016/S0167-4781(00)00010-5

Assawamongkholsiri T, Reungsang A, Pattra S (2013) Effect of acid, heat and combined acid-heat pretreatments of anaerobic sludge on hydrogen production by anaerobic mixed cultures. Int J Hydrogen Energy 38:6146–6153

Azman NF, Abdeshahian P, Kadier A, Nasser Al-Shorgani NK, Salih NKM, Lananan I, Kalil MS (2016) Biohydrogen production from de-oiled rice bran as sustainable feedstock in fermentative process. Int J Hydrog Energy 41(1):145–156. https://doi.org/10.1016/j.ijhydene.2015.10.018

Balat M, Balat M (2009) Political, economic and environmental impacts of biomass-based hydrogen. Int J Hydrog Energy 34(9):3589–3603. https://doi.org/10.1016/j.ijhydene.2009.02.067

Beckers L, Hiligsmann S, Lambert SD, Heinrichs B, Thonart P (2013) Improving effect of metal and oxide nanoparticles encapsulated in porous silica on fermentative biohydrogen production by clostridium butyricum. Bioresour Technol 133:109–117. https://doi.org/10.1016/j.biortech.2012.12.168

Bensah EC, Mensah M (2013) Chemical pre-treatment methods for the production of cellulosic ethanol: technologies and innovations. Int J Chem Eng 2013:719607. https://doi.org/10.1155/2013/719607

Bohara RA, Thorat ND, Pawar SH (2016) Immobilization of cellulase on functionalized cobalt ferrite nanoparticles. Korean J Chem Eng 33:216–222

Bunker CE, Smith MJ (2005) Nanoparticles for hydrogen generation. J Mater Chem 21:12173–12180

Chandrasekhar K, Venkata Mohan S (2014) Induced catabolic bio-electrohydrolysis of complex food waste by regulating external resistance for enhancing acidogenic biohydrogen production. Bioresour Technol 165:372–382

Chen HZ, Liu ZH (2017) Enzymatic hydrolysis of lignocellulosic biomass from low to high solids loading. Eng Life Sci 17:489–499

Cheng JJ, Timilsina GR (2011) Status and barriers of advanced biofuel technologies: a review. Renew Energy 36:3541–3549. https://doi.org/10.1016/j.renene.2011.04.031

Chozhavendhan S, Rajamehala M, Karthigadevi G, Praveenkumar R, Bharathiraja B (2020) A review on feedstock, pretreatment methods, influencing factors, production and purification processes of bio-hydrogen production. CSCEE 2:100038

Cianchetta S, Maggio BD, Burzi PL, Galletti S (2014) Evaluation of selected white-rot fungal isolates for improving the sugar yield from wheat straw. Appl Biochem Biotechnol 173:609–623

Cipolatti EP, Silva MJA, Klein M, Feddern V, Feltes MMC, Oliveira JV, Ninow JL, de Oliveira D (2014) Nanomaterials for biocatalyst immobilization-state of the art and future trends. J Mol Cata B Enzym 99:56–67

Das D, Veziroğlu TN (2001) Hydrogen production by biological processes: a survey of literature. Int J Hydrog Energy 26(1):13–28

Datta S, Christena LR, Rajaram YRS (2013) Enzyme immobilization: an overview on techniques and support materials. 3 Biotech 3:1–9

Ding J, Wang X, Zhou XF, Ren NQ, Guo WQ (2010) CFD optimization of continuous stirred-tank (CSTR) reactor for biohydrogen production. Bioresour Technol 101(18):7005–7013. https://doi.org/10.1016/j.biortech.2010.03.146

Dutta N, Saha MK (2019) Nanoparticle induced enzyme pre-treatment method for increased glucose production from lignocellulosic biomass under cold conditions. J Sci Food Agric 99(2):767–780

Engliman NS, Abdul PM, Wu SY, Jahim JM (2017) Influence of iron (II) oxide nanoparticle on biohydrogen production in thermophilic mixed fermentation. Int J Hydrog Energy 42(45): 27482–27493. https://doi.org/10.1016/j.ijhydene.2017.05

Fu J, Reinhold J, Woodbury NW (2011) Peptide-modified surfaces for enzyme immobilization. PLoS One 6(4):e18692

Gadhe A, Sonawane SS, Varma MN (2015) Enhancement effect of hematite and nickel nanoparticles on biohydrogen production from dairy wastewater. Int J Hydrog Energy 40: 4502–4511

Gill CS, Price BA, Jones CW (2007) Sulfonic acid-functionalized silica coated magnetic nanoparticles catalysts. J Catal 251(1):145–152. https://doi.org/10.1016/j.jcat.2007.07.007

Gong X, Chen L, Lin B (2020) Analyzing dynamic impacts of different oil shocks on oil price. Energy 198:117306

Gonzalez G, Urrutia H, Roeckel M, Aspe E (2005) Protein hydrolysis under anaerobic, saline conditions in presence of acetic acid. J Chem Technol Biotechnol 157:151–157. https://doi.org/10.1002/jctb.1165

Hallenbeck P (2012) Hydrogen production by cyanobacteria. In: Hallenbeck PC (ed) Microbial technologies in advanced biofuels production. Springer, New York, pp 15–28

Han J, Rong J, Wang Y, Liu Q, Tang X, Li C, Ni L (2018) Immobilization of cellulase on thermo-sensitive magnetic microspheres: improved stability and reproducibility. Bioprocess Biosyst Eng 41:1051–1060

Harmoko C, Sucipto KI, Ery RS, Hartono SB (2016) Vinyl functionalized cubic mesoporous silica nanoparticles as supporting material to enhance cellulase enzyme stability. J Eng Appl Sci 11: 2981–2992

Hartmann M, Kostrov X (2013) Immobilization of enzymes on porous silicas—benefits and challenges. Chem Soc Rev 42:6277–6289

Heinrichs B, Rebbouh L, Geus JW, Lambert S, Abbenhuis HCL, Grandjean F, Long GJ, Pirard JP, van Santen RA (2008) Iron (III) species dispersed in porous silica through sol-gel chemistry. J Non-Cryst Solids 354(2–9):665–672

Hu C, Wang N, Zhang W, Zhang S, Meng Y, Yu X (2015a) Immobilization of Aspergillus terreus lipase in self-assembled hollow nanospheres for enantioselective hydrolysis of ketoprofen vinyl ester. J Biotechnol 194:12–18

Hu S, Jiang F, Hsieh YL (2015b) 1D lignin-based solid catalysts for cellulose hydrolysis to glucose and nanocellulose. ACS Sustain Chem Eng 3:2566–2574

Huang YB, Fu YJ (2013) Hydrolysis of cellulose to glucose by solid acid catalysts. Green Chem 15(5):1095–1111

Huang PJ, Chang KL, Hsieh JF, Chen ST (2015) Catalysis of rice straw hydrolysis by the combination of immobilized cellulase from Aspergillus niger on β-cyclodextrin-Fe_3O_4 nanoparticles and ionic liquid. Biomed Res Int 2015:1–9. https://doi.org/10.1155/2015/409103

Ingle AP, Rathod J, Pandit R, da Silva SS, Rai M (2017) Comparative evaluation of free and immobilized cellulase for enzymatic hydrolysis of lignocellulosic biomass for sustainable bioethanol production. Cellulase 24:5529–5540

Ismail I, Hassan MA, Soon CS (2011) Effect of retention time on biohydrogen production by microbial consortia immobilised in polydimethylsiloxane. Afr J Biotechnol 10:601–609

Jafari O, Zilouei H (2014) Enhanced biohydrogen and subsequent biomethane production from sugarcane bagasse using nano-titanium dioxide pretreatment. Bioresour Technol 2014:670–678

Jariyaboon R, Sompong O, Kongjan P (2015) Bio-hydrogen and bio-methane potentials of skim latex serum in batch thermophilic two-stage anaerobic digestion. Bioresour Technol 198:198–206

Kapdan IK, Kargi F (2006) Biohydrogen production from waste materials. Enzyme Microb Technol 38:569–582

Khan MM, Lee J, Cho MH (2013) Electrochemically active biofilm mediated bio-hydrogen production catalyzed by positively charged gold nanoparticles. Int J Hydrog Energy 38:5243–5250

Kim DH, Jang S, Yun YM, Lee MK, Moon C, Kang WS et al (2014) Effect of acid pretreatment on hydrogen fermentation of food waste: microbial community analysis by next generation sequencing. Int J Hydrog Energy 39:16302–16329

Kirli B, Kapdan IK (2016) Selection of microorganism immobilization particle for dark fermentative biohydrogen production by repeated batch operation. Renew Energy 87:697–702

Kumar SG, Gobinath C, Karpagam K, Hemamalini V, Premkumar K, Sivaramakrishnan S (2012) Phyto-synthesis of silver nanoscale particles using Morinda citrifolia L. and its inhibitory activity against human pathogens. Colloids Surf B: Biointerfaces 95:235–240. https://doi.org/10.1016/j.colsurfb.2012.03.001

Lai DM, Deng L, Guo QX, Fu Y (2011) Hydrolysis of biomass by magnetic solid acid. Energy Environ Sci 4(9):3552–3557

Lambert S, Alié C, Pirard JP, Heinrichs B (2004) Study of textural properties and nucleation phenomenon in Pd/SiO_2, Ag/SiO_2 and Cu/SiO_2 cogelled xerogel catalysts. J Non-Cryst Solids 342(1–3):70–81

Lee I, Wang W, Ji S (2012) Device and method for pretreatment of biomass. US Patent No. US20120036765 A1

Lenihan P, Orozco A, O'Neil E, Ahmad MNM, Rooney DW, Walker GM (2010) Dilute acid hydrolysis of lignocellulosic biomass. Chem Eng J 156:395–403

Li Y, Wang XY, Jiang XP, Ye JJ, Zhang YW, Zhang XY (2015) Fabrication of graphene oxide decorated with Fe_3O_4@SiO_2 for immobilization of cellulase. J Nanopart Res 17:8

Lima JS, Araújo PHH, Sayer C, Souza AAU, Viegas AC, de Oliveira D (2017) Cellulase immobilization on magnetic nanoparticles encapsulated in polymer nanospheres. Bioprocess Biosyst Eng 40:511–518

Lin CY, Lay CH, Sen B, Chu CY, Kumar G, Chen CC, Chang JS (2012) Fermentative hydrogen production from wastewaters: a review and prognosis. Int J Hydrog Energy 37:15632–15642

Lin R, Cheng J, Ding L, Song W, Liu M, Zhou J, Cen K (2016) Enhanced dark hydrogen fermentation by addition of ferric oxide nanoparticles using Enterobacteraerogenes. Bioresour Technol 207:213–219. https://doi.org/10.1016/j.biortech.2016.02.009

Lineweaver H, Burk DJ (1934) The determination of enzyme dissociation constants. Am Chem Soc 56(3):658–666

Logan BE (2004) Biologically extracting energy from wastewater: bio-hydrogen production and microbial fuel cells. Environ Sci Technol 38:160–167

Matijevic E, Scheiner P (1978) Ferric hydrous oxide sols. III. Preparation of uniform particles by hydrolysis of Fe (III)- chloride, -nitrate and -perchlorate solutions. J Colloid Interface Sci 63: 509–524

Mittal A, Black SK, Vinzant TB, O'Brien M, Tucker MP, Johnson DK (2017) Production of furfural from process-relevant biomass-derived pentoses in a biphasic reaction system. ACS Sustain Chem Eng 5:5694–5701

Mohanraj S, Anbalagan K, Kodhaiyolii S, Pugalenthi V (2014) Comparative evaluation of fermentative hydrogen production using Enterobacter cloacae and mixed culture: effect of Pd (II) ion and phytogenic palladium nanoparticles. J Biotechnol 192:87–95

Mullai P, Yogeswari MK, Sridevi K (2013) Optimisation and enhancement of biohydrogen production using nickel nanoparticles—a novel approach. Bioresour Technol 141:212–219

Nath D, Manhar AK, Gupta K, Saikia D, Das SK, Mandal M (2015) Phytosynthesized iron nanoparticles: effects on fermentative hydrogen production by Enterobacter cloacae DH-89 *Bull.* Mater Sci 38:1533–1538

Navlani-Garcia M, Mori K, Kuwahara Y, Yamashita H (2018) Recent strategies targeting efficient hydrogen production from chemical hydrogen storage materials over carbon-supported catalysts. NPG Asia Mater 10:277–292

Onda A, Ochi T, Yanagisawa K (2008) Selective hydrolysis of cellulose into glucose over solid acid catalysts. Green Chem 10(10):1033–1037

Orts WJ, McMahan CM (2016) Biorefinery developments for advanced biofuels from a sustainable array of biomass feedstocks: survey of recent biomass conversion research from agricultural research service. Bioenergy Res 9(2):430–446

O-Thong S, Prasertsan P, Karakashev D, Angelidaki I (2008) Thermophilic fermentative hydrogen production by the newly isolated Thermoanaerobacterium thermosaccharolyticum PSU-2. Int J Hydrog Energy 33(4):1204–1214. https://doi.org/10.1016/j.ijhydene.2007.12.015

Otto M, Floyd M, Bajpai S (2008) Nanotechnology for site remediation. Remediation J 19(1): 99–108

Pedersen JV, Iverseb FK (2009) Devices and methods for discharging pretreated biomass from higher to lower pressure regions. Patent No. 2009147512 A2

Pena L, Ikenberry M, Hohn KL, Wang D (2012) Acid functionalized nanoparticles for pretreatment of wheat straw. J Biomater Nanobiotechnol 3:342–352

Periyasamy K, Santhalembi L, Mortha G, Aurousseau M, Boyer A, Subramanian S (2018) Bioconversion of lignocellulosic biomass to fermentable sugars by immobilized magnetic cellulolytic enzyme cocktails. Langmuir 34:6546–6555

Phan NTS, Jones CW (2006) Highly accessible catalytic sites on recyclable organosilane functionalized magnetic nanoparticles: an alternative to functionalized porous silica catalysts. J Mol Catal 253(1–2):123–131

Qi W, He C, Wang Q, Liu S, Yu Q, Wang W et al (2018) Carbon-based solid acid pretreatment in corncob saccharification: specific xylose production and efficient enzymatic hydrolysis. ACS Sustain Chem Eng 6(3):3640–3648

Rai M, Ingle AP, Gaikwad S, Dussan KJ and da Silva SS (2017) Role of nanoparticles in enzymatic hydrolysis of lignocellulose in ethanol. In: Nanotechnology for bioenergy and biofuel production. Springer, pp 153–171

Raita M, Arnthong J, Champreda V, Laosiripojana N (2015) Modification of magnetic nanoparticles lipase designs for biodiesel production from palm oil. Fuel Process Technol 134:189–197

Reddy K, Nasr M, Kumari S, Kumar S, Gupta SK, Enitan AM, Bux F (2017) Biohydrogen production from sugarcane bagasse hydrolysate: effects of pH, S/X, Fe^{2+}, and magnetite nanoparticles. Environ Sci Pollut Control Ser 24:8790–8804

Reynaldo PB, Adriano E, Marcelo M, Silvia N (2018) Enzymatic hydrolysis of sugarcane biomass and heat integration as enhancers of ethanol production. J Renew Mater 6:183–194

Rizwan M, Buthe A, Hamid M, Wang P (2016) Cost-efficient entrapment of β-glucosidase in nanoscale latex and silicone polymeric thin films for use as stable biocatalysts. Food Chem 190: 1078–1085

Rocha GJM, Martin C, da Silva VFN, Gomez EO, Goncalves AR (2012) Mass balance of pilot-scale pretreatment of sugarcane bagasse by steam explosion followed by alkaline delignification. Bioresour Technol 111:447–452. https://doi.org/10.1016/j.biortech.2012.02.005

Shang CY, Li WX, Zhang RF (2015) Immobilization of Candida rugosa lipase on ZnO nanowires/ macroporous silica composites for biocatalytic synthesis of phytosterol esters. Mater Res Bull 68:336–342

Shanmugam S, Hari A, Pandey A, Mathimani T, Felix L, Pugazhendhi A (2020) Comprehensive review on the application of inorganic and organic nanoparticles for enhancing biohydrogen production. Fuel 270:117453

Sheldon RA (2007) Cross-linked enzyme aggregates (CLEAs): stable and recyclable biocatalysts. Biochem Soc Trans 35:1583–1587

Singh N, Dhanya BS, Verma ML (2020) Nano-immobilized biocatalysts and their potential biotechnological applications in bioenergy production. Mater Sci Energy Technol 3(2020): 808–824

Srivastava N, Srivastava M, Kushwaha D, Gupta VK, Manikanta A, Ramteke PW, Mishra PKO (2017) Efficient dark fermentative hydrogen production from enzyme hydrolyzed rice straw by Clostridium pasteurianum (MTCC116). Bioresour Technol 238:552–558

Stober W, Fink A (1968) Controlled growth of monodisperse silica spheres in the micron size range. J Colloid Interf Sci 26:62–69

Su TC, Fang Z, Zhang F, Luo J, Li XK (2015) Hydrolysis of selected tropical plant wastes catalyzed by a magnetic carbonaceous acid with microwave. Sci Rep 5:17538. https://doi.org/10.1038/srep17538

Suhara H, Kodama S, Kamei I, Maekawa N, Meguro S (2012) Screening of selective lignin-degrading basidiomycetes and biological pretreatment for enzymatic hydrolysis of bamboo culms. Int Biodeter Biodegr 75:176–180

Sun Y, He J, Yang G, Sun G, Sage V (2019) A review of the enhancement of bio-hydrogen generation by chemicals addition. Catalysts 9(4):353

Swarnalatha V, Esther RA, Dhamodharan R (2013) Immobilization of α-amylase on gum acacia stabilized magnetic nanoparticles, an easily recoverable and reusable support. J Mol Catal B Enzym 96:6–13

Taha M, Shahsavari E, Al-Hothaly K, Mouradov A, Smith AT, Ball AS, Adetutu EM (2015) Enhanced biological straw saccharification through co-culturing of lignocellulosic degrading microorganisms. Appl Biochem Biotechnol 175:3709–3728

Taherdanak M, Zilouei H, Karimi K (2016) The effects of FeO and NiO nanoparticles versus Fe^{2+} and Ni^{2+} ions on dark hydrogen fermentation. Int J Hydrog Energy 41:167–173

Taherzadeh MJ, Karimi K (2007) Enzyme-based hydrolysis processes for ethanol from lignocelulosic materials: a review. Bioresources 2:707–738

Takagaki A, Tagusagawa C, Domen K (2008) Glucose production from saccharides using layered transition metal oxide and exfoliated nanosheets as a water tolerant solid acid catalyst. Chem Commun 42:5363–5365. https://doi.org/10.1039/B810346A

Thandavan K, Gandhi S, Sethuraman S, Rayappan JBB, Krishnan UM (2013) A novel nano-interfaced superoxide biosensor. Sens Actuators B 176:884–892
Verma ML, Chaudhary R, Tsuzuki T, Barrow CJ, Puri M (2013) Immobilization of β-glucosidase on a magnetic nanoparticle improves thermostability: application in cellobiose hydrolysis. Bioresour Technol 135:2–6
Verma ML, Kumar S, Das A, Randhawa JS, Chamundeeswari M (2019) Chitin and chitosan-based support materials for enzyme immobilization and biotechnological applications. Environ Chem Lett 18(2):1–9. https://doi.org/10.1007/s10311-019-00942-5
Villeneuve P, Muderhwa JM, Graille J, Haas MJ (2000) Customizing lipases for biocatalysis: a survey of chemical, physical and molecular biological approaches. J Mol Catal B Enzym 9:113–148
Wang J, Wan W (2009) Factors influencing fermentative hydrogen production: a review. Int J Hydrog Energy 34(2):799–811. https://doi.org/10.1016/j.ijhydene.2008.11.015
Wang W, Ji S, Lee I (2013) Fast and efficient nanoshear hybrid alkaline pretreatment of corn stover for biofuel and materials production. Biomass Bioenergy 51:35–42
Wang H, Covarrubias J, Prock H, Wu X, Wang D, Bossmann SH (2015) Acid functionalized magnetic nanoparticle as heterogeneous catalysts for biodiesel synthesis. J Phys Chem C 119: 26020–26028
Wimonsong P, Nitisoravut R (2014) Biohydrogen enhancement using highly porous activated carbon. Energy Fuels 28:4554–4559
Wimonsong P, Nitisoravut R (2015) Comparison of different catalyst for fermentative hydrogen production. J Clean Energy Technol 3:128–131
Xu J, Huo S, Yuan Z, Zhang Y, Xu H, Guo Y, Liang C, Zhuang X (2011) Characterization of direct cellulase immobilization with superparamagnetic nanoparticles. Biocatal Biotransform 29:71–76
Yallappa S, Manjanna J, Sindhe MA, Satyanarayan ND, Pramod SN, Nagaraja K (2013) Microwave assisted rapid synthesis and biological evaluation of stable copper nanoparticles using T. arjuna bark extract. Spectrochim Acta A Mol Biomol Spectrosc 110:108–115
Yang N, Li WH (2013) Mango peel extract mediated novel route for synthesis of silver nanoparticles and antibacterial application of silver nanoparticles loaded onto nonwoven fabrics. Ind Crop Prod 48:81–88. https://doi.org/10.1016/j.indcrop.2013.04.001
Yang G, Wang J (2018) Improving mechanisms of biohydrogen production from grass using zero-valent iron nanoparticles. Bioresour Technol 266:413–420
Yang B, Dai Z, Ding SY, Wyman CE (2011) Enzymatic hydrolysis of cellulosic biomass. Biofuels 2(4):421–449
Yang D, Wang X, Shi J, Wang X, Zhang S, Han P, Jiang Z (2015) In situ synthesized Rgo-Fe_3O_4 nanocomposites as enzyme immobilization support for achieving high activity recovery and easy recycling. Biochem Eng J 105:273–228
Yang C, Mo H, Zang L, Chen J, Wang Z, Qiu J (2016) Surface functionalized natural inorganic nanorod for highly efficient cellulase immobilization. RSC Adv 6:76855–76860
Zang L, Qiu J, Wu X, Zhang W, Sakai E, Wei Y (2014) Preparation of magnetic chitosan nanoparticles as support for cellulase immobilization. Ind Eng Chem Res 53:3448–3454
Zhang Y, Shen J (2007) Enhancement effect of gold nanoparticles on biohydrogen production from artificial wastewater. Int J Hydrog Energy 32(1):17–23. https://doi.org/10.1016/j.ijhydene.2006.06.004
Zhang F, Deng X, Fang Z, Tian X, Kozinski J (2011) Hydrolysis of crystalline cellulose over Zn–Ca–Fe oxide catalyst. Petrochem Technol 40:43–48
Zhang BL, Zhang HP, Fan XL, Li XJ, Yin DZ, Zhang QY (2013) Preparation of thermoresponsive Fe_3O_4/P(acrylic acid-methylmethacrylate-N-isopropylacrylamide) magnetic composite microspheres with controlled shell thickness and its releasing property for phenolphthalein. J Colloid Interfaces Sci 398:51–58

Zhang WJ, Qiu JH, Feng HX, Wu XL, Zang LM, Yi W, Sakai E (2014) Preparation and characterization of functionalized magnetic silica nanospheres with the immobilized cellulase. Appl Mech Mater 543:3892–3895

Zhang W, Qui J, Feng H, Zang L, Sakai E (2015) Increase in stability of cellulase immobilized on functionalized magnetic nanospheres. J Magn Magn Mater 375(1):117–123

Zhang J, Fan C, Zhang H, Wang Z, Zhang J, Song M (2018) Ferric oxide/carbon nanoparticles enhanced bio-hydrogen production from glucose. Int J Hydrog Energy 43:8729–8738

Zhao W, Zhang Y, Du B, Wei D, Wei Q, Zhao Y (2013) Enhancement effect of silver nanoparticles on fermentative biohydrogen production using mixed bacteria. Bioresour Technol 142:240–245

Zheng XJ, Yu HQ (2004) Biological hydrogen production by enriched anaerobic cultures in the presence of copper and zinc. J Environ Sci Health A 39(1):89–101. https://doi.org/10.1081/ese-120027370

Zheng JH, Wu X, Wang MQ, Ran DH, Xu W, Yang JH (2008) Study on the interaction between silver nanoparticles and nucleic acids in the presence of cetyltrimethylammonium bromide and its analytical application. Talanta 74:526–532

Zheng P, Wang J, Lu C, Xu Y, Sun Z (2013) Immobilized β-glucosidase on magnetic chitosan microspheres for hydrolysis of straw cellulose. Process Biochem 48:683–687

Zhou Z, Hartmann M (2013) Progress in enzyme immobilization in ordered mesoporous materials and related applications. Chem Soc Rev 42:3894–3912. https://doi.org/10.1039/c3cs60059a

Zhu S, Xu X, Rong R, Li B, Wang X (2016) Evaluation of zinc-doped magnetite nanoparticle toxicity in the liver and kidney of mice after sub-chronic intragastric administration. Toxicol Res 5:97–106

Chapter 7
Microbial-Mediated Synthesis of Nanoparticles and Their Role in Bioethanol Production

Sreedevi Sarsan, Shanthipriya Ajmera, Sindhura Podduturi, and Sai Prashanthi Govumoni

Abstract The rising population and the advent of globalization and industrialization has trumped the world's energy production due to the exhaustive utilization of fossil fuels. This in turn will lead to sustainability problems such as huge fluctuation in prices, environmental impacts, and declining quantity of fossil fuels. To meet the intense energy demands of future generations, economical and sustainable alternate energy sources should be explored. Biofuels, particularly biodiesel and bioethanol, have gained importance because of their inherent advantages such as renewability, sustainability, and lower emission of greenhouse gases. Commercially, vegetable oils, animal fats, carbohydrates, lignocellulosic biomass, and microbial biomass are being employed in biofuel production using conventional processes like fermentation and transesterification. However, these processes offer challenges in the form of the higher cost of production and other technological barriers. Hence, efforts are required to develop more efficient and low-cost processes employing various strategies of production and optimization. Nanotechnology offers a promising alternative that contributes efficiently to biofuel production industry using nanoparticles. Nanoparticles display unique physicochemical and biological characteristics owing to the larger surface area: volume, high energy absorption, more chemical reactivity and stability, and greater mechanical strength. These properties led to many applications and various nanomaterials have been explored potentially for their capabilities in biofuel production. Nanotechnology also aims to offer remarkable solutions by altering the properties of biomass constituents and biocatalysts employed in the production of biofuels. Nanomaterials have a significant role in bioenergy sectors

S. Sarsan (✉) · S. P. Govumoni
Department of Microbiology, St. Pious X Degree and PG College for Women, Hyderabad, Telangana, India
e-mail: drsreedevi@stpiouscollege.org

S. Ajmera
Department of Microbiology, Palamuru University, Mahabubnagar, Telangana, India

S. Podduturi
Telangana Social Welfare Residential Degree College for Women, Warangal, Telangana, India

© The Author(s), under exclusive license to Springer Nature Singapore Pte Ltd. 2023
M. Srivastava, P. K. Mishra (eds.), *NanoBioenergy: Application and Sustainability Assessment*, Clean Energy Production Technologies,
https://doi.org/10.1007/978-981-19-6234-9_7

because of their exclusive structure, high specific area, high energy electrical storage capacity, and better heating and lighting efficiency. This chapter will review the phenomenal characteristics of nanoparticles and their methods of synthesis especially using microbes. An overview of the nanotechnology in biofuel production with a special emphasis on various approaches in production of bioethanol has been presented in this chapter. This chapter also intends to discuss the potential applications of nanoparticle-based biofuels and their future prospects in fulfilling the energy needs of the world.

Keywords Biofuel production · Bioethanol · Nanoparticles · Microbial synthesis · Alternate fuels

7.1 Introduction

The advent of globalization and industrialization along with a rapidly growing population led to trumping of the production of energy all over the world, by the excessive consumption of fossil fuels. To encounter the intense energy demands of future generations, eco-friendly and cost-effective alternative energy sources should be explored and given considerable attention. Recently, biofuels have gained a lot of interest and also garnered attention from the government, general public, and scientific researchers because of the increased fuel costs, need to improve energy security, and concern over excessive greenhouse gas emissions from fossil fuels resulting in increased pollution. Biofuels have been in production for the last few decades and are more advantageous than fossil fuels and hence are considered as potential alternative fuels for future generations. Recently, the use of biofuels as transport fuels has been of great importance due to their ability to reduce the dependence on petroleum and other fossil fuels as well as in creating new jobs and improving the rural economy and also reducing greenhouse gas emissions and regulating global warming effect (Ahorsu et al. 2018). Thus, apart from creating energy security, they have huge positive impacts on the world economy and environment.

Biofuel is defined as any fuel that is a derivative of renewable biomass and is broadly categorized into primary and secondary biofuels. Primary biofuels involve direct biomass combustion, whereas secondary biofuels are derived from biomass. Several Conversion processes are available for unlocking energy from biomass. Biofuels are categorized into three types based on the source of biomass (Fig. 7.1).

- I Generation biofuels: First-generation biofuels employ energy rich crops like wheat, corn, and sugarcane and examples include Bioethanol, biodiesel, and biogas.
- II Generation biofuels: Second-generation biofuels prefer nonedible, nonfood resources as the raw materials for production and include agricultural residues, wood, and energy crops that are known as lignocellulosic biomass and examples include cellulosic ethanol, biodiesel.

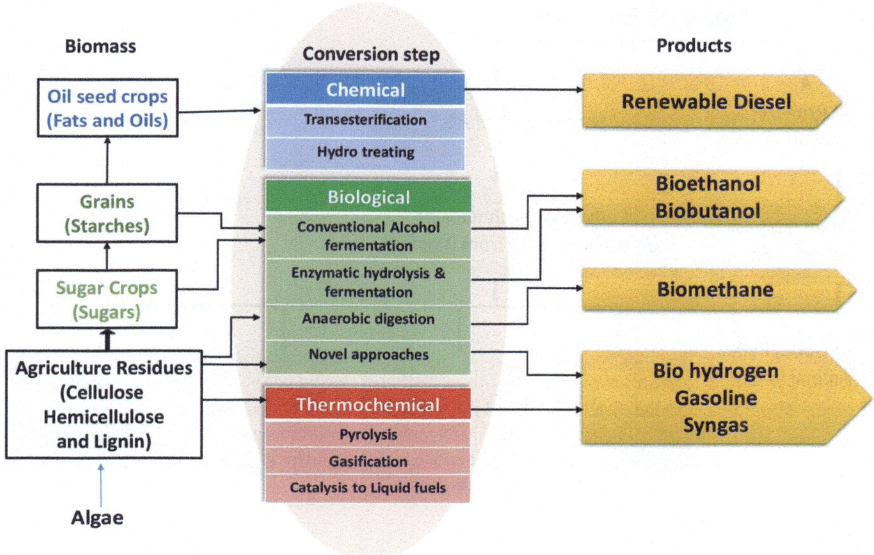

Fig. 7.1 Types of Biofuels

- III Generation biofuels: Third-generation biofuels use algal biomass as a substrate for production and examples include biodiesel, biomethane, biobutanol, biohydrogen, gasoline, and jet fuel.

The new emerging fourth-generation biofuels employ genetically modified algae and other microbes using noncultivable lands for biofuel production. This category includes electro fuels and photobiological solar fuels and is characterized by zero carbon emissions (Aro 2016; Ozdenkci et al. 2017; Parada et al. 2017; Silva et al. 2018).

7.1.1 Nanotechnology in Biofuels

Biofuels are commercially produced from vegetable oils, animal fats, carbohydrates, lignocellulosic biomass and microbes, and other biomass constituents using processes like fermentation, transesterification, pyrolysis, gasification, etc. (Fig. 7.2) (Bradley et al. 2009; Limayem and Ricke 2012; Eggert and Greaker 2014). However, these conventional processes offer challenges in the form of higher cost production and other technological barriers. Hence, to make these processes more efficient in terms of energy and cost, efforts are required to develop various production and optimization strategies. For example, processes employing proper Pretreatment and enzymes and fermentation could result in high yield and quality of biofuels (Patumsawad 2011). Nanotechnology offers a promising alternative that

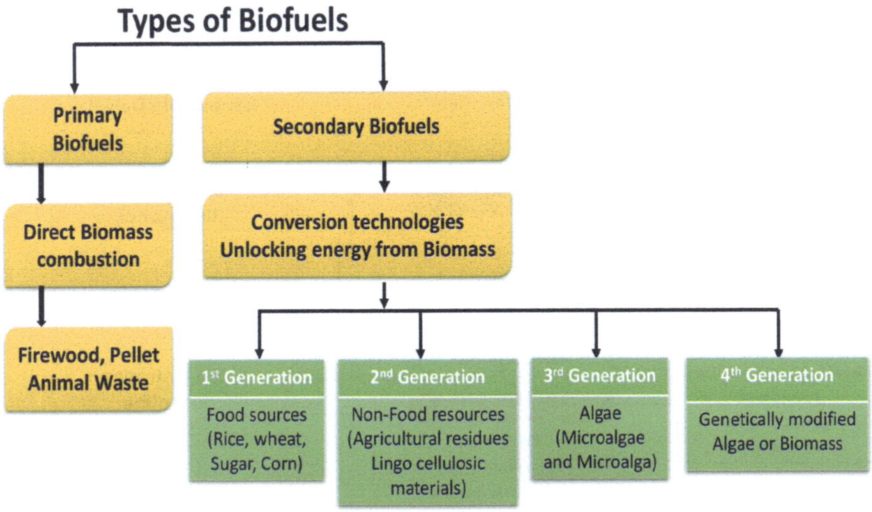

Fig. 7.2 Conversion of Biomass into Biofuels

contributes efficiently to the biofuel production industry. Nanotechnology is an efficient way of enhancing the production of renewable biofuels and offers a viable solution by altering the properties of feedstock and biocatalysts utilized in the production of biofuels. Nanoparticles display unique physical, chemical, biological, mechanical, magnetic, electronic, and optical properties which distinguish them from their large-scaled materials (Biswas et al. 2012). Nanoparticles have received increasing attention in many areas from enzymatic biocatalysis to biomedical and environmental applications because of their small size, high surface area: volume, greater mechanical strength, high energy absorption, chemical reactivity and stability, and low toxicity (Ariga et al. 2014; Malgras et al. 2015). Nanoparticles play a vital role and have been implemented in various processes of renewable energy due to their exclusive structure, high specific area, high energy electrical storage capacity, and better heating and lighting efficiency (Serrano et al. 2009; Ansari and Husain 2012). Nanobiotechnology offers many probable roles in the production of sustainable and alternate fuels and various nanomaterials have been explored potentially for their capabilities in biofuel production.

7.1.2 Nanoparticles

According to ASTM standard terminology relating to nanotechnology nanoparticle is defined as a small particle with >1 nm in length and smaller than 100 nm and rarely exhibits size-related intensive property. Nanoparticles are divided into many types based on their structural morphology, dimensionality composition, uniformity, and agglomeration. Based on dimensionality, nanoparticles are of different types: 0D (Nano dots), 1D(Graphene), 2D (Carbon nanotubes), and 3D (gold nanoparticles). Nanoparticles can be flat, cylindrical, spiral, spherical, and crystalline in nature and exist as single constituent material or arranged as nanocomposites. Based on the nature of material fabrication, nanoparticles may be either inorganic or organic type. Inorganic nanoparticles comprise of carbon-based NP, ceramic Np, semiconducting NP, and metal and metal oxides NP. Organic nanoparticles are categorized as polymeric NP and Biomolecule derived NP. Organic nanoparticles such as polymerosomes, liposomes, polymer constructs, and micelles are used for nanoimaging as well as drug and gene delivery in the target cells or tissues (Qiu and Bae 2006). Meantime, inorganic nanoparticles have received considerable significance owing to unique size, material, and physicochemical characteristics such as optical properties, magnetic properties, inert nature, ease of functionalization, and stability (Lohse and Murphy 2012). Inorganic nanoparticles like carbon nanotubes, metal oxide NPs, magnetic nanoparticles, etc., have potential applications in bioenergy production.

7.1.3 Synthesis of Nanoparticles

Several methods can be employed for the synthesis of nanoparticles (NP) including physical, chemical, and biological approaches. Top-down approach and bottom-up approach are the two main strategies employed in the production of nanoparticles. The methods of NP synthesis using top-down approach are simple and are involved in breaking down of bulk materials such as silicate and gold into nanoscale-sized materials. Some of the commonly used methods of this approach include mechanical milling, laser ablation, anodization, photolithography, ion and plasma etching nanolithography, sputtering and thermal decomposition. In Bottom-up method or constructive method, the synthesis of nanoparticles occurs by the buildup of materials from atoms to clusters to nanoscale components. Examples include electrochemical reduction method, microemulsions, sol–gel processing, laser pyrolysis, hydrothermal approaches, microbial reduction, precipitation, and are the most common methods employed in nanoparticles synthesis (Chan and Kwok 2011; Ealias and Saravanakumar 2017).

7.1.3.1 Physical Methods

Physical methods apply mechanical pressure, thermal energy, high energy radiations, or electrical energy resulting in the abrasion, melting, evaporation, or condensation of materials to form nanoparticles. Top-down approach is followed for the operation of these methods and produces uniform monodisperse nanoparticles and these methods are beneficial due to the lack of solvent contamination. Physical methodologies are less economical as they produce abundant waste during synthesis.

Ball milling, inert gas condensation, laser ablation/pulse laser deposition, melt mixing, sputtering, laser, and Flash spray pyrolysis are the most frequently employed methods for the formation of nanoparticles. Ball milling is the widely used method in which a powdered material is exposed to high-energy collisions with the balls. This method is simple and cost-effective and is one of the well-known methods of preparing nanoparticles (De Carvalho et al. 2013; Manawi et al. 2018).

7.1.3.2 Chemical Methods

Chemical methods used in the synthesis of nanoparticles include electrochemical reduction method, sol–gel method, microemulsions, plasma enhanced chemical vapor deposition, and hydrothermal synthesis. (1) Sol–gel method is the most popular bottom-up approach method used in the synthesis of nanoparticles. It is a wet chemical process that depends on hydrolysis and condensation of metal oxides or metal alkoxide solution. Hydrolysis step leads to the formation of colloidal solution known as Sol, which is then changed to a semisolid phase known as scenarios gel through the process of condensation. In later stages, gel is subjected to high-temperature drying to obtain desired nanoparticles (Gu et al. 2004; Xu et al. 2007; Pradeep et al. 2008). This is an extremely beneficial method because of simple, inexpensive, and homogenous technique and purity of the nanoparticles obtained. (2) Chemical vapor deposition method involves deposition of inorganic nanoparticles on suitable substrate surface. There are three steps involved in this method. In the first stage, a volatile precursor through a carrier gas is introduced into the reaction chamber. A chemical reaction then occurs in which vapors adsorb onto the surface of the heated substrate and leads to intermediate product development. In the last stage, solid grains and nanoparticles are formed through the decomposition of intermediate products (Manawi et al. 2018). (3) Hydrothermal synthesis is another approach for the preparation of nanomaterials and usually depends on synthesis in the solution phase. This method offers significant advantages over other methods such as generation of nanoparticles that are stable at high temperatures and preparation of nanomaterials with high vapor pressure (Gan et al. 2020). (4) Coprecipitation method is a simple technique that allows simultaneous nucleation, growth, coarsening and/or agglomeration reactions. It is the most convenient way used in the synthesis of a large number of magnetic nanoparticles. In coprecipitation reactions, metals are precipitated in the form of hydroxide from a solvent mixture containing salt precursor and base. The main advantage of coprecipitation method lies in obtaining homogenous small size nanomaterials (Maaz et al. 2009; Petcharoen and Sirivat 2012; Mascolo et al. 2013). (5) Microemulsions provide a new approach in the process of synthesizing different nanoparticles since they have large interfacial areas, high thermodynamic stability, and larger interfacial areas. Microemulsions generally comprise of three components such as water, a polar phase, oil a nonpolar phase, and a surfactant to stabilize the droplets of water in oil or oil in water. This method provides control on the size of the particle and other properties namely geometry, homogeneity, and surface area of nanoparticles (López-Quintela et al. 2004). (6) Reduction method of nanoparticle synthesis involves layer by layer formation of atoms and is suitable for the synthesis of hybrid nanoparticles (Kalska-Szostko 2012). For example, graphene AuNPS synthesis can be done by deposition of graphene sheets onto an electrode followed by suspension of the electrode in metallic precursor electrolytic solution.

7.1.3.3 Biological Methods

Physicochemical methods of nanoparticle synthesis have become obsolete as these processes employ high temperature and pressures and use toxic and hazardous chemicals during their synthesis and are also expensive (Narayanan and Sakthivel 2010). Therefore, green nanotechnology has emerged as a promising approach to design novel nanoparticles using a green chemistry approach. Green methods promote environmental sustainability and are a safe and effective approach for nanoparticles synthesis as they are economical and offer easy characterization and high success rates (Singh et al. 2016; Das et al. 2017; Abdelghany et al. 2018; Aman and Narendra 2019). It is an alternative approach that provides a new possibility for nanoparticle synthesis using natural reducing agents and/or stabilizing agents. The synthesis of nanoparticles by biological means is a bottom-up approach that includes biological entities like plants and microbes (bacteria, yeast, and molds) or their metabolic byproducts. The stability of nanoparticles may be improved when capped with molecule derivatives of the organisms employed (Ballotin et al. 2016). The nanoparticles produced using biological methods display great biocompatibility and are simple, nontoxic, and cost-effective (Gholami-Shabani et al. 2014). Biogenic methods are extremely helpful in the synthesis of sustainable, safer, and environment-friendly nanoparticles and are also easy to scale up.

Plant-Based Nanoparticles

Green methods of synthesis using vitamins, amino acids, and plant extracts are significantly attractive and greatly popularized due to their great potential for reducing the toxicity of nanoparticles (Baruwati et al. 2009). Many plant extracts are considered as novel resources in green route synthesis of safe and nontoxic Nps. Owing to the presence of important phytochemicals used as reducing agents in plant extracts, particularly in leaves, diverse plant species are being employed in the green route synthesis of metal/metal oxide Nps. In this process of green synthesis, bio-molecules of plant extracts (aldehydes, phenols, amides, ascorbic acids, flavones, ketones, and terpenoids) bring about the reduction of metal salts into metal Nps. Plants are important biological entities that have exemplary potential to reduce metallic nanoparticles and stabilize them via one-pot synthesis. Reports suggest that various plants such as aloe vera (Aloe barbadensis), tulasi (Ocimum sativum), alfalfa (Medicago sativa), oats (Avenasativa), coriander (Coriandrum sativum), lemon grass (Cymbopogon flexuosus), lemon (Citrus limon), and mustard (Brassica juncea) are involved in the production of silver Nps as well as gold Nps (Singh et al. 2018). Plants like pinus species are thought to be the top candidates for nanoparticle synthesis, as these have large amounts of phenolic compounds containing hydroxyl and carboxyl groups by which iron ions are inactivated (Iravani 2011).

The production of silver NPs (AgNPs) from plants is widely investigated and is known to be produced by oxidation of Ag^{+} to Ag^{0} by different phytocompounds like flavonoids, aldehydes, terpenoids, ketones, tannins, polyphenols, phenolic acids, and protein of plant extracts (Ovais et al. 2018). Noruzi et al. described the production of AuNps utilizing the bioreduction potential of aqueous extracts of Thuja orientalis (Cypress) leaves (Noruzi et al. 2012). Biosynthesis of silver and gold Nps using *Szyygiumaromaticum* bud extract and *Murrayakoenigii* leaf extract that contain natural reducing agent eugenol has been explained (Singh et al. 2010; Christensen et al. 2011). Silver nanoparticles are synthesized utilizing latex of the plant *Plumeria rubra* and were explained by Patil et al. (2012). The use of fruit extracts of *Emblica officinalis* in the synthesis of silver and gold nanoparticles was investigated and were found to be highly stable (Ankamwar et al. 2005). Zinc, nickel, copper, and cobalt nanoparticles were synthesized intracellularly in alfalfa, mustard, and sunflower plants. Various plant extracts such as Curcuma longa, Cinnamomum camphora, Cinnamomum zeylanicum, Gardenia jasminoides, Anogeisus latifolia, Pinus resinosa, Glycine max, Ocimum sanctum, Musa paradisiac, Pulicaria glutinosa, and Doipyros kaki are employed in the green synthesis of Pd and Pt Nps (Siddiqi and Husen 2016).

Microbe-Based Nanoparticles

There are several benefits of microbial-mediated nanoparticle synthesis such as high yielding, economical, and environmentally safe (Mandal 2006; Thakkar et al. 2010). This process of nanoparticle synthesis employs microbial culture filtrates (extracellular and intracellular) as reducing agents. Microbes like bacteria and fungi are extensively utilized as bioreactors for the production of nanoparticles owing to their faster growth rate, easy to cultivate, and their potential to grow using optimum environmental conditions. Microorganisms have inherent ability to synthesize nanoparticles of inorganic origin by reduction mechanism via intracellular and extracellular ways (Fig. 7.3). Microorganisms possess the ability to trap and transport metal ions present in the environment and convert into their elemental forms mediated by enzymatic action which are then either accumulated or secreted (Li et al. 2011b).

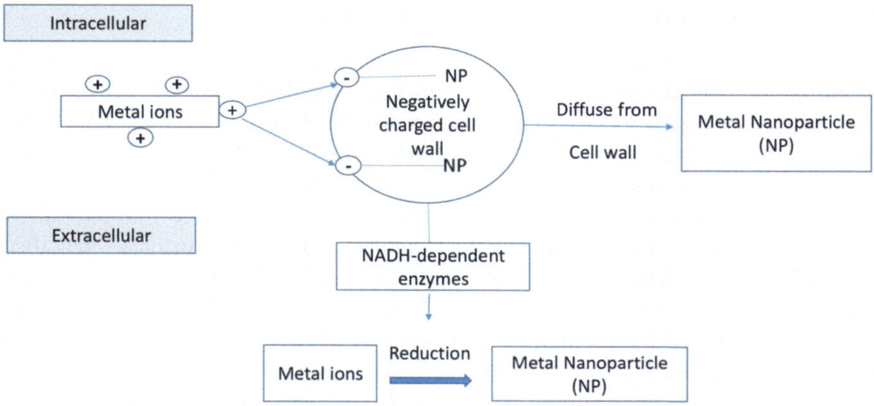

Fig. 7.3 Mechanism of Synthesis of Nanoparticles by Microorganisms

7.2 Microbial-Mediated Synthesis of Nanoparticles

The biological synthesis of metallic nanoparticles employing microbial flora is considered to be eco-friendly as well as low-cost process. Various prokaryotic and eukaryotic microbial species could be used in the production of nanoparticles that are high yielding, economically safe, and eco-friendly (Mandal 2006; Thakkar et al. 2010). Different metal Nps such as gold, silver, copper, lead, iron, cadmium, platinum as well as metal oxide Nps such as titanium oxide and zinc oxide are reported to be synthesized by several microbial strains of bacteria, fungi, yeast and algae (Hulkoti and Taranath 2014). These nanoparticles are highly useful with a lot of applications in agriculture, medicine, textile industry, electricity industry, cosmetic industry, drug delivery, and biochemical sensors (Pramila and Sushil 2016). The microorganisms provide a diverse environment intended for the biological route of nanoparticle synthesis. Microorganisms trap the metal ions present in the environment and then convert them into their elemental forms by catalytic action of various metabolic enzymes. The cell wall carries a negative charge and it has a very important role in the process. The positively charged metal ion encounters an electrostatic reaction with the negatively charged cell wall. The metal ions are reduced into nanoparticles by the action of the localized cell wall enzymes and then get secreted out. The Biosynthesis of nanoparticles may be regarded as intracellular or extracellular depending on the site of nanoparticle formation (Li et al. 2011a, b). In the intracellular synthesis method, the metal ions are being transported mediated by enzymes into the cell of microorganisms resulting in the formation of nanoparticles. Whereas in the extracellular synthesis method, the metal ions are trapped onto the cell surface followed by their reduction mediated by enzymes (Zhang et al. 2015). Due to its easy recovery process, the extracellular method is the method of choice.

7.2.1 Bacteria in Nanoparticles Synthesis

Bacteria are considered as the most suitable candidates for the synthesis of nanoparticles. Diverse species of bacteria are explored for potential applications in green nanotechnology, particularly in the production of nanoparticles. Bacteria have gained considerable attention in biosynthesis of metallic nanoparticles. Bacteria has the potential to synthesize unique nanostructured materials such as biomineralized nanostructures which include magnetosomes, silicified frustules, calcified coccoliths, organic nanomaterials nano cellulose, and bacterial nanowires (Kroger and Poulsen 2008; Nwodo et al. 2012; Gama et al. 2012; Malvankar and Lovley 2012; Yan et al. 2017; Skeffington and Scheffel 2018). Bacteria are regarded as potential biofactories for the biosynthesis of silver, palladium, gold, titanium, magnetite, titanium dioxide, cadmium sulfide, and other nanoparticles. Several studies reported the biosynthesis of Nps from different microbial species such as *Marinobacter Pelagius* producing gold Nps (Sharma et al. 2012), *Pseudomonas stutzeri* AG259 producing silver nanoparticles (Slawson et al. 1994), and *Lactobacillus strains* synthesizing titanium nanoparticles (Prasad et al. 2007).

Some bacterial strains have developed resistance to most toxic heavy metals and can even survive at high metal ion concentrations. Microbial detoxification can be made by intracellular bioaccumulation or by extracellular biomineralization or reduction or precipitation which converts toxic, soluble inorganic metal ions into nontoxic, insoluble metal nanoclusters. These organisms employ different mechanisms for the production of nanomaterials intracellularly and extracellularly. In the intracellular mechanism, the metal ions are transported as a result of interaction between the positive metal ion and negative ions of the bacterial cell wall and then reduced to nanoparticles catalyzed by cell wall enzymes, and finally diffused out of the bacterial cell. The extracellular method of synthesis of gold nanoparticles from *Rhodopseudomonas capsulate* involves the reduction of gold metal ions by accepting electrons from NADH by reductase enzyme secreted by *R. capsulate* and the resultant nanoparticles size ranged between 1 and 200 nm (Hulkoti and Taranath 2014).

Pseudomonas stutzeri AG259 has been successfully used in the production of silver nanocrystals (Menon et al. 2017). AgNPs were synthesized by utilizing cell-free culture extracts of *Ps. proteolytica*, *Ps. antarctica*, *Ps. meridian*, *Bacillus cecembensis, Arthrobacter gangotriensi, and Arthrobacter kerguelensis* (Shivaji et al. 2011). Iron oxidizing bacteria, for example, *Gallionella, Mariprofundus, and Leptothrix* form iron oxide nanoparticles extracellularly (Hashimoto et al. 2007). Iron oxidizing bacteria such as *Geobacter bemidjiensis*is play a role in the synthesis of ferric oxyhydroxide nanoparticle aggregates (Luef et al. 2013). Moreover, Magnetotactic bacteria such as *Aquaspirillum magnetotacticum* and *Magnetospirillum magnetotacticum* are able to synthesize iron oxide nanoparticles. Moreover, bacterial species like *Shewanellaoneidensis* and *Desulfosporosinus* sps. are reportedly known to synthesize uranium dioxide nanoparticles (Jeevanandam et al. 2016). Magnetic nanoparticles offer exceptional applications in the production of bioethanol from lignocellulosic components through enzyme immobilization (Alftren and Hobley 2013). Besides, the synthesis of inorganic nanomaterials, few organisms belonging to genera such as *Gluconacetobacter* have shown the ability to biosynthesize bacterial nano cellulose that shows high purity, mechanical stability, and crystallinity over nanocrystalline cellulose and nano fibrillated cellulose (Golmohammadi et al. 2017).

7.2.2 Fungi in Nanoparticles Synthesis

Fungi is one of the significant microbial agents employed in the biosynthesis of nanoparticles and provide many advantages due to their higher bioaccumulation, economic viability, easy cultivation, and scale-up processes resulting from simple techniques of biomass handling and downstream and recovery processes. Fungi has a vital role in the synthesis of nanoparticles by reducing metal ions into insoluble complexes such as metal sulfides (Mehra and Winge 1991). Fungi possess good bio reductant potential and hence can be used for the synthesis of nanoparticles in a quick and sustainable manner, due to the release of large quantities of proteins, enzymes (hydrogenases, nitrate reductase), and organic acids. Fungi are commonly utilized as stabilizing and reducing agents owing to properties such as high tolerance to heavy metals as well as the capability of internalizing and accumulating metal ions. Moreover, they are easy to be cultivated on a commercial scale and could be used to form nanoparticles of desired shapes and sizes (Gade et al. 2008; Ahluwalia et al. 2014; Azmath et al. 2016; Khan et al. 2017). Fungi are known to synthesize nanoparticles by intracellular and extracellular modes. In the intracellular mechanism of synthesis, metal precursors are added to the mycelial culture followed by its internalization. Thereafter the synthesis, nanoparticles are extracted by using methods such as centrifugation and filtration for the disruption of biomass with a subsequent release of nanoparticles (Castro-Longoria et al. 2011; Rajput et al. 2016; Molnar et al. 2018). A distinct method of ion transportation is seen in *Verticillium sp.* where the bioreduction takes place in the cell wall. In Actinomycetes, metal ions undergo reduction on the mycelia surface along with the cytoplasmic membrane. By intracellular mode of synthesis 1–40 nm sized nanoparticles were developed. Extracellular method is a highly preferable method wherein the metal precursors are added to the aqueous filtrate containing fungal biomolecules resulting in the formation of nanoparticles in the dispersion without requiring additional procedures for the release of Nps from the fungal cells (Azmath et al. 2016; Sabri et al. 2016; Costa Silva et al. 2017; Gudikandula et al. 2017). However, purification of nanoparticles dispersion is required for the removal of fungal residues and other impurities by employing methods like dialysis, filtration, and ultracentrifugation techniques (Ashrafi et al. 2013; Qidwai et al. 2018; Yahyaei and Pourali 2019).

Different fungi such as *Aspergillus fumigatus, Asergillus niger, Cladosporium cladosporioides, Fusarium semitectum, Fusarium oxysporum, Fusarium solani, Trichothecium sps., Phoma glomerata, Phaenerochaete chrysosporium, Penicillium fellutanum*, *Penicillium brevicompactum*, *Trichoderma viridae*, *Trichoderma asperellum*, and *Volvariella volvaceae* have been explored for nanoparticle synthesis both extracellularly and intracellularly. Many authors have described the process of gold and silver nanoparticles synthesis on the surface of fungal mycelium as well as on their cell membrane (Pighi et al. 1989; Mukherjee et al. 2001b; Chen et al. 2003; Ahmad et al. 2005; Gericke and Pinches 2006a, b; Vigneshwaran et al. 2007; Singh et al. 2017). *Penicillium* is considered a superior genera among all fungi intended for the production of silver nanoparticles via extracellular mode (Sadowski et al. 2008). However, syntheses of Ag nanocrystals are also reported by *Aspergillus* species (Binupriya et al. 2010; Gade et al. 2008) and *Fusarium solani* (Ingle et al. 2009). The synthesis of nanoparticles by extracellular method is reported in *Pleurotussajorcaju* (Nithya and Ragunathan 2009). Fungi such as *Verticillium* and *Fusarium* have been reportedly known to produce gold nanoparticles (Mukherjee et al. 2001a; Mukherjee 2002). Moreover, these were also found to synthesize magnetic nanoparticles extracellularly (Bharde et al. 2006). *Fusarium oxysporum* is reported to produce Ag Nps which were very stable and sizes ranging between 5 and 15 nm (Ahmad et al. 2003). *Verticillium* was also reported to produce uniform $CaCO_3$ nanocrystallites of 70–100 nm size (Rautaray et al. 2004). The biological method of synthesis of silver nanoparticles of approximately 8.92 nm size was demonstrated in *Aspergillus flavus*. The "sil" gene encoded by plasmids was found responsible for reducing silver ions in large-scale processes (Vigneshwaran et.al 2007). *Trichoderma viride* was reported to synthesize AgNPs extracellularly by using solution of silver nitrate (Fayaz et al. 2010).

7.2.3 Yeasts in Nanoparticle Synthesis

Yeasts offer several benefits over bacteria in mass production of metal nanoparticles because of their ability to grow rapidly, easy cultivation, utilization of simple nutrients, and ease in controlling production. For the same, *Saccharomyces pombe* and *Candida glabrata* have been reportedly known for the production of cadmium sulfide, selenium, titanium, silver, and gold nanoparticles intracellularly (Boroumand Moghaddam et al. 2015; Feng et al. 2017; Subashini and Bhuvaneswari 2018; Rana et al. 2020). Kowshik et al. (2003) isolated the yeast strain MKY3, capable of producing silver nanoparticles in huge quantities when inoculated with aqueous silver nitrate. *Candida glabrata* and *Schizosaccharomyces pombe* are demonstrated to synthesize cadmium nanoparticles (Dameron et al. 1989). The intracellular method of synthesis of lead sulfide Nps was explored in the marine yeast *Rhodosporidium diobovatum* (Seshadri et al. 2011). *Candida versicolor* has the potential to synthesize silver nanoparticles in the absence of surfactants and can be widely applied as water-soluble catalysts for living cells (Sanghi and Verma 2009). The yeast cells involve the presence of membrane-bound oxidoreductases and quinones during the synthesis of nanoparticles. These enzymes' function is influenced by the fluctuations in pH. For instance, increase in pH of yeast cell internal environment leads to the activation of reductases and eventually results in metal ions reduction and nanoparticles formation (Salunke et al. 2015).

7.2.4 Algae in Nanoparticle Synthesis

Different species of cyanobacteria and eukaryotic green algae like Spirulina, Chlorella, Lyngbya, etc. may be employed for biorecovery of metals and green synthesis of nanoparticles at an affordable cost (Bakir et al. 2018). In S. platensis, the gold particles are reduced from Au^{3+} to Au^0 resulting in the synthesis of gold nanoparticles (Uma Suganya et al. 2015).

7.3 Role of Nanotechnology in Production of Biofuels

There are different kinds of biofuels generated which are differentiated on the basis of biomass source and the techniques employed in their production.

- Transesterification of lipid fraction yields biodiesel
- Fermentation of the carbohydrate components generates Bioethanol
- Anaerobic digestion yields biogas
- Gasification and anaerobic digestion results in Biomethanol
- Photolysis or photo fermentation yields Biohydrogen
- Direct combustion yields Biosyngas

The biofuel industries are moving ahead with a focus on sustainable energy and renewable fuel systems thereby overcoming environmental pollution and climate change issues. Their attention has been on various components involved in biofuel production such as biomass pretreatment, overall process optimization, and cost. Nanotechnology is considered as a very potent method in the processing and manufacture of biofuels in a process efficient and cost-effective manner (Sekhon 2014; Rai and Da Silva 2017). Some of the exceptional characteristics of nanoparticles that prove to be of great use in biofuel production include high surface area; high degree of crystallinity; ability to catalyze reactions; durability; their high recovery rate; reusability; recycling; and many other properties (García-Martínez 2010). Different nanomaterials like nanofiber, nanotubes, nanosheets, and other metal nanoparticles are reported to possess many applications as nanocatalysts during biofuel production. These biofuels are considered as best alternative energy resources because they are renewable as well as environment friendly (Antunes et al. 2017). Nanoparticles have been considered to be significant in the production of biofuels due to their distinct physical and chemical properties. The chief important property is their easy recovery from reaction mixture. There are many nanomaterials possessing exceptional properties which are employed in the production of biofuels, eg., TiO_2, ZnO, Fe_3O_4, carbon, graphene, and fullerene (Ahmed and Douek 2013).

Nanotechnology has different applications in the biofuels and bioenergy sector like in feedstock modification or development of more efficient catalysts, etc. Conventionally, enzymes were employed to hydrolyze biomass to generate bioethanol, biodiesel, and other biofuels (Michalska et al. 2015; Verma and Barrow 2015; Terán-Hilares et al. 2016). Nanostructures are considered as a replacement to the enzymes thus resulting in more efficient catalysis and biofuel production. They may also be used for immobilizing enzymes so that biocatalysts can be recovered from the medium and reused for more efficient production. Also, magnetic properties may be added to immobilized systems thus improving the technology (Verma and Barrow 2015; Rai et al. 2016). Magnetic nanoparticles are often employed as carrier molecules to immobilize enzymes and thus find wide applications in biofuel production. These magnetic nanoparticles are of small size and have a high surface area: volume along with quantum as well as immobilizing properties which make them largely employable in the biofuel sector.

7.3.1 Biodiesel Production

Biodiesel is a mixture of esters, usually resulting from transesterification of vegetable oils and animal fats. Biodiesel is an alternative fuel and provides many advantages over fossil fuels such as they are safe and biodegradable. They possess superior lubricant properties, high combustion efficiency, and result in reduced emission of CO_2 and other particulate matter (Hossain et al. 2008; Feyzi and Norouzi 2016). There are many functionalized nanomaterials which function as nanocatalysts and are reported to be employed in biofuel production from different feedstocks (Table 7.1). These nanocatalysts were synthesized using different methods such as coprecipitation method, sol–gel method, aerogel method, and impregnation method.

7.3.2 Biohydrogen Production

A wide range of anaerobic bacteria and microalgae generates molecular hydrogen from various organic materials using various metabolic routes such as steam reformation, dark and photo microbial fermentation, and biophotolysis (Kapdan and Kargi 2006; Ran et al. 2006; Wang et al. 2007a, b; Shaishav et al. 2013; Tiwari and Kiran 2017). Several operational conditions that include substrate concentration, hydraulic retention time, pH, temperature, etc., must be optimally maintained to enhance the performance of microbes involved in biohydrogen production (Das et al. 2008). Nanoparticles increase the rate of transfer of electrons in anaerobic conditions and its faster reaction with electron donors improve the kinetics of biohydrogen production and enhances the activity of microorganisms involved (Serrano et al. 2009; Ali et al. 2017). Nanoparticles aid in overcoming certain process barriers like fermentation inhibitors, maintenance of the acidogenic phase, lower yield, and

Table 7.1 Different Nanocatalysts and feedstock used for the production of biodiesel

S. no.	Feedstock	Nanocatalyst	Yield (%)	References
1.	Sunflower oil	$Cs/Al/Fe_3O_4$	94.8	Feyzi et al. (2013)
		$Ca/Fe_3O_4@SiO_2$	97.0	Feyzi and Norouzi (2016)
2.	Rapeseed oil	$K_2O/\gamma\text{-}Al_2O_3$	94.0	Han and Guan (2009)
3.	Sunflower oil and rapeseed oil	MgO	98.0	Verziu et al. (2008)
4.	Soybean oil	ZrO2 loaded with $C_4H_4O_6HK$	98.03	Qiu et al. (2011)
		MgO nanoparticles on TiO_2 support	95.0	Mguni et al. (2012)
		Fe/Cd and Fe/Sn oxide nanoparticles	84.0	Alves et al. (2014)
5.	Jatropha oil	Hydrotalcite-derived particles with Mg/Al oxides	95.2	Deng et al. (2011)
		CaO	98.54	Reddy et al. (2016)
6.	Palm oil	TiO2-ZnO	92.2	Madhuvilakku and Piraman (2013)
		$KF/\gamma\text{-}Al_2O_3$/honeycomb ceramic (HC) monolithic catalyst	96.0	Gao et al. (2015)
7.	Olive oil	ZnO nanorods	94.8	Molina (2013)
8.	Glyceryl trioleate	Fe/Fe_3O_4 magnetic nanoparticles	95.0	Wang et al. (2015)
9.	Karanja oil and Jatropha oil	Li-Cao	99.0	Kaur and Ali (2011)
10.	Cooking oils	CaO and MgO nanoparticles	98.95	Tahvildari et al. (2015)
11.	Chinese tallow seed oil	KF/CaO	96.8	Wen et al. (2010)

Source: Adapted from Antunes et al. (2017)

substrate conversion thereby increasing the production. However, it should be noted that nanoparticles also possess antimicrobial characteristics as they inhibit microbial growth at higher concentrations.

Nanoparticles have proved to effectively enhance the biohydrogen production via dark fermentative process (Zhang and Shen 2007). In the Dark fermentative process, biohydrogen is produced from renewable resources such as feedstocks and microbes via fermentation process. The main biohydrogen-producing pathway is the acetic reaction. Nanoparticles help the microorganisms involved in biohydrogen production by reducing the lag phase of growth and favoring the acetic reaction. The gold nanoparticles have a large surface area:volume which allows the binding of bacteria to their active site. Nano-based additives help in improving the activity of proteins like iron and nickel hydrogenases and ferredoxins which facilitate the electron transfer in the microbes producing biohydrogen (Ramsurn and Gupta 2013). Metallic nanoparticles are highly stable and can be used in various approaches of dark

Table 7.2 Explicit properties of nanoparticles leading to higher biohydrogen yield

S. no.	Nanoparticles	Properties leading to enhanced biohydrogen production	References
1.	Gold nanoparticle	Improved substrate utilization by 56%	Zhang and Shen (2007)
2.	Silver nanoparticle	Enhanced glucose conversion	Zhao et al. (2013)
3.	FeO nanoparticles	Iron acted as a cofactor in active sites of the hydrogenase enzyme and enhanced the activity of hydrogenases	Ali et al. (2017), Yang and Wang (2018), Beckers et al. (2013)
4.	TiO_2 nanoparticles	Remarkable photocatalyst	Pandey et al. (2015), Salgado et al. (2016)
5.	Silica nanoparticles	High thermal stability	Harish et al. (2018)

fermentation employed for enhancing biohydrogen production (Huirache et al. 2013). The porous and morphological feature of silica nanoparticles has outstanding properties like high thermal stability that facilitate biohydrogen production (Harish et al. 2018). Iron in nanoparticles acted as a cofactor of the hydrogenase enzyme and eventually enhance the activity of biohydrogen-producing organisms (Ali et al. 2017). Thus, higher biohydrogen yield using nanoparticles is possible because, nanoparticles possess explicit properties like smaller size, high porosity, and high ratio of surface area:volume and they also help to maintain pH and stimulate the activity of hydrogenase enzyme, etc. (Table 7.2).

Microalgae are involved in photosynthetic/photo fermentative biohydrogen production. Nanoparticles tend to act by increasing their growth (by favoring the production of carbohydrates); improving photosynthetic activity, protein level, nitrogen metabolism, etc., by serving as catalytic agents or boosting the activity of key enzymes that are crucial for the metabolism of microalgal species (Yang et al. 2006; Ahmad et al. 2018). Silver nanoparticles and Gold nanorods increased the photosynthetic activity of *Chlorella vulgaris.* An increase in chlorophyll and carotenoid pigments was facilitated by Fe^0 nanoparticles, promoting growth (Eroglu et al. 2013). Gold nanoparticles possess a large ratio of surface area:volume that enable their binding to active sites of biomolecules thereby causing a stimulatory effect on the process of biohydrogen production. The microbial processes are also enhanced by enhancing the activity of enzymes involved in biohydrogen production (Sekoai et al. 2019). TiO_2 nanoparticles increased the production by 50% (Pandey et al. 2015). Photocatalytic hydrogen production occurs using a photocatalyst and an illuminating source to split water molecules to form hydrogen and oxygen. Nanoparticles serve as a photocatalyst in biohydrogen production. TiO_2 nanoparticles are nontoxic and express chemical stability making it a remarkable photocatalyst (Salgado et al. 2016).

7.3.3 *Biogas Production*

Biogas energy is a renewable source whose production can be increased by using nanotechnology and the phenomenal characteristics of nanoparticles (Hussein's 2015). Organic wastes rich in carbon and nitrogen sources include plant, animal, and human wastes that are anaerobically digested in order to form biogas and its constituents (Romero et al. 2016). Addition of certain metal ions (iron, cobalt, and nickel) in trace amounts act as catalysts and increase the activity of methanogenic bacteria thereby increasing energy production. Researchers have confirmed that nanomaterials are beneficial to be used in biofuel production rather than atomic and bulky constituents (Feng et al. 2010, 2014). In the anaerobic digestion process, the nanoparticles offer a symbiotic relationship. The nanoparticle enables the microbes to act as catalytic agents and in turn the microbes can alter the oxidation state of nanoparticles. Magnetic nanoparticles are useful in the methanogenesis process due to their high coercivity and strong paramagnetic property (Yang et al. 2015). Various nanoparticles (Fe, Co, Ni) enhanced the methane gas production resulting in higher biogas yields (Casals et al. (2014); Abdelsalam et al. (2016). Fe^0 nanoparticles enhanced methane concentration in biogas and enhanced the rate of biogas production by 30.4% using waste-activated sludge. It also acted as a promising adsorbent by reducing H_2S impurity by 98% which otherwise would be corrosive to the equipments used during biogas production, harmful to humans and reduces the density and colorific value of methane (Song et al. 2017). ZVI nanoparticle were also reported to increase the yield of biogas (Su et al. 2013; Karri et al. 2005; Wang et al. 2016; Yang et al. 2013).

7.4 Nanoparticles in Bioethanol Production

Bioethanol is the most commonly produced and the most extensively used alternative fuel in the transportation sector due to its low production cost and environment-friendly technology (Saini et al. 2015). The advantages of bioethanol are high octane number, low boiling point, higher heat of vaporization, better combustion, and lower exhaust emissions. It has comparable energy content and can be blended with gasoline and can be used in vehicles without modification of the existing engines (Waqas et al. 2016). Bioethanol produced from food crops including sugar and starch-rich feedstocks like corn, wheat, sugar cane, potato, cassava, etc., is known as first-generation biofuel (Bertrand et al. 2016). Starch stored in these crops serves as a high-yield feedstock for the production of bioethanol. First-generation bioethanol is more effective and widely produced bioethanol for commercial use and produced on a large scale. Although first-generation ethanol appears to be a promising substitute, it cannot sufficiently meet the global energy needs and is not adequate enough to replace fossil fuels, especially in the transportation sector. The main drawback of first-generation biofuel like bioethanol is an inherent competition between food and biofuel feedstocks over the utilization of cultivable lands mainly used for food crops, subsequently causing an upsurge of food prices and ultimately leading to food insecurity (Dutta et al. 2014; Manochio et al. 2017; Bastos 2018; Hirani et al. 2018; Branco et al. 2019). Therefore, many other processes like second-generation processes to produce bioethanol have been explored.

Biofuels which are produced from nonedible crops, non-food crops, and waste biomass including lignocellulosic biomass residues of food crops (e.g., corn stalks, sugarcane bagasse, wheat, and rice straw) or trees and grasses grown specifically for energy are known as second-generation (2G) biofuels (Antunes et al. 2014; Zabed et al. 2016). The cultivation of these feedstocks neither requires extra land nor is expensive and thus will not raise concerns over food sustainability. The residual biomass of forest, agricultural, industrial, or municipal wastes are used in the processing and production of bioethanol. Among these raw materials, lignocellulosic substrates are considered as the best resources for bioethanol production and are hence considered as the best option to lower the usage of conventional fuels and thereby reduce environmental pollution (Balat and Balat 2009).

7.4.1 Basic Steps in the Process of Bioethanol Production

Bioethanol production using biomass sources is a complicated process. There are four major steps in the large-scale process of bioethanol production which include (1) Pretreatment, (2) Hydrolysis/Saccharification, (3) Fermentation, (4) Recovery and dehydration (Fig. 7.4). The pretreatment step is an energy-consuming and expensive step which involves cleaning and reduction of the size of particles by various means such as milling and grinding. The recalcitrant nature of

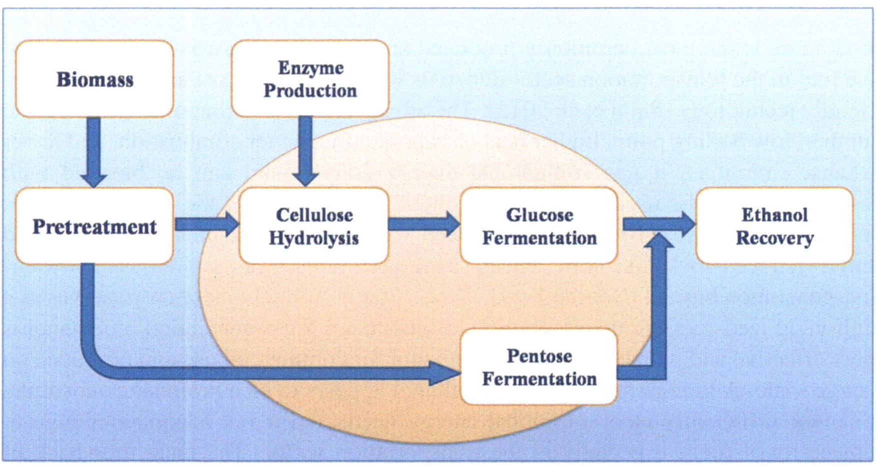

Fig. 7.4 Basic steps in the production of Bioethanol

lignocellulosic biomass makes the pretreatment steps essential and is intended to make the cellulose chains free from lignin and make the cellulose accessible to enzymatic hydrolysis (Kumari and Singh 2018; Dimos et al. 2019). After pretreatment, biomass undergoes acid hydrolysis or enzymatic hydrolysis which causes the breakdown of polysaccharides into monomer sugars such as glucose and xylose. Later microorganisms, especially yeasts are involved in the fermentation of these sugars to yield ethanol, CO_2, and some organic compounds. Next is the distillation step, which is an energy-consuming step and helps in the separation of ethanol from alcohol-water solution. The process consists of two parts—primary distillation which yields 95% concentrated ethanol and dehydration which results in 99 % concentration of ethanol.

Lignocellulosic biomass is a complicated structure consisting of 35–50% of cellulose, 20–35% of hemicellulose, 15–20% of lignin, and other minor constituents (Balan 2014). The lignocellulosic material thus contains cellulose and hemicelluloses that has to be broken down by pretreatment methods in order to be used for bioethanol generation. The pretreatment method selected must lead to a high yield of fermentable sugars, avoid their degradation, prevent formation of inhibitory or toxic compounds, economical and also recover lignin and hemicelluloses for production of valuable by-products(Galbe and Zacchi 2012; Mafe et al. 2015; Seidl and Goulart 2016; Kumar and Sharma 2017). There are various pretreatment methods available:

- Physical methods (milling, microwaves, ultrasonication, pyrolysis)
- Chemical methods (acid hydrolysis, alkaline hydrolysis, organosolv process, ozonolysis)
- Physicochemical methods (steam explosion, carbondioxide explosion, ammonia fiber explosion, wet oxidation)
- Biological methods (whole cells or enzymatic treatment).

There are several constraints for utilizing lignocellulosic feedstocks in the process of bioethanol production and hence it is commercially not a viable process. Pretreatment is a necessary step in order to convert the lignocellulosic biomass (LCB) into ethanol, nevertheless difficult because of the recalcitrant nature of LCB, formation of several inhibitors, and expensive process (Zabed et al. 2016). Biological pretreatment of LCB using lignin and hemicellulose degrading microorganisms or enzymes such as cellulases is a viable alternative to physical and chemical treatments owing to less energy consumption, no inhibitors formation during the process, and an environment-friendly process (Zabed et al. 2016). Production of enzymes is one of the most expensive steps during the entire procedure of biomass conversion. Approximately 18% cost of the overall bioethanol production is occupied by the cellulases enzymes used in treating (hydrolyzing) lignocellulosic biomass. The development of technology to use nanoparticles for the recovery of enzymes as well as their reuse will extensively diminish the production cost and thus will impact and benefit the overall process. Many metal nanomaterials act as a catalyst in bioethanol production during fermentation by influencing the enzymatic activity or gas-liquid mass transfer rate. Surface modification of nanoparticles provides a larger surface area for gas-liquid mass transfer and also improves the properties of active sites thereby increasing the yield of biofuel (Zhao et al. 2000).

7.4.2 Pretreatment of Lignocellulosic Biomass using Nanoparticles

Recently, research studies have demonstrated the use of nanoparticles to pretreat and hydrolyze lignocellulosic biomass for better conversion into biofuels. Nanoparticles and their suspensions exhibit similar hydrolytic action to that achieved by using chemicals during the processing of lignocellulosic biomass (Wang et al. 2012). The chief benefit of using nanoparticles in the pretreatment of substrates is that it requires only minimum amounts as well as they can be reused and recycled for further use. The cell wall of raw biomass is easily penetrated by the metal nanoparticles due to their phenomenal physical properties. In addition, they also release carbohydrates to be used in bioethanol production on interaction with biomolecules (Razack et al. 2016). A research study stated that the yield of carbohydrate increased up to 15.26% from *Chlorella vulgaris* biomass on 40 min of incubation using silver nanoparticles. The incubation time was found to be decreased by an increase in the concentration of nanoaparticles (Razack et al. 2016; Kushwaha et al. 2018). Ferrite nanoparticles resulted in generation of higher amounts of carbohydrates from wheat straw when compared to the control without nanoparticles. Magnetic nanoparticles have also been employed in the pretreatment process of lignocellulosic substrates as well as in the recovery of enzymes or nanobiocatalysts thus ensuring a cost-effective process. The Fe_3O_4 magnetic nanoparticles when used for pretreatment of lignocellulosic biomass have shown to significantly improve the biogas production (Khalid et al. 2018). Nanocrystals comprising silica functionalized with sulfonates combined with ferrous oxide magnetic nanoparticles were developed and used as a catalyst for the hydrolysis of cellobiose biomass (Verma 2017). The use of alkyl sulfonate treated nanoparticle catalysts resulted in the hydrolysis of 78% cellobiose and can also be recovered and further reused in the hydrolysis of biomass. The silica-coated magnetic nanoparticles obtained from the reaction mixture showed a substantial improvement in hemicellulose hydrolysis at 80 °C. Protease-associated magnesium nanoparticles when used for pretreatment of a biomass sample, resulted in more amount of aminoalkanoic acid generated at 95 °C along with an enhanced removal of lignin (18-fold) than when cellulase enzymes were employed for pretreatment. These products on further hydrolysis with xylanase exhibited much increase in biogas production when compared to untreated samples (Singhvi and Kim 2020).

7.4.3 *Enzyme Immobilization on Nanoparticles to Improve Bioethanol Yield*

Enzymes are crucial in the pretreatment of Lignocellulosic biomass substrates (LB) resulting in the generation of simple sugars. Different enzymes like cellulases, hemicellulases, lignases, xylanases, pectinases, etc., play an important role in hydrolyzing lignocellulosic substrates resulting in the production of biofuels. This pretreatment process is highly expensive and contributes to the higher cost of the entire process of biofuel production (Fig. 7.5). Nanotechnology proposes to immobilize enzymes on diverse nanomaterials to improve bioethanol production. Hence, the enzymes are frequently fixed onto nanoparticles or other support materials thereby causing an improvement in their properties and thus making the pretreatment of lignocellulosic biomass an economically viable process. Novel immobilization techniques are being developed to enhance the performance of immobilized enzymes, which are mainly focused on their reusability, stability, and to make the separation process easier. Enzymes like cellulases and hemicellulases are immobilized onto the surface of nanomaterials and have been shown to display high catalytic activity and are also reusable for many cycles.

The technique of immobilization involves the confinement of enzymes in a fixed region, so as to retain their activity and reuse the enzymes repeatedly. Various methods of enzyme immobilization are available. Physical methods include attachment through membranes to retain enzymes whereas chemical methods involve crosslinking and binding on supports. Different inorganic and organic substrates

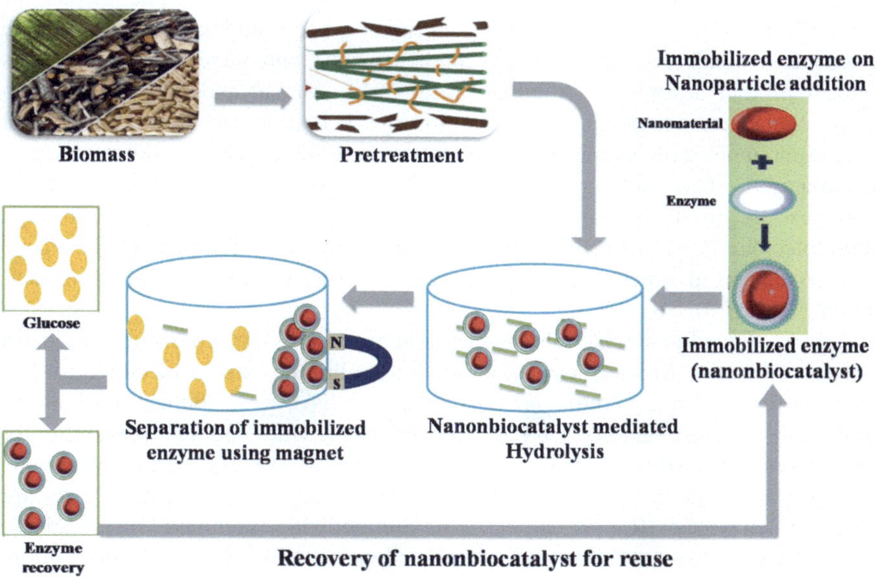

Fig. 7.5 Immobilized Enzymes for Lignocellulosic Biomass conversion to Biofuel

Table 7.3 Nanomaterials used for immobilization of enzymes used in bioethanol production

S. no.	Enzyme	Nanomaterials	References
1.	β-glucosidase	Polymer magnetic nanofibers	Lee et al. (2010)
2.	Cellulase	Single-walled carbon nanotubes	Goh et al. (2012)
3.	Lipase	Magnetic chitosan microspheres	Xie and Wang (2012)
4.	*S. cerevisiae* cells	Silica nanoparticles	Beniwal et al. (2018)
5.	Cellulase (*Aspergillus niger)*	TiO_2 nanoparticles	Lupoi and Smith (2012)
6.	Cellulase (*Aspergillus fumigatus)*	MnO_2 nanoparticles	Cherian et al. (2015)
7.	β-glucosidase (*Aspergillus niger)*	Magnetic nanoparticles	Verma et al. (2013a, b)

are used as support materials for enzyme immobilization. Many enzymes having different size, morphology, and porosity are frequently immobilized on natural inorganic support materials like clays, pumice, silica, zeolite, or manmade substances like glass, metal oxides, ceramics, and magnetic materials. Many natural organic polymers such as cellulose, starch, dextran, chitosan, agar, and alginate are also typically used for enzyme immobilization.

Several studies have indicated that immobilization of enzymes can be employed to address inhibitor difficulties in bioethanol synthesis and thus increases ethanol yield. Enzyme immobilization on nanoparticles is typically accomplished through adsorption or binding by means of ionic and covalent bonds. The method of immobilization of enzymes using covalent binding is considered more appropriate as covalent bonds are formed between the enzyme and nanoparticles and thus reduces protein desorption (Abraham et al. 2014; Ahmad and Sardar 2014). To supply the functional group for enzyme linkage, it is necessary to modify or coat these compounds with chemically active polymers for their stable immobilization on nanomaterials. Crosslinking methods employing aldehydes or diiminoesters or diamines are used for linking support nanomaterials with enzymes through intermolecular bond formation between them. Numerous research studies have been conducted in which different nanomaterials were used for enzyme immobilization (Table 7.3).

Among several nanomaterials, magnetic nanoparticles are promising candidates for immobilization of several enzymes such as cellulases and hemicellulases. The enzymes immobilized onto magnetic nanoparticles have wide applications because they result in higher stability and easy recovery of enzymes and their reusability for numerous cycles (Alftren and Hobley 2013; Rai et al. 2016). The method of enzyme immobilization onto magnetic nanoparticles helps in later recovery as well as further reuse of enzymes (Rai et al. 2016). Immobilizing cellulase on magnetic nanoparticles causes the cell wall of microalgae to hydrolyze upon interaction. This causes the release of lipid components (lipid extraction). This method yielded 93.56% biodiesel under optimal conditions (Duraiarasan et al. 2016). The β-glucosidase enzyme which is being immobilized by entrapment method onto polymer magnetic nanofibers was successfully used in the production of bioethanol from lignocellulosic materials (Lee et al. 2010). In one of the research studies, fungal β-glucosidase enzyme immobilized on magnetic nanoparticles was effectively employed as a nanocatalyst in bioethanol production (Verma et al. 2013a, b). In another study, the enzyme immobilized onto magnetic Fe_3O_4 nanoparticles incorporated in single-walled carbon nanotubes was used in bioethanol production (Goh et al. 2012). The immobilization of lipase enzyme onto magnetic chitosan microspheres was achieved using glutaraldehyde by chemical coprecipitation method (Xie and Wang 2012).

In addition to magnetic nanoparticles, many other nanomaterials like silica and titanium oxide, fullerene, graphene, and carbon nanotubes have been preferred for enzyme immobilization (Huang et al. 2011; Cho et al. 2012; Verma et al. 2013a, b). Beniwal et al. (2018) used the nano-biocatalytic system along with immobilized *S. cerevisiae*, which improved the ethanol production from cheese whey. The use of cellulase enzyme immobilized onto silica-based nanoparticles in hydrolysis of cellulose, resulted in production of more glucose than when free enzyme was used, proving the employability of immobilized enzymes in simultaneous process of saccharification and fermentation. The method of adsorption was used for enzyme immobilization on titanium oxide Nps and was effectively used to hydrolyze lignocellulosic biomass for bioethanol production processes (Lupoi and Smith 2012). A study demonstrated that the cellulase enzyme isolated from fungus *Aspergillus fumigatus* was immobilized by covalent binding technique onto MnO_2 nanoparticles and was reported to potentially enhance its thermostability property up to 70 °C (Cherian et al. 2015).

The technique of immobilization provides enzymes with a high affinity towards lignocellulosic substrates, resulting in high rates of biomass hydrolysis. Furthermore, the technique also helps improve their pH and temperature stability, benefitting the fermentations involving simultaneous saccharification and fermentation reactions. The structure formed as a result of nanoparticles and enzymes interaction depends upon the type of nanoparticles as well as their concentration employed in the process of immobilization. The structure formed bring about many changes in its surface properties thereby causing altered protein stability. Many research studies demonstrated an improvement in the stability of cellulase enzyme when iron oxide and nickel oxide like nanomaterials are used for immobilization (Srivastava et al. 2015). The immobilization of enzyme cellulases using glutaraldehyde-based iron oxide nanoparticles exhibited more stability at varied pH and temperature than free enzymes. Immobilization of cellulases by using $NiCo_2O_4$ demonstrated improved thermal stability at 80 °C over free enzymes. Thus, the immobilization of enzymes on nanoparticles possesses improved catalytic efficiency towards lignocellulosic substrates and hydrolyze them to yield simple sugars.

Apart from enzymes, there is a possibility for immobilization of even microbial cells on nanoparticles and used in the bioethanol fermentation process (Antunes et al. 2017). The synthesis of bioethanol using microbes immobilized on nanoparticles has been documented in the literature. For example, *S. cerevisiae* cells immobilized on the magnetic nanoparticles produced a high amount of bioethanol approximately 264 g/L.h (Ivanova et al. 2011). Many studies indicate that the approaches concerning immobilization of either enzymes or whole microbial cells on different nanomaterials are appropriate for safe and cost-effective production of bioethanol.

7.5 Factors Influencing the Role of Nanoparticles in Production of Bioethanol

Several factors influence the features of nanoparticles which would in turn affect their performance during the process of bioethanol and other biofuel production.

7.5.1 *Size and Shape*

Nanoparticles are of huge interest as they are of smaller size and have a larger ratio of surface area:volume. Nanoparticles show wide variations in their chemical and physical properties depending on the differences in their sizes and shapes (Bogunia-Kubik and Sugisaka 2002; Daniel and Astruc 2004; Zharov et al. 2005). Therefore, new and innovative applications may be accomplished by regulating the size and shape to nm scale in order to obtain desired particles. Because of their extremely small size, nanoparticles provide a larger surface area so that more enzymes are attached leading to high enzyme loading per unit mass of particles (Jia et al. 2003). Nanomaterials have large surface area which will provide the matrix for the immobilized enzymes thus contributing to improved stability. The enzyme stability may also be enhanced through the multiple attachments of the enzyme on the nanomaterials surface and thus reducing the unfolding of proteins (Khoshnevisana et al. 2011; Zhu et al. 2014). Furthermore, the enzymes which are bound with nanomaterials possess many advantages like easy separation, less product contamination, continuous processing, and higher yield of product (Zhang et al. 2015). Different approaches are made to achieve desired nanoparticles of varied sizes and shapes. For example, novelty could be achieved in the physical properties of nanocomposites solutions resulting from blending of nanofibers with nanoparticles (Wang et al. 2007a, b; He et al. 2009; Saquing et al. 2009; Lee et al. 2010).

7.5.2 *Temperature*

Temperature is the most significant parameter to be considered during the process of nanoparticle synthesis. Different methods of nanoparticle synthesis employ different temperatures, for instance, physicochemical methods generally employ a temperature above 300 °C whereas biological methods employ moderate temperatures usually below 100 °C temperature. The pore size, morphology/shape as well as stability of nanoparticles are usually affected by temperature. The effect of temperature on enzyme stability is explained through several experimental research studies. Nanomaterials display a crucial role in the improvement of pH and temperature stability of cellulase and other enzymes involved in bioethanol production. The improved enzyme stability is attributed to many distinctive physicochemical properties like large ratio of surface area:volume, strong adsorption capacity, high surface reactivity, and catalytic efficiency (Willner et al. 2006; Ansari and Husain 2012; Pandurangan and Kim 2015). Many researchers have proposed that the immobilizing property of nanoparticles also aids in enhancing the stability of enzymes. Immobilization of laccase was done onto the cellulose nanofiber produced by the bacteria *Gluconacetobacter xylinus* (Saravana Kumar et al. 2016). In this study, the lignocellulosic inhibitory derivatives like furfural, acetosyringone, and coniferyl aldehyde were completely degraded at 40 °C after 36 h of incubation by immobilized laccase. Also, it showed a better pH and temperature stability than free enzyme. A crude thermostable cellulase (*Aspergillus fumigatus* AA001) upon immobilization with ZnO nanoparticles showed thermal stability, even at 65 °C temperature (Srivastava and Jaiswal 2016). Cherian et al. (2015) reported that when cellulase enzyme immobilized on MnO_2 nanoparticles were used for bioethanol production, it resulted in a higher concentration of ethanol around 21.96 g/L at 70 °C for 2 h under solid state fermentation. Another study reported 90% conversion of biodiesel at 30 °C within 24 h using electrospun polyacrylonitrile nanofibrous membrane produced by *Pseudomonas cepacia* (Li et al. 2011a).

7.5.3 pH

The performance of metallic nanoparticles is greatly affected by the pH during synthesis. At >7 pH values, nanoparticles show improved stability as a result of aggregation among the particles. The alterations in pH during nanoparticle synthesis controls the variations in the sizes and shapes of nanoparticles. Herrera-Becerra et al. (2008) utilized alfalfa biomass to synthesize iron oxide nanoparticles. During synthesis at a low level of pH, i.e., pH 5 larger nanoparticles were formed whereas pH 10 yielded smaller nanoparticles. At low pH, there may be an agglomeration of gold nanoparticles leading to a smaller number of nucleation as shown in research studies by Ahmed and Ikram (2016). The rod-shaped gold nanoparticles synthesized from wheat biomass exhibited various geometries like tetrahedral, hexagonal, decahedral, icosahedral, multi-twinned, and irregular between pH 2 and 6 at room temperature (Armendariz et al. 2004). Similar studies reported that most of the nanoparticles synthesized from different substrates like alfalfa, wheat, and oat were of irregular shape at pH 2 (Gardea-Torresdey 2002; Armendariz et al. 2004). Thus, pH impacts the nanoparticles to a great extent, especially on their shape and size.

7.5.4 Concentration of Nanoparticles

The concentration of nanoparticles plays a significant role in the synthesis of desired products as well as in controlling the reaction rate. Silver Nps have the potential to lyse the cell wall of microalgae and more silver nanoparticles can rupture the microalgal cell membrane effectively. A study on various concentrations of AgNPs employed for saccharification showed that a concentration of 150 mg/g was needed for maximum Saccharification (Razack et al. 2016). Similar results were obtained in other studies indicating that the nanoparticles are more effective at higher concentrations (Hazani et al. 2013; Palomares et al. 2011).

7.5.5 Method of Synthesis of Nanoparticles

Various physical, chemical, and biological methods can be used in the production of nanoparticles (Chen et al. 2008). Previously, the techniques like evaporation and condensation led to the formation of different nanoparticles including gold, silver, lead, fullerene, and other materials (Gurav et al. 1994, Schmidt-Ott (1988); Magnusson et al. 1999, Kruis et al. 2000). Later many other physical methods like ball milling, thermal evaporation, pulsed laser desorption, spray pyrolysis, sputter deposition, lithographic techniques, etc., are being employed for nanoparticles synthesis. Chemical methods of nanoparticle synthesis include chemical solution/vapor deposition, soft chemical method, electrodeposition, hydrolysis, sol–gel method, coprecipitation, and wet chemical methods. These chemical methods may result in production of enormous amount of nanoparticles in short term but these processes should provide the capping agents meant to stabilize the size of nanoparticles. An extensive variety of materials like proteins, polysaccharides, synthetic polymers, polylactic acid, poly-D-L-glycolide, methacrylic acid, etc., are used for the preparation of nanoparticles (Sauto et al. 2012). Propyleneimine, polyetherimide, etc., are used as copying agents whereas ethylene glycol, borohydride, citrate, ascorbate, etc., are used as reducing agents (Abou El-Nour et al. 2010; Duan et al. 2015). The reagents which are generally used in the process of synthesis and stabilization of nanoparticles are toxic chemicals and result in harmful derivatives (Vithiya and Sen 2011). Biological methods have been recently developed for the synthesis of nanoparticles in order to provide simple, environment-friendly, and less expensive protocols. The biological synthesis of nanoparticles is generally simple, clean, single step, nontoxic, and environment-friendly approach (Ahmed and Ikram 2016). A number of microorganisms are found to be competent enough to synthesize inorganic nanoparticles composite intracellularly and extracellularly. The biological sources for nanoparticle synthesis are unicellular and multicultural organisms like bacteria, molds, yeasts, and plants. The intracellular and extracellular synthesis of nanoparticles by these organisms occur at physiological temperature, pH, and pressure. Moreover, these methods have other advantages such as they exhibit controlled toxicity, have control over size characteristics, and are easy, rapid, less expensive, and eco-friendly (Ingale and Chaudhari 2013).

7.5.6 Immobilization of Nanoparticles

There are several advantages of utilizing nanomaterials in processing and production of biofuels. Many studies have focused on developing enzyme immobilization methods to overcome disadvantages of free enzyme including short lifetime, instability, and separation. In Particular, the immobilized nanomaterial helps in increasing the surface-to-volume ratio, so as to obtain enzymes with high loading capacity and also maintain enzyme stability (Mathew et al. 2009; Rao 2015). Lately, nano-sized materials have been developed and introduced which can be employed as enzyme stabilization carriers (Kim et al. 2006). Novel immobilization techniques are being developed to enhance the performance of immobilized enzymes, which are mainly focused on their reusability, stability, and to make the separation process easier.

Nanofibrous membranes are made by electro-spinning method. These membranes possess a large area on their surface for enzymatic attachment and high porosity for diffusion of substrates. Normally Electro spun polyacrylonitrile fibers are used to immobilize lipases in the biodiesel synthesis whereas in this technique the nanofibers are used to adsorb lipase on the surface (Sakai et al. 2010; Li et al. 2011a). When compared to other techniques, the stability of immobilized lipase on nanofibers was good as it retained 91% of its initial conversion activity even after repeated reaction cycles (Li et al. 2011a). The immobilized laccase from cellulose nanofiber which was produced by the bacteria *Gluconacetobacter xylinus* showed a better pH and thermal stability when compared to free laccase and also it completely degrades the major inhibitors like furfural, acetosyringone, and coniferyl aldehyde within 36 h (Saravana Kumar et al. 2016).

The purpose of using magnetic nanoparticles is easy separation from the reaction medium during processing and also providing a larger surface area thus aiding the high loading capacity of target molecules (Pan et al. 2012; Singamaneni et al. 2011). Magnetic nanoparticles have many advantages such as they are highly dispersible in aqueous solutions, size-tunable, capable of forming nanocrystals, and have excellent thermal stability and reusability (Zhu et al. 2013). Aluminum oxide is a potential nanomaterial that can be employed for the immobilization of different enzymes utilized in the production of biofuels. It plays an important role in the production of platinum or silica catalyst via electrostatic adsorption phenomenon and also the loading capacity of the immobilized enzyme is moderated (Biswas 2019; Siepmann et al. 2008).

7.5.7 Stability and Reusability of Nanoparticles

Enzymes are utilized in the production of biofuel mainly in two phases (1) The pretreatment stage where lignocellulosic substrates are hydrolyzed resulting in the production of fermented sugars and (2) The transesterification stage where biodiesel is produced from plants. The half-life of many enzymes is very short (few minutes to some hours) which need to be increased. Thus, the main difficulty in handling enzymes is their shelf life which could be overcome by using advanced knowledge of nanotechnology and modern biotechnology to generate a solution for producing eco-friendly fuels (Whitcombe et al. 2014). Nanomaterials are extremely desirable due to factors like high surface areas enabling higher enzyme interactions, enhanced immobilization efficiency, improved long-term storage as well as reusability and recycling stability of the enzymes which will reduce the cost of the enzyme. For industrial applications, the reusability of enzymes is considered to be of high importance. This can be accomplished by immobilizing the cellulases and other enzymes with nanostructured materials like multi-walled nanotubes and graphene so that they can be utilized for repeated cycles of reactions (Gao et al. 2014).

Gold nanoparticles formed from leaf extracts of *Coriandrum sativum* were stable at room temperature even up to 1 month (BadriNarayanan and Sakthivel 2008). The Ag and Au Nps produced from the broth of neem leaves were reported to be stable for even 4 weeks after their synthesis (Shankar et al. 2004). The activity as well as the stability of the enzyme cellulase were improved on physical adsorption with the magnetic Nps (Khoshnevisana et al. 2011). The immobilization of enzymes onto glass beads and other nanostructures enhanced their stability by almost thousand times (Wang et al. 2001). The immobilization of laccase enzymes with nanomaterials not only increases their half-life but also stabilizes the enzyme (Xu 2015). Research studies were made on the stability of both free and immobilized cellulase and finally concluded that the enzyme cellulase of *Aspergillus fumigatus* JCF on immobilization with MnO_2 nanoparticles was more stable than free enzyme and was found to retain 89% of its initial activity after 2 h of reaction time. He also observed that even after the fifth cycle, the enzyme maintained around 60% of its original activity (Cherian et al. 2015). Similarly, β-glucosidase enzyme immobilized on Fe_3O_4 nanoparticles retained their activity up to 16 cycles during bioethanol production (Verma et al. 2013a, b). A crude cellulase enzyme complex when immobilized onto magnetic Fe_3O_4 nanoparticles by carbodiimide activation showed an increase in its thermal stability and was effectively active even after six repeated cycles (Jordan et al. 2011). A research study confirmed that co-immobilization on Au-magnetic silica nanoparticles improved the reusability of cellulases and also suggested that the loss in enzyme activity after many reuses might be due to the weakening of the bond between the matrix and the enzyme with resulting leaching of the enzyme from the system (Cho et al. 2012).

7.6 Future Prospects of Nanoparticles in Bioethanol Production

Nanomaterials are the most popular catalysts particularly in the field of bioethanol production because of their potential catalytic efficacy. A variety of nanocatalysts like carbon nanotubes, magnetic nanoparticles, metal oxide Nps, engineered nanostructures, etc., are considered to be promising among sustainable fuel resources owing to their beneficial properties. Particularly, the technique of immobilization of enzymes with nanoparticles is considered as an outstanding approach as it favors the reusability and recycling of cellulases and other enzymes or whole microbial cells as biocatalysts in the production of biofuels.

In spite of many advantages of nanotechnology in biofuel production, it also has several limitations. During the enzyme immobilization processes on nanomaterials, the size of some enzyme catalysts makes the coating process difficult. Due to bigger pore size than nanomaterials, catalysts are not embedded properly in the channels while others still stay on the surface and even obstruct the diffusion process, leading to lower enzyme activity or their loss during the operation. In addition to the technical challenges, there are also environmental concerns regarding the excessive use of heavy metal embedded nanoscale materials that could negatively impact human health (Singh et al. 2011). Owing to their extremely small size, nanoparticles may be easily incorporated by the organisms in the food web leading to biomagnification. They are inhaled or ingested or enter via dermal contact in humans and cause severe cytotoxic effects upon interaction with DNA, lipids, and proteins like biomolecules in various tissues of kidneys, lungs, and liver (Fischer and Chan 2007; Stander and Theodore 2011). The studies on ecotoxicity of nanoparticles have to be made using animal models, in order to properly understand the damage and actual risk caused by the accumulation of particles. Many research experiments are designed to study the important safety features and thus limit the exposure to nanoparticles and thereby preventing their harmful effects on the ecosystem. Such studies on nanoparticles are crucial in prevention or reduction of toxicity to humans and environment while upholding a balance between the applications and safety aspects.

However, biofuels are considered as the future fuel and nanoparticles play a vital role in the production of next-generation biofuels. The technique of employing nano-based biocatalysts is significant leading to the economic viability of several production methods. Yet, many recommendations need to be proposed for further studies in order to accelerate their applications in various bioprocesses. The behavior of various shapes and sizes of nanoparticles to perform bioprocesses should be thoroughly understood. The screening of effects of different nanoparticles and the range of concentrations on microbial activity ought to be made in order to determine the optimum process conditions. Nanoparticles used should not be toxic to the microorganisms and should be less expensive and environment friendly. Moreover, there is a need to focus on the improvements to be made in computational power and algorithms so as to aid in the screening and detection of suitable nanoparticles to be utilized in the process of biofuel production. Finally, the techno-economic assessments should be appropriately designed in order to conduct up-scaled biofuel production processes. Thus, nano-based biofuels are promising and are expected to replace soon the conventional systems at commercial scale in the future.

7.7 Conclusion

Exhaustion of existing fossil fuels and the associated global ecological problems have demanded the researchers to explore new alternative and renewable energy sources. Nanotechnology and nanoparticles tend to have a very vital role in the process of manufacturing and production of future generation biofuels namely, bioethanol and biodiesel. Among the various processes employed, nanotechnology employing nanoparticles has gained importance in the production of biofuels, owing to their beneficial properties. Different approaches can be developed to tackle the problems associated with bioenergy processes such as employing nanocatalysts or feedstock modifications will improve the production processes. Different nanomaterials like carbon nanotubes, metal oxide nanoparticles, and magnetic nanoparticles, possessing unique properties can be employed in the production of bioethanol and other biofuels. In addition to their incredible benefits and advantages, there could be deleterious effects of nanoparticles when used in excess and in uncontrolled ways thereby negatively impacting human health and ecosystem, especially when applied in large-scale production processes. Although nanotechnology has proved to be highly beneficial for the production of biofuels and may be recommended for bioethanol production processes, certain safety concerns need to be addressed meticulously before being commercialized.

References

Abdelghany TM, Al-Rajhi MH, Abboud M (2018) Recent advances in green synthesis of silver nanoparticles and their applications: about future directions. A review. Bionanoscience 8:5–16

Abdelsalam E, Samer M, Attia YA, Abdel-Hadi MA, Hassan HE, Badr Y (2016) Comparison of nanoparticles effects on biogas and methane production from anaerobic digestion of cattle dung slurry. Renew Energy 87:592–598

Abou El-Nour KMM, Eftaiha A, Al-Warthan A, Ammar RAA (2010) Synthesis and applications of silver nanoparticles. Arab J Chem 3:135–140

Abraham RE, Verma ML, Barrow CJ, Puri M (2014) Suitability of magnetic nanoparticle immobilised cellulases in enhancing enzymatic saccharification of pre-treated hemp biomass. Biotechnol Biofuel 7(1):1–2

Ahluwalia V, Kumar J, Sisodia R, Shakil NA, Walia S (2014) Green synthesis of silver nanoparticles by Trichoderma harzianum and their bioefficacy evaluation against Staphylococcus aureus and Klebsiella pneumonia. Ind Crops Prod 55:202–206

Ahmad R, Sardar M (2014) Immobilization of cellulase on TiO2 nanoparticles by physical and covalent methods: a comparative study. Indian J Biochem Biophys 51:314–320

Ahmad A, Mukherjee P, Senapati S (2003) Extracellular biosynthesis of silver nanoparticles using the fungus *Fusarium oxysporum*. Colloid Surf B Biointerfaces 28:313–318

Ahmad A, Senapati S, Khan MI, Kumar R, Sastry M (2005) Extra-/intracellular biosynthesis of gold nanoparticles by an alkalotolerant fungus, Trichothecium sp. J Biomed Nanotechnol 1:47–53

Ahmad B, Shabbir A, Jaleel H, Khan MMA, Sadiq Y (2018) Efficacy of titanium dioxide nanoparticles in modulating photosynthesis, peltate glandular trichomes and essential oil production and quality in Mentha piperita. L. Curr Plant Biol 13:6–15

Ahmed M, Douek M (2013) The role of magnetic nanoparticles in the localization and treatment of breast cancer. Biomed Res 2013:1–11

Ahmed S, Ikram S (2016) Biosynthesis of gold nanoparticles: a green approach. J Photochem Photobiol B 161:141–153

Ahorsu R, Medina F, Constantí M (2018) Significance and challenges of biomass as a suitable feedstock for bioenergy and biochemical production: a review. Energies 11(12):3366

Alftren J, Hobley TJ (2013) Covalent immobilization of β-glucosidase on magnetic particles for lignocellulose hydrolysis. Appl Biochem Biotechnol 169:2076–2087

Ali A, Mahar RB, Soomro RA, Sherazi STH (2017) Fe3O4 nanoparticles facilitated anaerobic digestion of organic fraction of municipal solid waste for enhancement of methane production. Energy Sources Part A Recov Util Environ Eff 39(16):1815–1822

Alves MB, Medeiros F, Sousa MH, Rubim JC, Suarez PA (2014) Cadmium and tin magnetic nanocatalysts useful for biodiesel production. J Braz Chem Soc 25:2304–2313

Aman G, Narendra KJ (2019) Advances in green synthesis of nanoparticles. Artif Cell Nanomed Biotechnol 47(1):844–851

Ankamwar B, Damle C, Ahmad A, Sastry M (2005) Biosynthesis of gold and silver nanoparticles using Emblica Officinalis fruit extract, their phase transfer and transmetallation in an organic solution. J Nanosci Nanotechnol 5:1665–1671

Ansari SA, Husain Q (2012) Potential applications of enzymes immobilized on/in nano materials: a review. Biotechnol Adv 30(3):512–523

Antunes FAF, Chandel AK, Mmilessi TSS, Santos JC, Rosa CA, Da Silva SS (2014) Bioethanol production from sugarcane bagasse by a novel Brazilian pentose fermenting yeast Scheffersomyces shehatae UFMG-HM 52:2: evaluation of fermentation medium. Int J Chem Eng 2014:180681

Antunes FAF, Gaikwad S, Ingle AP, Pandit R, dos Santos JC, Rai M, da Silva SS (2017) Bioenergy and biofuels: nanotechnological solutions for sustainable production. In: Nanotechnology for bioenergy and biofuel production. Springer, Cham, pp 3–18

Ariga K, Yamauchi Y, Rydzek G, Ji Q, Yonamine Y, Wu KCW, Hill JP (2014) Layer-by-layer nanoarchitectonics: invention, innovation, and evolution. Chem Lett 43(1):36–68

Armendariz V, Herrera I, Peralta-Videa JR, Jose-Yacaman M, Troiani H, Santiago P, Gardea-Torresdey JL (2004) Size controlled gold nanoparticles formation by Avena sativa biomass: use of plants in nanobiotechnology. J Nanopart Res 6:377–382

Aro EM (2016) From first generation biofuels to advanced solar biofuels. Ambio 45(1):S24–S31

Ashrafi SJ, Rastegar MF, Ashrafi M, Yazdian F, Pourrahim R, Suresh AK (2013) Influence of external factors on the production and morphology of biogenic silver nanocrystallites. J Nanosci Nanotechnol 13:2295–2301

Azmath P, Baker S, Rakshith D, Satish S (2016) Mycosynthesis of silver nanoparticles bearing antibacterial activity. Saudi Pharm J 24:140–146

BadriNarayanan K, Sakthivel N (2008) Coriander leaf mediated biosynthesis of gold nanoparticles. Mater Lett 62:4588–4590

Bakir EM, Younis NS, Mohamed ME (2018) Cyanobacteria as nanogold factories: chemical and anti-myocardial infarction properties of gold nanoparticles synthesized by *Lyngbyamajuscula*. Mar Drugs 16(6):217

Balan V (2014) Current challenges in commercially producing biofuels from lignocellulosic biomass. ISRN Biotechnol: 463074

Balat M, Balat H (2009) Recent trends in global production and utilization of bio-ethanol fuel. Appl Energy 86(11):2273–2282
Ballotin D, Fulaz S, Souza ML, Corio P, Rodrigues AG, Souza AO (2016) Elucidating protein involvement in the stabilization of the biogenic silver nanoparticles. Nanoscale Res Lett 11:313
Baruwati B, Polshettiwar V, Varma RS (2009) Glutathione promoted expeditious green synthesis of silver nanoparticles in water using microwaves. Green Chem 11:926–930
Bastos RG (2018) Biofuels from microalgae: bioethanol. In: Jacob-Lopes E, Zepka LQ, Queiroz MI (eds) Energy from microalgae. Springer, Cham, pp 229–246
Beckers L, Hiligsmann S, Lambert SD, Heinrichs B, Thonart P (2013) Improving effect of metal and oxide nanoparticles encapsulated in porous silica on fermentative biohydrogen production by Clostridium butyricum. Bioresour Technol 133:109–117
Beniwal A, Saini P, Kokkiligadda A, Vij S (2018) Use of silicon dioxide nanoparticles for β-galactosidase immobilization and modulated ethanol production by co-immobilized K. marxianus and S. cerevisiae in deproteinized cheese whey. LWT Food Sci Technol 87: 553–561
Bertrand E, Vandenberghe LPS, Soccol CR, Sigoillot JC, Faulds C (2016) First generation bioethanol. In: Green fuels technology. Springer International Publishing, Cham, pp 175–212
Bharde A, Rautray D, Bansal V (2006) Extracellular biosynthesis of magnetite using fungi. Small 2: 135–141
Binupriya AR, Sathishkumar M, Yun SI (2010) Myco-crystallization of silver ions to nanosized particles by live and dead cell filtrates of Aspergillus oryzae var. viridis and its bactericidal activity toward Staphylococcus aureus KCCM 12256. Ind Eng Chem Res 49:852–858
Biswas A (2019) Nanotechnology in biofuels production: a novel approach for processing and production of bioenergy. Springer International Publishing AG, part of Springer Nature. https://doi.org/10.1007/978-3-319-94797-6_10
Biswas A, Bayer IS, Biris AS, Wang T, Dervishi E, Faupel F (2012) Advances in top-down and bottom-up surface nanofabrication: techniques, applications & future prospects. Adv Colloid Interface Sci 170(1–2):2–27
Bogunia-Kubik K, Sugisaka M (2002) From molecular biology to nanotechnology and nanomedicine. Biosystems 65:123–138
Boroumand Moghaddam A, Namvar F, Moniri M, Azizi S, Mohamad R (2015) Nanoparticles biosynthesized by fungi and yeast: a review of their preparation, properties, and medical applications. Molecules 20(9):16540–16565
Bradley D, Cuypers D, Pelkmans L (2009) Second generation biofuels and trade: an exploratory study. Canada IEA Task 40:1–55
Branco RH, Serafim LS, Xavier AM (2019) Second generation bioethanol production: on the use of pulp and paper industry wastes as feedstock. Fermentation 5(1):4
Casals E, Barrena R, García A, González E, Delgado L, Busquets-Fité M et al (2014) Programmed iron oxide nanoparticles disintegration in anaerobic digesters boosts biogas production. Small 10(14):2801–2808
Castro-Longoria E, Vilchis-Nestor AR, Avalos-Borja M (2011) Biosynthesis of silver, gold and bimetallic nanoparticles using the filamentous fungus Neurosporacrassa. Colloid Surf B Biointerfaces 83:42–48
Chan HK, Kwok PCL (2011) Production methods for nanodrug particles using the bottom-up approach. Adv Drug Deliv Rev 63:406–416
Chen JC, Lin ZH, Ma XX (2003) Evidence of the production of silver nanoparticles via pretreatment of Phoma sp. 3·2883 with silver nitrate. Lett Appl Microbiol 37:105–108
Chen H, Roco MC, Li X, Lin Y (2008) Trends in nanotechnology patents. Nat Nanotechnol 3:123–125
Cherian E, Dharmendirakumar M, Baskar G (2015) Immobilization of cellulase onto MnO2 nanoparticles for bioethanol production by enhanced hydrolysis of agricultural waste. Chinese J Catal 36(8):1223–1229

Cho E, Jung S, Kim HJ, Lee YG, Nam KC, Leed HJ, Bae HJ (2012) Co-immobilization of three cellulases on Au-doped magnetic silica nanoparticles for the degradation of cellulose. Chem Commun 48:886–888

Christensen L, Vivekanandhan S, Misra M, Mohanty A (2011) Biosynthesis of silver nanoparticles using Murraya koenigii (curry leaf): an investigation on the effect of broth concentration in reduction mechanism and particle size. Adv Mat Lett 2:429–434

Costa Silva LP, Oliveira JP, Keijok WJ, Silva AR, Aguiar AR, Guimaraes MCC (2017) Extracellular biosynthesis of silver nanoparticles using the cell-free filtrate of nematophagus fungus Duddingtoniaflagans. Int J Nanomed 12:6373–6381

Dameron CT, Reese RN, Mehra RK, Kortan AR, Carroll PJ (1989) Biosynthesis of cadmium sulphide quantum semiconductor crystallites. Nature 338:596–597

Daniel MC, Astruc D (2004) Gold nanoparticles: assembly, supramolecular chemistry, quantum-size-related properties, and applications toward biology, catalysis, and nanotechnology. Chem Rev 104:293–346

Das D, Khanna N, Veziroğlu NT (2008) Recent developments in biological hydrogen production processes. Chem Ind Chem Eng Quart 14:57–67

Das RK, Pachapur VL, Lonappan L, Naghdi M, Pulicharla R, Maiti S, Brar SK (2017) Biological synthesis of metallic nanoparticles: plants, animals and microbial aspects. Nanotechnol Environ Eng 2(1):1–21

De Carvalho JF, De Medeiros SN, Morales MA, Dantas AL, Carriço AS (2013) Synthesis of magnetite nanoparticles by high energy ball milling. Appl Surf Sci 275:84–87

Deng X, Fang Z, Liu Y-h, Chang-Liu Y (2011) Production of biodiesel from Jatropha oil catalyzed by nanosized solid basic catalyst. Energy 36(2):777–784

Dimos K, Paschos T, Louloudi A, Kalogiannis KG, Lappas AA, Papayannakos N et al (2019) Effect of various pretreatment methods on bioethanol production from cotton stalks. Fermentation 5(1):5

Duan H, Wang D, Li Y (2015) Green chemistry for nanoparticle synthesis. Chem Soc Rev 44(16): 5778–5792

Duraiarasan S, Razack SA, Manickam A, Munusamy A, Syed MB, Ali MY, Mohiuddin MS (2016) Direct conversion of lipids from marine microalga C. salina to biodiesel with immobilised enzymes using magnetic nanoparticle. J Environ Chem Eng 4(1):1393–1398

Dutta K, Daverey A, Lin JG (2014) Evolution retrospective for alternative fuels: first to fourth generation. Renew Energy 69:114–122

Ealias AM, Saravanakumar MP (2017) A review on the classification, characterisation, synthesis of nanoparticles and their application. IOP Conf Ser Mater Sci Eng 263:032019

Eggert H, Greaker M (2014) Promoting second generation biofuels: does the first generation pave the road? Energies 7:4430–4445

Eroglu E, D'Alonzo NJ, Smith SM, Raston CL (2013) Vortex fluidic entrapment of functional microalgal cells in a magnetic polymer matrix. Nanoscale 5(7):2627–2631

Fayaz M, Tiwary CS, Kalaichelvan PT (2010) Blue orange light emission from biogenic synthesized silver nanoparticles using *Trichoderma viride*. Colloid Surf B Biointerfaces 75:175–178

Feng XM, Karlsson A, Svensson BH, Bertilsson S (2010) Impact of trace element addition on biogas production from food industrial waste-linking process to microflora. FEMS Microbiol Ecol 74:226–240

Feng Y, Zhang Y, Quan X, Chen S (2014) Enhanced anaerobic digestion of waste activated sludge digestion by the addition of zero valent iron. Water Resour 52:242–250

Feng H, Liu SY, Huang XB, Ren R, Zhou Y, Song CP, Qian DH (2017) Green biosynthesis of CdS nanoparticles using yeast cells for fluorescence detection of nucleic acids and electrochemical detection of hydrogen peroxide. Int J Electrochem Sci 12(1):618–628

Feyzi M, Norouzi L (2016) Preparation and kinetic study of magnetic Ca/Fe3O4@ SiO2 nanocatalysts for biodiesel production. Renew Energy 94:579–586

Feyzi M, Hassankhani A, Rafiee H (2013) Preparation and characterization of Cs/Al/Fe3O4 nanocatalysts for biodiesel production. Energy Convers Manag 71:62–68

Fischer HC, Chan WC (2007) Nanotoxicity: the growing need for in vivo study. Curr Opin Biotechnol 18:565–571
Gade AK, Bonde P, Ingle AP, Marcato PD, Durán N, Rai MK (2008) Exploitation of Aspergillus niger for synthesis of silver nanoparticles. J Biobased Mater Bioenergy 2:243–247
Galbe M, Zacchi G (2012) Pretreatment: the key to efficient utilization of lignocellulosic materials. Biomass Bioenergy 46:70–78
Gama M, Gatenholm P, Klemm D (eds) (2012) Bacterial nanocellulose: sophisticated multifunctional material. CRC Press, Boca Raton
Gan YX, Jayatissa AH, Yu Z, Chen X, Li M (2020) Hydrothermal synthesis of nanomaterials. J Nanomater: 1–3
Gao W, Huang H, Shi H, Feng X, Song W (2014) Nitrogen-rich graphene from small molecules as high performance anode material. Nanotechnology 25:415402
Gao L, Wang S, Xu W, Xiao G (2015) Biodiesel production from palm oil over monolithic KF/γ-Al2O3/honeycomb ceramic catalyst. Appl Energy 146:196–201
García-Martínez JE (2010) Nanotechnology for the energy challenge. Wiley, Hoboken. https://doi.org/10.1002/9783527665105
Gardea-Torresdey JL, Tiemann KJ, Parsons JG, Gamez G, Jose-Yacaman M (2002) Characterization of trace level Au(III) binding to alfalfa biomass (Medicago sativa) by GFAAS. Adv Environ Res 6(3):313–323
Gericke M, Pinches A (2006a) Biological synthesis of metal nanoparticles. Hydrometallurgy 83: 132–140
Gericke M, Pinches A (2006b) Microbial production of gold nanoparticles. Gold Bulletin 39:22–28
Gholami-Shabani M, Akbarzadeh A, Norouzian D, Amini A, GholamiShabani Z, Imani A (2014) Antimicrobial activity and physical characterization of silver nanoparticles green synthesized using nitrate reductase from Fusariumoxysporum. Appl Biochem Biotechnol 172:4084–4098
Goh WJ, Makam VS, Hu J, Kang L, Zheng M, Yoong SL et al (2012) Iron oxide filled magnetic carbon nanotube–enzyme conjugates for recycling of amyloglucosidase: toward useful applications in biofuel production process. Langmuir 28(49):16864–16873
Golmohammadi H, Morales-Narvaez E, Naghdi T, Merkoci A (2017) Nanocellulose in sensing and biosensing. Chem Mater 29:5426–5446
Gu F, Wang SF, Lü MK, Zhou GJ, Xu D, Yuan DR (2004) Photoluminescence properties of SnO2 nanoparticles synthesized by Sol−Gel method. J Phys Chem B 108:8119–8123
Gudikandula K, Vadapally P, Charya MS (2017) Biogenic synthesis of silver nanoparticles from white rot fungi: their characterization and antibacterial studies. OpenNano 2:64–78
Gurav A, Kodas T, Wang L, Kauppinen E, Joutsensaari J (1994) Generation of nano-meter size fullerene particles via aerosol routes. Chem Phys Lett 218:304–308
Han H, Guan Y (2009) Synthesis of biodiesel from rapeseed oil using K2O/γ-Al2O3 as nano-solid-base catalyst. Wuhan Univ J Nat Sci 14:75–79
Harish K, Nagasamy V, Himangshu B, Anuttam K (2018) Metallic nanoparticle: a review. Biomed J Sci Tech Res 4:1–11
Hashimoto H, Yokoyama S, Asaoka H (2007) Characteristics of hollow microtubes consisting of amorphous iron oxide nanoparticles produced by iron oxidizing bacteria, Leptothrixochracea. J Magn Magn Mater 310:2405–2407
Hazani AA, Ibrahim MM, Shehata AI, El-Gaaly GA, Daoud M, Fouad D, Rizwana H, Nadine MS (2013) Moubayed ecotoxicity of Ag-nanoparticles on twomicroalgae, Chlorella Vulgaris and Dunaliellatertiolecta. Arch Biol Sci 65:1447–1457
He D, Hu B, Yao QF, Wang K, Yu SH (2009) Large-scale synthesis of flexible free-standing SERS substrates with high sensitivity: electrospun PVA nanofibers embedded with controlled alignment of silver nanoparticles. ACS Nano 3(12):3993–4002
Herrera-Becerra R, Zorrilla C, Rius JL, Ascencio JA (2008) Electron microscopy characterization of biosynthesized iron oxide nanoparticles. Appl Phys A Mater Sci Process 91:241–246

Hirani AH, Javed N, Asif M, Basu SK, Kumar A (2018) A review on first-and second-generation biofuel productions. In: Kumar A, Ogita S, Yau YY (eds) Biofuels: greenhouse gas mitigation and global warming. Springer, New Delhi, pp 141–154

Hossain AS, Salleh A, Boyce AN, Chowdhury P, Naqiuddin M (2008) Biodiesel fuel production from algae as renewable energy. Am J Biochem Biotechnol 4(3):250–254

Huang R, Su R, Qi W, He Z (2011) Bioconversion of lignocellulose into bioethanol: process intensification and mechanism research. Bioenergy Res 4(4):225–245

Huirache AR, Nava R, Peza-Ledesma CL, Lara-Romero J, Alonso-Núñez G, Pawelec B (2013) SBA-15 mesoporous silica as catalytic support for hydrodesulfurization catalysts—review. Mater 6:4139–4167

Hulkoti NI, Taranath TC (2014) Biosynthesis of nanoparticles using microbes—A review. Colloid Surf B Biointerfaces 121:474–483

Hussein AK (2015) Applications of nanotechnology in renewable energies—A comprehensive overview and understanding. Renew Sustain Energy Rev 42:460–476

Ingale AG, Chaudhari AN (2013) Biogenic synthesis of nanoparticles and potential applications: an eco-friendly approach. J Nanomed Nanotechol 4:165

Ingle A, Rai M, Gade A, Bawaskar M (2009) Fusarium solani: a novel biological agent for the extracellular synthesis of silver nanoparticles. J Nanopart Res 11:2079–2085

Iravani S (2011) Green synthesis of metal nanoparticles using plants. Green Chem 13:2638

Ivanova V, Petrova P, Hristov J (2011) Application in the ethanol fermentation of immobilized yeast cells in matrix of alginate/magnetic nanoparticles, on chitosan-magnetite microparticles and cellulose-coated magnetic nanoparticles. arXiv preprint arXiv:1105.0619

Jeevanandam J, Chan YS, Danquah MK (2016) Biosynthesis of metal and metal oxide nanoparticles. ChemBioEng Rev 3(2):55–67

Jia H, Zhu G, Wang P (2003) Catalytic behaviors of enzymes attached to nanoparticles: the effect of particle mobility. Biotechnol Bioeng 84:406–414

Jordan J, Kumar CS, Theegala C (2011) Preparation and characterization of cellulase-bound magnetite nanoparticles. J Mol Catal B: Enzyme 68(2):139–146

Kalska-Szostko B (2012) Electrochemical methods in nanomaterials preparation. In: Recent trend in electrochemical science and technology, pp 261–280

Kapdan IK, Kargi F (2006) Bio-hydrogen production from waste materials. Enzyme Microb Technol 38(5):569–582

Kaur M, Ali A (2011) Lithium ion impregnated calcium oxide as nano catalyst for the biodiesel production from karanja and jatropha oils. Renew Energy 36:2866–2871

Karri S, Sierra-Alvarez R, Field JA (2005) Zero valent iron as an electron-donor for methanogenesis and sulfate reduction in anaerobic sludge. Biotechnol Bioeng 92(7):810–819

Khalid MK, Asad M, Henrich-Noack P, Sokolov M, Hintz W, Grigartzik L et al (2018) Evaluation of toxicity and neural uptake in vitro and in vivo of superparamagnetic iron oxide nanoparticles. Int J Mol Sci 19(9):2613

Khan NT, Khan MJ, Jameel J, Jameel N, Rheman SUA (2017) An overview: biological organisms that serves as nanofactories for metallic nanoparticles synthesis and fungi being the most appropriate. Bioceram Dev Appl 7:101

Khoshnevisana K, Bordbar AK, Zare D, Davoodi D, Noruzi M, Barkhi M, Tabatabaei M (2011) Immobilization of cellulase enzyme on superparamagnetic nanoparticles and determination of its activity and stability. Chem Eng J 171:669–673

Kim J, Grate JW, Wang P (2006) Nanostructures for enzyme stabilization. Chem Eng Sci 61(3): 1017–1026

Kowshik M, Ashtaputre S, Kharrazi S, Vogel W, Urban J (2003) Extracellular synthesis of silver nanoparticles by a silver-tolerant yeast strain MKY3. Nanotechnology 14:95–100

Kroger N, Poulsen N (2008) Diatoms—From cell wall biogenesis to nanotechnology. Annu Rev Genet 42:83–107

Kruis FE, Fissan H, Rellinghaus B (2000) Sintering and evaporation characteristics of gas-phase synthesis of size-selected PbS nanoparticles. Mater Sci Eng B 69:329–334

Kumar AK, Sharma S (2017) Recent updates on different methods of pretreatment of lignocellulosic feedstocks: a review. Bioresour Bioprocess 4(1):7
Kumari D, Singh R (2018) Pretreatment of lignocellulosic wastes for biofuel production: a critical review. Renew Sustain Energy Rev 90:877–891
Kushwaha D, Upadhyay SN, Mishra PK (2018) Nanotechnology in bioethanol/biobutanol production. Biofuel Bioref Technol 5:115–127
Lee SM, Jin LH, Kim JH, Han SO, Na HB, Hyeon T, Koo YM, Kim J, Lee JH (2010) Beta-glucosidase coating on polymer nanofibers for improved cellulosic ethanol production. Bioprocess Biosyst Eng 33(1):141–147
Li SF, Fan YH, Hu RF, Wu WT (2011a) Pseudomonas cepacia lipase immobilized on to the electro spun PAN nanofibrous membranes for biodiesel production from soybean oil. J Mol Catal B: Enzyme 72:40–45
Li X, Xu H, Chen ZS, Chen G (2011b) Biosynthesis of nanoparticles by microorganisms and their applications. J Nanomater 2011:1–16
Limayem A, Ricke SC (2012) Lignocellulosic biomass for bioethanol production: current perspectives, potential issues and future prospects. Prog Energy Combust Sci 38(4):449–467
Lohse SE, Murphy CJ (2012) Applications of colloidal inorganic nanoparticles: from medicine to energy. J Am Chem Soc 134:15607–15620
López-Quintela MA, Rivas J, Blanco MC, Tojo C (2004) Synthesis of nanoparticles in microemulsions. In: Nanoscale materials, pp 135–155
Luef B, Fakra SC, Csencsits R (2013) Iron-reducing bacteria accumulate ferric oxyhydroxide nanoparticle aggregates that may support planktonic growth. ISME J 7:338–350
Lupoi JS, Smith EA (2012) Near-infrared Raman spectroscopy measurements of ethanol detection limits in simultaneous saccharification and fermentation liquors. In: Developments in enzyme immobilization and near-infrared Raman spectroscopy with downstream renewable energy applications, vol. 1001, pp 124
Maaz K, Karim S, Mumtaz A, Hasanain SK, Liu J, Duan JL (2009) Synthesis and magnetic characterization of nickel ferrite nanoparticles prepared by co-precipitation route. J Magn Magn Mater 321:1838–1842
Madhuvilakku R, Piraman K (2013) Biodiesel synthesis by TiO2–ZnO mixed oxide nanocatalyst catalyzed palm oil transesterification process. Bioresour Technol 150:55–59
Mafe OA, Davies SM, Hancock J, Du C (2015) Development of an estimation model for the evaluation of the energy requirement of dilute acid pretreatments of biomass. Biomass Bioenergy 72:28–38
Magnusson MH, Deppert K, Malm JO, Bovin JO, Samuelson L (1999) Gold nanoparticles: production, reshaping, and thermal charging. J Nanopart Res 1:243–251
Malgras V, Ji Q, Kamachi Y, Mori T, Shieh FK, Wu KCW, Ariga K, Yamauchi Y (2015) Templated synthesis for nanoarchitectured porous materials. Bull Chem Soc Jpn 88(9): 1171–1200
Malvankar NS, Lovley DR (2012) Microbial nanowires: a new paradigm for biological electron transfer and bioelectronics. ChemSusChem 5:1039–1046
Manawi Y, Ihsanullah AS, Al-Ansari T, Atieh M, Manawi YM (2018) A review of carbon nanomaterials' synthesis via the chemical vapor deposition (CVD) method. Materials 11(5):822
Mandal D (2006) The use of microorganisms for the formation of metal nanoparticles and their application. Appl Microbiol Biotechnol 69(5):485–492
Manochio C, Andrade BR, Rodriguez RP, Moraes BS (2017) Ethanol from biomass: a comparative overview. Renew Sustain Energy Rev 80:743–755
Mascolo M, Pei Y, Ring T (2013) Room temperature Co-precipitation synthesis of magnetite nanoparticles in a large pH window with different bases. Materials 6:5549–5567
Mathew AP, Laborie MP, Oksman K (2009) Cross-linked chitosan-chitin whiskers nanocomposites with improved permeation selectivity and pH stability. Biomacromolecules 10(6):1627–1632
Mehra RK, Winge DR (1991) Metal ion resistance in fungi: molecular mechanisms and their regulated expression. J Cell Biochem 45:30–40

Menon S, Shanmugam R, Kumar V (2017) A review on biogenic synthesis of gold nanoparticles, characterization, and its applications. Resour Technol 3:516–527

Mguni LL, Meijboom R, Jalama K (2012) Biodiesel production over nano-MgO supported on titania. World Acad Sci Eng Technol 64:894–898

Michalska K, Bizukojć M, Ledakowicz S (2015) Pretreatment of energy crops with sodium hydroxide and cellulolytic enzymes to increase biogas production. Biomass Bioenergy 80: 213–221

Molina CMM (2013) ZnO nanorods as catalyst for biodiesel production from olive oil. M.Sc. Thesis. University of Louisville

Molnar Z, Bodai V, Szakacs G, Erdelyi B, Fogarassy Z, Safran G (2018) Green synthesis of gold nanoparticles by thermophilic filamentous fungi. Sci Rep 8:3943

Mukherjee P (2002) Extracellular synthesis of gold nanoparticles by the fungus Fusariumoxysporum. ChemBioChem 3(5):461–463

Mukherjee P, Ahmad A, Mandal D (2001a) Bioreduction of AuCl 4 ions by the fungus Verticilliumsp and surface trapping of the gold nanoparticles formed. Angew Chem Int Ed 40:3585–3588

Mukherjee P, Ahmad A, Mandal D, Senapati S, Sainkar SR, Khan MI, Parishcha R (2001b) Fungus-mediated synthesis of silver nanoparticles and their immobilization in the mycelial matrix: a novel biological approach to nanoparticle synthesis. Nano Lett 1:515–519

Narayanan KB, Sakthivel N (2010) Biological synthesis of metal nanoparticles by microbes. Adv Coll Interface Sci 56(1):1–13

Nithya R, Ragunathan R (2009) Synthesis of silver nanoparticles using Pleurotus sajor caju and its antimicrobial study. Digest J Nanomater Bios 4:623–629

Noruzi M, Zare D, Davoodi D (2012) A rapid biosynthesis route for the preparation of gold nanoparticles by aqueous extract of cypress leaves at room temperature. Spectrochimica Acta Part A Mol Biomol Spectrosc 94:84–88

Nwodo U, Green E, Okoh A (2012) Bacterial exopolysaccharides: functionality and prospects. Int J Mol Sci 13:14002–14015

Ovais M, Khalil AT, Islam NU, Ahmad I, Ayaz M, Saravanan M et al (2018) Role of plant phytochemicals and microbial enzymes in biosynthesis of metallic nanoparticles. Appl Microbiol Biotechnol 102(16):6799–6814

Ozdenkci K, De Blasio C, Muddassar HR, Melin K, Oinas P, Koskinen J, Sarwar G, Jarvinen M (2017) A novel biorefinery integration concept for lignocellulosic biomass. Energy Convers Manag 149:974–987

Palomares IR, Boltes K, Pinas FF, Leganes F, Calvo EG, Santiago JJ, Rosal R (2011) Physicochemical characterization and ecotoxicological assessment of CeO2 nanoparticles using two aquatic microorganisms. Toxicol Sci 119:135–145

Pan Y, Du X, Zhao F, Xu B (2012) Magnetic nanoparticles for the manipulation of proteins and cells. Chem Soc Rev 41(7):2912–2942

Pandey A, Gupta K, Pandey A (2015) Effect of nanosized TiO2 on photofermentation by Rhodobactersphaeroides NMBL-02. Biomass Bioenerg 72:273–279

Pandurangan M, Kim DH (2015) ZnO nanoparticles augment ALT, AST, ALP and LDH expressions in C2C12 cells. Saudi J Biol Sci 22(6):679–684

Parada MP, Osseweijer P, Duque JAP (2017) Sustainable biorefineries, an analysis of practices for incorporating sustainability in biorefinery design. Ind Crops Prod 106:105–123

Patil CD, Patil SV, Borase HP, Salunke BK, Salunkhe RB (2012) Larvicidal activity of silver nanoparticles synthesised using Plumeria rubra plant latex against Aedes aegypti and Anopheles stephensi. Parasitol Res 110:1815–1822

Patumsawad S (2011) Second generation biofuels: technical challenge and R and D opportunity in Thailand. J Sustain Energy Environ: 47–50

Petcharoen K, Sirivat A (2012) Synthesis and characterization of magnetite nanoparticles via the chemical co-precipitation method. Mater Sci Eng B 177:421–427

Pighi L, Pumpel T, Schinner F (1989) Selective accumulation of silver by fungi. Biotechnol Lett 11: 275–280
Pradeep A, Priyadharsini P, Chandrasekaran G (2008) Sol–gel route of synthesis of nanoparticles of MgFe2O4 and XRD, FTIR and VSM study. J Magn Magn Mater 320:2774–2779
Pramila K, Sushil KS (2016) Microbes mediated synthesis of metal nanoparticles: current status and future prospects. Int J Nanomater Bios 6(1):1–24
Prasad K, Jha AK, Kulkarni AR (2007) Lactobacillus assisted synthesis of titanium nanoparticles. Nanoscale Res Lett 2:248–250
Qidwai A, Pandey A, Kumar R, Shukla SK, Dikshit A (2018) Advances in biogenic nanoparticles and the mechanisms of antimicrobial effects. Ind J Pharm Sci 80:592–603
Qiu LY, Bae YH (2006) Polymer architecture and drug delivery. Pharm Res 23:1–30
Qiu F et al (2011) Biodiesel production from mixed soybean oil and rapeseed oil. Appl Energy 88:2050–2055
Rai M, dos Santos JC, Soler MF, Franco Marcelino PR, Brumano LP, Ingle AP, da Silva SS (2016) Strategic role of nanotechnology for production of bioethanol and biodiesel. Nanotechnol Rev 5(2):231–250
Rai M, Da Silva SS (2017) Nanotechnology for bioenergy and biofuel production. Springer
Rajput S, Werezuk R, Lange RM, Mcdermott MT (2016) Fungal isolate optimized for biogenesis of silver nanoparticles with enhanced colloidal stability. Langmuir 32:8688–8697
Ramsurn H, Gupta RB (2013) Nanotechnology in solar and biofuels. ACS Sustain Chem Eng 1(7): 779–797
Rao A (2015) Nanotechnology: applications in biofuel productions. Conference paper, VIT University. https://www.reserachgate.net/publication/271902586
Ran CQ, Chen ZA, Zhang W, Yu XJ, Jin MF (2006) Characterization of photo-biological hydrogen production by several marine green microalgae. Wuhan Ligong Daxue Xuebao 28(2):258–263
Rana A, Yadav K, Jagadevan S (2020) A comprehensive review on green synthesis of nature-inspired metal nanoparticles: mechanism, application and toxicity. J Clean Prod 272:122880
Rautaray D, Ahmad A, Sastry M (2004) Biological synthesis of metal carbonate minerals using fungi and actinomycetes. J Mater Chem 14:2333–2340
Razack SA, Duraiarasan S, Mani V (2016) Biosynthesis of silver nanoparticle and its application in cell wall disruption to release carbohydrate and lipid from C. vulgaris for biofuel production. Biotechnol Rep 11:70–76
Reddy ANR, Saleh AA, Islam MS, Hamdan S, Maleque MA (2016) Biodiesel production from crude Jatropha oil using a highly active heterogeneous nanocatalyst by optimizing transesterification reaction parameters. Energy Fuel 30:334–343
Romero GMS, Vila J, Mata Alvarez J, Chimenos JM, Astals S (2016) The role of additives on anaerobic digestion: a review. Renew Sustain Energy Rev 58:1486–1499
Sabri MA, Umer A, Awan GH, Hassan MF, Hasnain A (2016) Selection of suitable biological method for the synthesis of silver nanoparticles. Nanomater Nanotechnol 6:1–20
Sadowski Z, Maliszewska IH, Grochowalska B, Polowczyk I, Kozlecki T (2008) Synthesis of silver nanoparticles using microorganisms. Mater Sci Poland 26:419–424
Saini JK, Saini R, Tewari L (2015) Lignocellulosic agriculture wastes as biomass feedstocks for second-generation bioethanol production: concepts and recent developments. 3 Biotech 5(4): 337–353
Sakai S, Liu Y, Yamaguchi T, Watanabe R, Kawabe M, Kawakami K (2010) Production of butyl-biodiesel using lipase physically-adsorbed on to electro spun poly- acrylonitrile fibers. Bioresour Technol 101:7344–7349
Salgado SYA, Zamora RMR, Zanella R, Peral J, Malato S, Maldonado MI (2016) Photocatalytic hydrogen production in a solar pilot plant using a Au/TiO2 photo catalyst. Int J Hydrog Energ 41:11933–11940
Salunke BK, Sawant SS, Lee SI, Kim BS (2015) Comparative study of MnO2 nanoparticle synthesis by marine bacterium Saccharophagus degradans and yeast Saccharomyces cerevisiae. Appl Microbiol Biotechnol 99(13):5419–5427

Sanghi R, Verma P (2009) Biomimetic synthesis and characterisation of protein capped silver nanoparticles. Bioresour Technol 100(1):501–504

Saquing CD, Manasco L, Khan SA (2009) Electrospun nanoparticle-nanofiber composites via a one-step synthesis. Small 5(8):944–951

Saravana Kumar T, Han-Sung P, Ae-Young M, Myoung-Suk C, Dae-Hyuk K, Seung-Moon P (2016) Detoxification of furanic and phenolic lignocellulose derived inhibitors of yeast using laccase immobilized on bacterial cellulosic nanofibers. J Mol Catal B: Enzyme 134:196–205

Sauto EB, Severino P, Santana MHA (2012) Preparation of polymeric nanoparticles by polymerization of monomers—Part I. Polímeros 22:96–100

Schmidt-Ott A (1988) New approaches to in-situ characterization of ultrafine agglomerates. J Aerosol Sci 5:553–563

Seidl PR, Goulart AK (2016) Pretreatment processes for lignocellulosic biomass conversion to biofuels and bioproducts. Curr Opin Green Sustain Chem 2:48–53

Sekhon BS (2014) Nanotechnology in agri-food production: an overview. Nanotechnol Sci Appl 7: 31–53

Sekoai PT, Ouma CNM, du Preez SP, Modisha P, Engelbrecht N, Bessarabov DG, Ghimire A (2019) Application of nanoparticles in biofuels: an overview. Fuel (London, England) 237:380–397

Serrano E, Rus G, García-Martínez J (2009) Nanotechnology for sustainable energy. Renew Sustain Energy Rev 13(9):2373–2384

Seshadri S, Saranya K, Kowshik M (2011) Green synthesis of lead sulfide nanoparticles by the lead resistant marine yeast, Rhodosporidiumdiobovatum. Biotechnol Prog 27:1464–1469

Shaishav S, Singh RN, Satyendra T (2013) Biohydrogen from algae: fuel of the future. Int Res J Environ Sci 2:44–47

Shankar S, Rai A, Ahmad A, Sastry M (2004) Rapid synthesis of Au, Ag, and bimetallic Au core-Ag shell nanoparticles using Neem (Azadirachta indica) leaf broth. J Colloid Interface Sci 275(2):496–502

Sharma N, Pinnaka AK, Raje M, Fnu A, Bhattacharyya MS (2012) Exploitation of marine bacteria for production of gold nanoparticles. Microb Cell Factories 11:86

Shivaji S, Madhu S, Singh S (2011) Extracellular synthesis of antibacterial silver nanoparticles using psychrophilic bacteria. Process Biochem 46:1800–1807

Siddiqi KS, Husen A (2016) Green synthesis, characterization and uses of palladium/platinum nanoparticles. Nanoscale Res Lett 11(1):1–13

Siepmann F, Siepmann J, Walther M, MacRae RJ, Bodmeier R (2008) Polymer blends for controlled release coatings. J Control Release 125:1–15

Silva FM, Mahler CF, Oliveira LB, Bassin JP (2018) Hydrogen and methane production in a two-stage anaerobic digestion system by co-digestion of food waste, sewage sludge and glycerol. Waste Manag 76:339–349

Singamaneni S, Bliznyuk VN, Binek C, Tsymbal EY (2011) Magnetic nanoparticles: recent advances in synthesis, self-assembly and applications. J Mater Chem 21(42):16819–16845

Singh AK, Talat M, Singh DP, Srivastava ON (2010) Biosynthesis of gold and silver nanoparticles by natural precursor clove and their functionalization with amine. J Nanopart Res 12:1667–1675

Singh J, Dutta T, Kim KH, Rawat M, Samddar P, Kumar P (2018) 'Green' synthesis of metals and their oxide nanoparticles: applications for environmental remediation. J Nanobiotechnol 16(1): 1–24

Singh T, Jyoti K, Patnaik A, Singh A, Chauhan R, Chandel SS (2017) Biosynthesis, characterization and antibacterial activity of silver nanoparticles using an endophytic fungal supernatant of Raphanus sativus. J Genet Eng Biotechnol 15(1):31–39

Singh P, Kim YJ, Zhang D, Yang DC (2016) Biological synthesis of nanoparticles from plants and microorganisms. Trends Biotechnol 34(7):588–599

Singh SK, Bansal A, Jha MK, Dey A (2011) Comparative studies on uptake of wastewater nutrients by immobilized cells of Chlorella minutissima and dairy waste isolated algae. Ind Chem Eng 53(4):211–219

Singhvi M, Kim BS (2020) Current developments in lignocellulosic biomass conversion into biofuels using nanobiotechology approach. Energies 13(20):5300
Skeffington AW, Scheffel A (2018) Exploiting algal mineralization for nanotechnology: bringing coccoliths to the fore. Curr Opin Biotechnol 49:57–63
Slawson RM, Lohmeier-Vogel EM, Lee H, Trevors JT (1994) Silver resistance in Pseudomonas stutzeri. Biometals 7:30–40
Song C, Liu Q, Ji N, Deng S, Zhao J, Li Y (2017) Reducing the energy consumption of membrane-cryogenic hybrid CO2 capture by process optimization. Energies 124:29–39
Srivastava N, Jaiswal P (2016) Production of cellulases using agriculture waste: application in biofuels production. In: Environmental biotechnology: a new approach. Daya Publishing House, New Delhi, pp 233–244
Srivastava N, Singh J, Ramteke PW, Mishra PK, Srivastava M (2015) Improved production of reducing sugars from rice straw using crude cellulase activated with Fe3O4/Alginate nanocomposite. Bioresour Technol 183:262–266
Stander L, Theodore L (2011) Environmental implications of nanotechnology—an update. Int J Environ Res Public Health 8:470–479
Su L, Shi X, Guo G, Zhao A, Zhao Y (2013) Stabilization of sewage sludge in the presence of nanoscale zero-valent iron (nZVI): abatement of odor and improvement of biogas production. J Mater Cycles Waste Manag 15(4):461–468
Subashini G, Bhuvaneswari S (2018) Nanoparticles from Fungi (Myconanoparticles). In: Fungi and their role in sustainable development: current perspectives. Springer, Singapore, pp 753–779
Tahvildari K, Anaraki YN, Fazaeli R, Mirpanji S, Delrish E (2015) The study of CaO and MgO heterogenic nanocatalyst coupling on transesterification reaction efficacy in the production of biodiesel from recycled cooking oil. J Environ Health Sci Eng 13
Terán-Hilares R, Reséndiz AL, Martínez RT, Silva SS, Santos JC (2016) Successive pretreatment and enzymatic saccharification of sugarcane bagasse in a packed bed flow-through column reactor aiming to support biorefineries. Bioresour Technol 203:42–49
Thakkar KN, Mhatre SS, Parikh RY (2010) Biological synthesis of metallic nanoparticles. Nanomed Nanotechnol Biol Med 6(2):257–262
Tiwari A, Kiran T (2017) Biofuels from microalgae. In: Advances in biofuels and bioenergy, pp 239–249
Uma Suganya KS, Govindaraju K, Ganesh Kumar V (2015) Blue green alga mediated synthesis of gold nanoparticles and its antibacterial efficacy against Gram positive organisms. Mater Sci Eng C 47:351–356
Verma ML (2017) Fungus-mediated bioleaching of metallic nanoparticles from agro-industrial by-products. In: Fungal nanotechnology. Springer, Cham, pp 89–102
Verma ML, Barrow CJ (2015) Recent advances in feedstocks and enzyme-immobilised technology for effective transesterification of lipids into biodiesel. In: Microbial factories, pp 87–103
Verma ML, Barrow CJ, Puri M (2013a) Nanobiotechnology as a novel paradigm for enzyme immobilization and stabilization with potential applications in biodiesel production. Appl Microbiol Biotechnol 97:23–39
Verma ML, Chaudhary R, Tsuzuki T, Barrow CJ, Puri M (2013b) Immobilization of β-glucosidase on a magnetic nanoparticle improves thermostability: application in cellobiose hydrolysis. Bioresour Technol 135:2–6
Verziu M, Cojocaru B, Hu J, Richards R, Ciuculescu C, Filip P, Parvulescu V (2008) Sunflower and rapeseed oil transesterification to biodiesel over different nanocrystalline MgO catalysts. Green Chem 10. https://doi.org/10.1039/b712102d
Vigneshwaran N, Ashtaputre NM, Varadarajan PV, Nachane RP, Paralikar KM, Balasubramanya RH (2007) Biological synthesis of silver nanoparticles using the fungus Aspergillus flavus. Mater Lett 61:1413–1418
Vithiya K, Sen S (2011) Biosynthesis of nanoparticles. Int J Pharm Sci Res 2:2781–2785

Wang P, Dai S, Waezsada SD, Tsao AY, Davison BH (2001) Enzyme stabilization by covalent binding in nanoporous sol–gel glass for nonaqueous biocatalysis. Biotechnol Bioeng 74:249–255

Wang D, Ikenberry M, Pe L, Hohn KL (2012) Acid-functionalized nanoparticles for pretreatment of wheat straw

Wang H, Covarrubias J, Prock H, Wu X, Wang D, Bossmann SH (2015) Acid-functionalized magnetic nanoparticle as heterogeneous catalysts for biodiesel synthesis. J Phys Chem C 119:26020–26028

Wang T, Zhang D, Dai L, Chen Y, Dai X (2016) Effects of metal nanoparticles on methane production from waste-activated sludge and microorganism community shift in anaerobic granular sludge. Sci Rep 6:25857

Wang YZ, Li YX, Sun G, Zhang GL, Liu H, Du S, Yang SA, Bai J, Yang QB (2007a) Fabrication of Au/PVP nanofiber composites by electrospinning. J Appl Polym Sci 105(6):3618–3622

Wang Z, Pan Y, Dong T, Zhu X, Kan T et al (2007b) Production of hydrogen from catalytic steam reforming of bio-oil using C12A7-O-based catalysts. Appl Catal A 320:24–34

Waqas M, Naser N, Sarathy M, Morganti K, Al-Qurashi K, Johansson B (2016) Blending octane number of ethanol in HCCI, SI and CI combustion modes. SAE Intl J Fuels Lubricant 9(3): 659–682

Wen L, Wang Y, Lu D, Hu S, Han H (2010) Preparation of KF/CaO nanocatalyst and its application in biodiesel production from Chinese tallow seed oil. Fuel 89:2267–2271

Whitcombe MJ, Kirsch N, Nicholls IA (2014) Molecular imprinting science and technology: a survey of the literature for the years from 2004 to 2011. J Mol 27:297–401

Willner I, Baron R, Willner B (2006) Growing metal nanoparticles by enzymes. Adv Mater 18(9): 1109–1120

Xie W, Wang J (2012) Immobilized lipase on magnetic chitosan microspheres for transesterification of soybean oil. Biomass Bioenerg 36:373–380

Xu J, Yang H, Fu W, Du K, Sui Y, Chen J (2007) Preparation and magnetic properties of magnetite nanoparticles by sol–gel method. J Magn Magn Mater 309:307–311

Xu R (2015) Enhancement of catalytic activity of immobilized laccase for diclofenac biodegradation by carbon nanotubes. Chem Eng J 262:88–95

Yahyaei B, Pourali P (2019) One step conjugation of some chemotherapeutic drugs to the biologically produced gold nanoparticles and assessment of their anticancer effects. Sci Rep 9. https://doi.org/10.1038/s41598-019-46602-0

Yan H, Chen X, Song H, Li J, Feng Y, Shi Z et al (2017) Synthesis of bacterial cellulose and bacterial cellulose nanocrystals for their applications in the stabilization of olive oil pickering emulsion. Food Hydrocoll 72:127–135

Yang G, Wang J (2018) Improving mechanisms of biohydrogen production from grass using zero-valent iron nanoparticles. Bioresour Technol 266:413–420

Yang F, Hong F, You W, Liu C, Gao F, Wu C (2006) Influence of nano-anatase TiO2 on the nitrogen metabolism of growing spinach. Biol Trace Elem Res 110:179–190

Yang Y, Guo J, Hu Z (2013) Impact of nano zero valent iron (NZVI) on methanogenic activity and population dynamics in anaerobic digestion. Water Res 47(17):6790–6800

Yang Z, Shi X, Wang C, Wang C, Wang L, Guo R (2015) Magnetite nanoparticles facilitate methane production from ethanol via acting as electron acceptors. Sci Rep 5:16118

Zabed H, Sahu JN, Boyce AN, Faruq G (2016) Fuel ethanol production from lignocellulosic biomass: an overview on feedstocks and technological approaches. Renew Sustain Energy Rev 66:751–774

Zhang W, Qiu J, Feng H, Zang L, Sakai E (2015) Increase in stability of cellulase immobilized on functionalized magnetic nanospheres. J Magn Magn Mater 375:117–123

Zhang Y, Shen J (2007) Enhancement effect of gold nanoparticles on biohydrogen production from artificial wastewater. Int J Hydrog Energy 32(1):17–23

Zhao W, Zhang Y, Du B, Wei D, Wei Q, Zhao Y (2013) Enhancement effect of silver nanoparticles on fermentative biohydrogen production using mixed bacteria. Bioresour Technol 142:240–245

Zhao XS, Lu GQ, Hu X (2000) Characterization of the structural and surface properties of chemically modified MCM-41 material. Micropor Mesopor Mater 41(1–3):37–47

Zharov VP, Kim JW, Curiel DT, Everts M (2005) Self-assembling nanoclusters in living systems: application for integrated photothermal nanodiagnostics and nanotherapy. Nanomedicine 1(4): 326–345

Zhu H, Hou C, Li Y, Zhao G, Liu X, Hou K, Li Y (2013) One-pot solvothermal synthesis of highly water-dispersible size-tunable functionalized magnetite nanocrystal clusters for lipase immobilization. Chem Asian J 8(7):1447–1454

Zhu Y, Wang G, Ni H, Xiao A, Cai H (2014) Cloning and characterization of a new manganese superoxide dismutase from deep-sea thermophile Geobacillus sp. EPT3. World J Microbiol Biotechnol 30(4):1347–1357

Chapter 8
Green Synthesis of Nanomaterials from Biomass Waste for Biodiesel Production

Zeenat Arif, Naresh K. Sethy, P. K. Mishra, and P. Kumar

Abstract An exponential increase in energy demand and associated problems with the use of fossil fuels in the environment provides a pathway to look for substitute for renewable energy source. In this idea, biodiesel seems to be a promising approach for being renewable and an alternative to fossil fuel due to its eco-friendly, nontoxic, biodegradable characteristics. Production of biodiesel process involves a chemical reaction that necessitates catalyst utilization. Although both types of catalyst (homogeneous/heterogeneous) have their own pros and cons, presently more attention is made to the use of green route synthesized nano-catalyst. Owing to the importance of fast recoverability and easy reusability of the catalysts after the reaction according to green chemistry principle, it becomes a significant factor for sustainable management. In this chapter, the advantages of biofuel over conventional fossil fuel, reaction mechanism, use of catalyst, green route catalyst synthesis from waste, and its efficiency in the synthesis of biodiesel are discussed. This chapter is concluded highlighting the existing challenges and the future prospects of catalysts synthesized using the green route for bioenergy application.

Keywords Biodiesel · Energy · Green · Nano-catalyst · Yield

8.1 Introduction to Biofuel

There is a continuous increase in energy consumption throughout the worldwide population. Satyanarayana et al. 2011 mentioned that the consumption rate of fossil fuel is 105 times faster than its creation and in the near future demand will be doubled and tripled for primary energy sources by 2035 and 2055, respectively. Other shortcomings of fossil fuel consumption are economic concerns and global warming resulting from greenhouse gases. Figure 8.1 depicts an era with different energy forms. Therefore, renewable fuels have attracted the attention of research and

Z. Arif (✉) · N. K. Sethy · P. K. Mishra · P. Kumar
Department of Chemical Engineering and Technology, Indian Institute of Technology (BHU), Varanasi, India

© The Author(s), under exclusive license to Springer Nature Singapore Pte Ltd. 2023

M. Srivastava, P. K. Mishra (eds.), *NanoBioenergy: Application and Sustainability Assessment*, Clean Energy Production Technologies,
https://doi.org/10.1007/978-981-19-6234-9_8

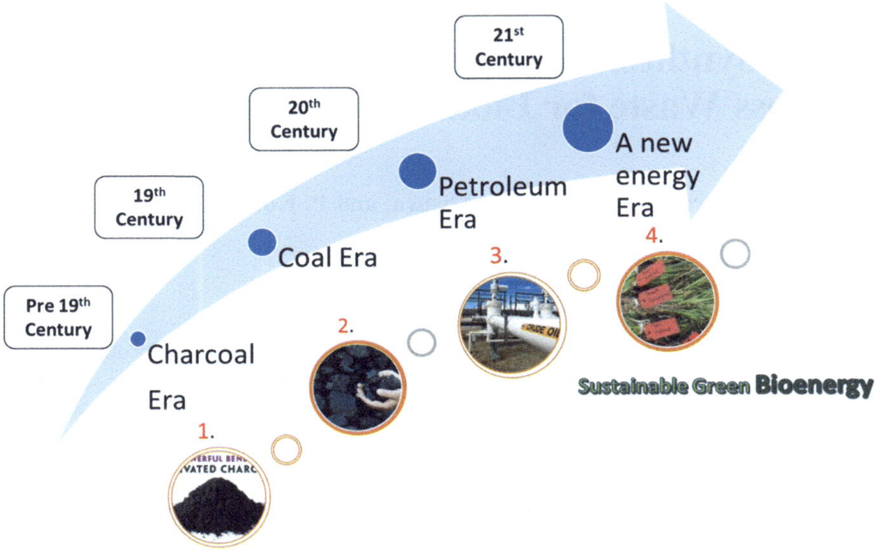

Fig. 8.1 Era showing existence of energy forms in different century

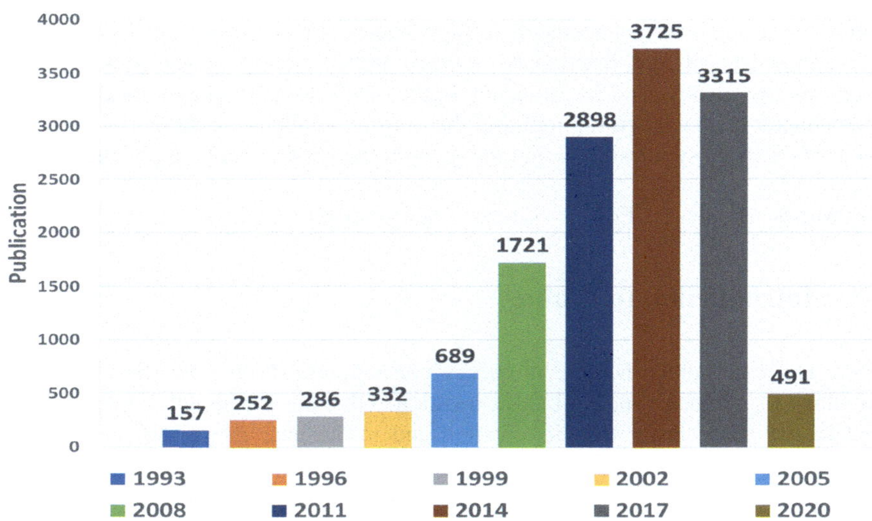

Fig. 8.2 Publication/Year on biodiesel production from 1993 to 2020 (Scifinder Database)

development toward their use in form of biodiesel, biogas, bioethanol, biohydrogen, etc. They drifted the attention of the researcher toward themselves because of their sustainability, low Green House Gases emission, and even reduction in carbon footprint (Hussein 2015). The data represented in Fig. 8.2 shows the research increasing trend in the area of biodiesel. The nature of the graph depicts an exponential rise in the publication from 157 in 1993 to 3725 in the year 2014. The data was obtained from the Scifinder database.

In the research field for last few decades, significant attention was given to making use of biofuels; an alternative fuel, e.g., biodiesel, owing to its sustainability and environment-friendly nature. There are many sources available such as mustard, sunflower, animal fat, and soybean for biofuel synthesis. Biodiesel is composed of fatty acid of methyl ester (FAME) or ethyl ester derived from recycled cooking and plant oils, animal fats, etc. Advantage of using biodiesel in conventionally designed diesel engines is reduction in the emission of carbon compounds such as monoxide (CO) and dioxide (CO_2), high hydrogen and oxygen content, and good lubricity of biodiesel enhances the engine life due to improved fuel injection system (Hussein 2015). Other significant properties which makes it prominent for fuel application is high combustion efficiency, flash point which makes it nonflammable and nonexplosive, cetane number, and low sulfur content (Changmai et al. 2020). Microorganism (algae), due to its high lipid content and good growth rate is considered as potential feedstock for biodiesel (Upadhyay et al. 2016). However, many challenges exist while using microalgae that need to be overcome particularly the cost factor and process intensification methods such as improvement in pretreatment process, process variable optimization to enhance the production rate and since other approaches like synthetic biology, genetic engineering nevertheless increases the yield but at the same time they were laborious, time-consuming, and led to increase in the production cost (Shanmugam et al. 2020) and were facing a problem of proper infrastructure for the production process (Patumsawad 2011). Thus, it urges to develop a technique which possess synergistic property of a simple/easy technique and is cost-effective to achieve the desired goal of a high production rate. Nanotechnology is a promising approach and can offer meaningful solutions to the challenges observed by changing the feed material property or using it in biocatalysts form for biofuel production. The potential applications of nanotechnology have encouraged researchers to investigate novel and build robust nano biocatalytic systems (Verma et al. 2013).

8.2 Introduction to Nanoparticles

Over the past three decades, nanotechnology has evolved as an independent field with numerous applications in diverse areas. A dimension less than 1/10th of micro is called a nano size dimension, for example, 0.1 μm is equal to 100 nm thus particle size (<100 nm) of materials are commonly defined as nanoparticles. These are smaller size particles, having a large surface area to volume ratio resulting in excellent catalytic activity, chemical steadiness, and nonlinear optical performance (Agarwal et al. 2017). There may be biological, chemical, and physical approaches to synthesize nanoparticles. The physical methods are costly and extreme processing conditions (e.g., high temperature and pressure) are needed to obtain particles in the

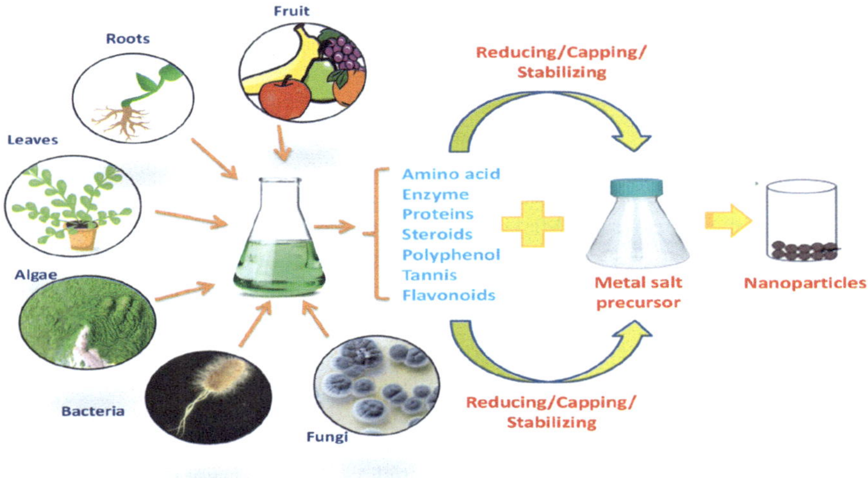

Fig. 8.3 Schematic representation of the green synthesis of nanoparticles

nano range. On the other hand, chemical methods need toxic and hazardous chemicals which are not eco-friendly for the environment and living creatures. Also, most of the toxic chemicals consumed cannot be recycled, reused, or degraded to harmless forms (Agarwal et al. 2017). The capping and stabilizing agents may be needed to prevent agglomeration of nanoparticles prepared through physical and chemical methods which add additional cost to the synthesis process. Synthesis of particles using plant extract or microorganism is termed as "Biosynthesis." It is cheap and easy to carry out process thereby being a widely used process (Agarwal et al. 2017). The schematic representation of nanoparticle formation using green approach is shown in Fig. 8.3.

This approach for particle synthesis is advantageous with respect to cost, biocompatibility, safety, and environmental friendliness. Use of different parts of plant extracts/agro-waste over microorganisms (bacteria and fungi) is preferred for nanoparticle synthesis because of the faster rate of biosynthesis, and the possibility of large-scale production without maintaining any stringent conditions which are necessary in case of microorganisms (Sharma et al. 2019). Considering the advantages of plant extract over microorganism, numerous nanoparticles were synthesized using physical and chemical methods from the extract of different parts of plants/agro-waste.

As mentioned earlier nanoparticles have found applications in diverse fields ranging from agriculture, chemical industry, etc., to healthcare and environmental remediation. Some typical applications also include as nanocarrier for drug delivery, as catalyst for dye reduction (Srikar et al. 2016), and in solar cell and photocatalytic systems (Pawar et al. 2019). Table 8.1 enlists different green route synthesized nanoparticles, properties, and their application.

Table 8.1 List of green routes synthesized nanoparticles: Shape, size, and application

Plant/Agro-waste	NPs	Results	References
Parthenium hysterophorus extract	TiO_2	Method: Microwave irradiation Average size: 20–50 nm Shape: spherical shape % Degradation of dye was >85%	Thandapani et al. (2018)
Azadirachta indica aqueous leaf extract	Ag	Average size: 34 nm Exhibit antimicrobial activities against gram-negative (*E. coli*) positive (*S. aureus*) bacteria	Ahmed et al. (2016)
Polygala tenuifolia	CdO	Shape: Trigonal Average size: 34 nm	Ghotekar (2019)
Moringa oleifera	ZnO	Spherical shaped particle Average size: 6–20 nm	Agarwal et al. (2017)
Jatropha curcas	TiO_2	Average crystalline size: 13 nm Average size: 10–20 nm Spherical shape particle Removal of COD and Cr is 82.26% and 76.48% respectively in tannery wastewater	Goutam et al. (2018)
Sesbania grandi flora	TiO_2	Average crystalline size: 42.58 nm Shape: Triangular, square, and spherical	Srinivasan et al. (2019)
Fraxinus rhynchophylla	ZnO_2	Average diameter 100–200 nm	Wang et al. (2020a, b)
Vitex agnus-castus	SnO_2	Average size: 4–13 nm Shape: spherical shape	Ebrahimian et al. (2020)
Cassia fistula and *Melia azadarach*	ZnO	Shape: Spherical Average size: 3–68 nm	Naseer et al. (2020)
Citrus Limon	Cu	Shape: Spherical Average size: 30 nm	Amer and Awwad (2021)
Watermelon rind	Pd	Shape: Spherical Average size: 96.4 nm	Lakshmipathy et al. (2015)
Wheat straw	SiO_2	Shape: NA Average size: 100 nm	Patel et al. (2017)
Rice husk	Silica	Shape: spherical Average size: 70 nm	Chen et al. (2012)

8.3 Benefits of Nanomaterials from Agro-Based Materials

The residue wastes in the farm field when burned directly on the open ground, composed of CH_4, CO, N_2, SO_2, and hydrocarbons with particulate matters of PM0.1 and PM2.5 which are directly released into atmosphere degrading the environment. Also, it results in fog formation and dense fine ash cloud due to low temperature and high humidity and dense fine ash cloud (Janta et al. 2022; Bray et al. 2019). Wheat straw and rice paddy residues comprise of oxides of nitrogen and nitrous oxides when burned produces dangerous levels of reactive nitrogen. Phairuang et al. 2019 mentioned in their study that in the month of October–November 2017, New Delhi possessed a threat of bad air quality and measured average concentration ranges from 22.43 to 718.95 μgm^{-3} which was very high when compared to average ambient air quality standards of 60 μgm^{-3}. An almost similar situation was reported for China, in the study of Liu et al. 2019, which highlights an increase in the concentration of PM10 particles during straw burning by 2–4 times and was having adverse health issues in 2015. This may also result in 75% in smoke formation which ultimately leads to a reduction in visibility and also affects the radiation balance. Exposure to fine particles is also health hazardous in terms of respiratory problem, asthma, chronic bronchitis, etc. It was also investigated that on the burning of rice straw there is an increase up to 14 times in semi-volatile organic compounds. Government can prohibit the burning of agro-commodities by imposing a ban, but there is no other alternative methodology to eradicate this pollution problem from the traditional method. So, the best way to utilize this waste is by converting them into a useful product and great demand of nanoparticle because of its unique characteristics could be one of the positive approaches in this field.

8.4 Synthesis of Different Nanomaterials from Agro-Wastes

8.4.1 Graphene

Graphene, a 2D substance made out of immaculate carbon is having hexagonal shape. Graphene is known for its good thermal conductivity, electronic properties (Schwierz 2010), high surface area (2600 m^2/g), modulus (1Tpa), and optical transparency (Xiao et al. 2014). Being expensive and having difficult production techniques but owing to its supreme property efforts are being made to search for inexpensive approaches to synthesize graphene. Graphene oxide (GO) a single atomic layered compound is one such material. It is formed from graphite oxidation. Hofmann's Hummer's and Staudenmaier's are well-known generalized synthesis methods for graphene (Novoselov et al. 2012). In recent years, chemical vapor deposition (CVD) (Singh and Vidyasagar 2014) and chemical exfoliation was adopted for graphene and its derivatives (Somanathan et al. 2015). These synthesis methods form high-quality graphene but high temperature, application of toxic chemical agents and expensive substrates are their main drawbacks. According to this study, the cost of accessible GO/gram is around $200. To overcome such issues, researchers were focusing to develop a new strategy for GO synthesis from agro-waste, which is eco-friendlier and more cost-effective (Du et al. 2012). Sugarcane bagasse was oxidized directly under the muffled atmosphere for the synthesis of GO and this method was termed as SOMA-GO (Somanathan et al. 2015). Group of researchers Omid et al. (2014) and Rajesh et al. (2016) reported the production of GO and reduced graphene oxide sheets of excellent quality from wastes of bagasse, fruit, leaf, and wood. Rice husk (as an agro-waste) was also used to produce graphene by adopting calcination and chemical activation due to its abundant existence (Muramatsu et al. 2014).

8.4.2 Silica Nanoparticles (SiO_2 NPs)

Silica (SiO_2) which is found in natural state exists in two forms (a) crystalline and (b) amorphous forms. Demand for amorphous silica is more as it has vast application in industrial and electronics field because it is a good semiconductor and they are nontoxic and possess a large surface area. Stober method, sol–gel, and thermal decomposition are different known methods for the formation of amorphous silica (Gu et al. 2015; Sun et al. 2016). Studies by Karimi and Mohsenzadeh (2016) show that SiO_2 NPs have the potential to increase crop yield thus can be used effectively in the agriculture field. The chemical synthesis procedure requires toxic, harmful, and expensive chemicals (Suriyaprabha et al. 2014). So, a number of methodologies are designed and developed for using agro-wastes for synthesizing SiO_2 NPs. Literature reports the production of SiO_2 NPs using different agro-wastes, e.g., rice husk, corn hub, and other plant sources (Kamath and Proctor 1998; Hariharan and Sivakumar 2013; Mohanraj et al. 2012). Wheat straw is a good substrate for silica synthesis as it is composed of silica greater than 70%.

Rice husk (a predominant agro-waste after rice processing) obtained from rice, a second major crop worldwide earlier was used as an additive for fuel and fertilizer (Mohammadinejad et al. 2016). SiO_2 composition ranges between 90 and 97% in rice husk hence it is considered as the best available source for the SiO_2 NPs synthesis. Hassan et al. (2014) used sol–gel technique for the synthesis of SiO_2 NPs. Similarly, Ghorbani et al. (2015) uses rice husk after pretreatment by hydrochloric acid for the production of SiO_2 NPs and the results show that 97% pure silicon was obtained.

After rice, corn ranks third worldwide in quantity generation after processing for silica nanoparticles production; Okoronkwo et al. (2016) study concludes that 47% of silica was composed in corn cob ash. Qadri et al. (2015) in their study used a residue of corn for the synthesis of nanoparticles and nanorods and concluded that the yield of pure silica was 97%.

Bagasse ash, an agro-industrial waste is used widely for power generation; leftover ash from burning is composed of large proportions of silica. Sana et al. (2014) for SiO_2 NPs formation used bagasse ash of sugar beet using laser ablation-mediated technique. Manjula-Rani et al. (2014) demonstrated the experiment for SiO_2 NPs formation by hydrolysis and alkali precipitation of cow dung ash and reported cent percent silica nanoparticles formation.

8.4.3 Carbon Nanomaterials (CNP)

Carbon is the sixth most abundant nonmetallic element of second period and belongs P block in periodic table. Graphite, diamond, and C-60 are its three elemental forms. CNPs are known for its applications in batteries and supercapacitors in form of good-performance electrode substances (Wen et al. 2014). CNPs have wide applications in the field of cancer treatment. An engineered CNP also has vast usage in electrical, mechanical, thermal, and optical sectors (Srivastava et al. 2015). CVD is a well-known technique to form Carbon nanomaterial at large-scale industrial scale. In India, nearly 400 tons of agro-waste are generated from bagasse, coconut shell, rice, etc., that could produce enormous amounts of carbon particles and so it is cost-effective potential source for the mass production of carbon nanomaterial. Table 8.2 enlists various types of nanoparticles derived from agro-waste.

8.5 Synthesis of Biofuel by Biomass: Renewable Technologies

Firstly, to convert biomass to biofuel, lignocellulose refined biomass used as a source is converted to hydrocarbon biofuel. Gasification and Liquefaction are two ways to bring conversion from biomass to biofuel. Thermochemical processes, an effective technology was followed to produce synthesis gas by gasification or bio-oils by pyrolysis, or by the hydrolysis route (Bridgwater 2012).

8.6 Reaction Mechanism for Biodiesel Production

Biodiesel is produced via esterification and transesterification/alcoholysis process. In transesterification process oils or triglyceride (TG) (edible or nonedible) along with alcohol undergoes nucleophilic reaction to form FAME and by-product as glycerol. The process includes three reversible reactions: Step 1: formation of diglyceride from triglyceride, Step 2: formation of monoglyceride from diglyceride, and Step 3: conversion of monoglyceride to glycerol. In each conversion step, an ester is formed thus it can simply say that three ester molecules are produced from one TG molecule. The transesterification reaction converts a triglyceride into biodiesel efficiently. However, in the esterification reaction, for a cent percent conversion of free fatty acids, it is essential to carry out a reaction using alcohol and carboxylic acids to form biodiesel. Both these reactions are carried out in two steps. Firstly, via esterification reaction conversion of high FFA content of vegetable to esters (FAME) takes place in ambience of acid catalyst, and then transesterification reaction to produce FAME from triglycerides using a basic catalyst (Changmai et al. 2020). Thus, the importance of catalysts in the synthesis of biodiesel was highlighted and the selection of suitable catalysts for this particular application will greatly enhance the reaction rate thereby increasing the biodiesel yield.

Table 8.2 List of different nanoparticles derived from agro-waste

Nanomaterial	Agro-waste source	Synthesis method	Size and crystallinity	References
Ag	*Coconut shell* 	Reduction of $AgNO_3$ using shell extract	14.5 nm and spherical	Sinsinwar et al. (2018)
Au	*Grape waste* 	–	20–25 nm and spherical	Krishnaswamy et al. (2014)
Silica	*Sugarcane bagasse* 	Flame synthesis, sol–gel	20 nm and spherical	Mohd et al. (2017)

(continued)

Table 8.2 (continued)

Nanomaterial	Agro-waste source	Synthesis method	Size and crystallinity	References
Calcium oxide	*Eggshell* 	Calcinations, hydration, dehydration	75 nm and spherical	Pandit and Fulekar (2017)
SiO_2, Al_2O_3, Fe_2O_3, CaO, MgO, Na_2O K_2O	*Groundnut shell* 	Ball milling	Size depends on ball milling hours and spherical	Yaro et al. (2017)
SiO_2	*Wheat straw* 	Pyrolysis, microwave process	100 nm and spherical	Patel et al. (2017)

Palladium	*Watermelon rind* 	$PdCl_2$, 10 mL of WRAE, blending at 150 rpm, 24 h, temperature: 30 °C followed by drying at 85 °C	96.4 nm and spherical	Palanisamy et al. (2021)
TiO_2	*Cajanus Cajan* 	Hydrolysis	30 nm and spherical	Arif et al. (2020a, b)
SiO_2	*Bamboo leaf* 	Sol-gel	20 nm and spherical	Sethy et al. (2019)

8.7 Enhancement in Biomass Conversion: Use of Nano-Catalysts

Waste edible oil, a known biomass source, an effective and useful substitute is used widely for the synthesis of biofuel. Transesterification is an effective process for biofuel synthesis by waste vegetable oil using different quantities of catalyst.

Heterogeneous catalysts are widely used over homogeneous catalysts for biodiesel production because of its reusability due to easy separation and eco-friendly nature (Arzamendi et al. 2007). Marciniuk et al. 2014, report CaO, Salinas et al. 2016 report SrO, and MgO was reported by Yahya et al. 2016 as heterogeneous base catalyst but major drawbacks associated with them is the requirement of high temperature and pressure, also they have low performance when compared to homogeneous catalysts. Moreover, catalyst deactivation due to coking, sintering, and poisoning is one of the main problems (Yatish et al. 2020b). In comparison to bulk catalysts, the use of nano-catalyst can produce biofuel under moderate operating conditions. Owing to the large ratio of surface to volume makes nanoparticles an excellent potential catalyst compared to bulk materials. Research studies report that to enhance the products' quality, the development of a new catalyst with excellent catalytic activity is necessary.

Use of catalysts in biofuel synthesis not only enhances the gasification methods to increase the production of syngas but at the same time retards the tar formation. Excellent catalytic activities due to high specific surface of nano-catalyst have circumvented the limitations faced while using heterogeneous catalysts and also the use of nano-catalysis reduces corrosion problems. The use of "a green" method for heterogeneous nano-catalysts leads to a reduction in production costs by increasing the reusability property of the catalyst. Akia et al. (2014) reported the synthesis of iron nano-catalyst from Pongamia pinnata and produces biodiesel from the same plant oil and methanol. The synergistic effect of using nano-catalysts includes tar content reduction and improving the product quality as well as conversion efficiency (Akia et al. 2014). Nanomaterials of Copper, Gold, Iron, Nickel, Silver, and Titanium had the potential to improve the yield of biohydrogen via different biological routes such as bio-photolysis, dark- and photo-fermentation (Lin et al. 2016a, b). Also, nanomaterials during fermentation, play the role of oxygen scavenger thus decreasing the oxidation-reduction potential thereby developing a suitable anaerobic environment so that hydrogenase enzyme can improve the biohydrogen yield (Taherdanak et al. 2015). The other advantageous feature of using a nanomaterial is that it improves the lignin removal from lignocellulosic biomass when employed in the pretreatment step and thus speeding up the entire process as well as improving the sugar yield (Ladole et al. 2017; Wei et al. 2015). Moreover, for efficient hydrolysis, nanomaterial's presence affects the pH and thermal stability of enzymes (Bilai et al. 2018).

8.8 Green Synthesized Nano-Catalyst for Biofuel Production: Application

Yadav et al. 2019 in their research adopted green approach for silver doped ZnO nanoparticles (NPs) preparation using an extract of turmeric root. Characterization of X-Ray Diffraction reveals 45 nm crystallite size particle and photo images of Scanning Electron Microscopy depict spherical shaped but agglomerated. X-Ray Photoelectron Spectroscopy analysis concludes the state of Ag is metallic and ZnO has a Wurtzite structure. Utilization of synthesized Ag-doped ZnO nanoparticles as catalyst in the esterification of *Terminalia belerica* oil for biodiesel synthesis shows 214 $mmolg^{-1}$ h^{-1} of hydrogen evolution hence reveals a prominence material toward biodiesel application. Production and purification of biodiesel was schematically represented in Fig. 8.4.

The article by Yadav et al. (2018a, b) demonstrated the experiment for doped ZnO nanoparticles with trivalent dysprosium ions (Dy^{3+}) formation using *E. tirucalli* plant (Euphorbiaceous family) latex by green combustion technique as it has the advantageous characteristics of being economical, more production. The characterization analysis results suggest that the particle size ranges bewteen 30 and 38 nm and were having a wurtzite hexagonal structure. The size may vary depending upon the concentration of latex taken. The synthesized ZnO: Dy^{3+} as catalytic agent resulting in 84.1% of biodiesel production was reported in the transesterification process.

ZnO nanoparticle without any doping also finds its suitability in biodiesel production. Raghavendra and his coauthor published the application of ZnO particle for biodiesel production in the year 2017. Their study reports the treatment of zinc

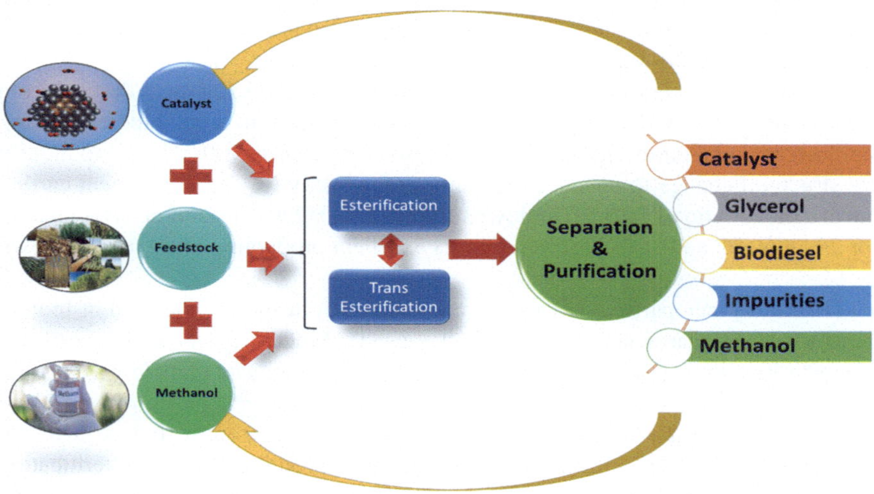

Fig. 8.4 Biodiesel pathway: Production and Separation

nitrate using extract obtained from the seed of *Garcinia gummi-gutta*, a reducing agent by combustion technique for the formation of multifunctional ZnO nanoparticles (NPs). The formed 10–20 nm sized ZnO NPs show photoluminescence (PL), and antioxidant properties. The reaction results reveal biodiesel yield was slightly above 80% and other significant properties such as acid value, carbon residue, density, copper strip corrosion, flash point kinematic viscosity, and calorific value when compared with the data of ASTM standards predicts an excellent product quality.

Yatish et al. (2020a) use a catalyst comprised of Copper oxide nanoparticles (CuO NPs) for biodiesel. The study demonstrated the leaves and seeds (reducing-cum-fuel agent) of *Terminalia chebula* plant commonly called as black- or chebulic myrobalan proves to be economical and cost-effective for CuO NPs formation, a heterogeneous catalyst via combustion method. The characterization results of CuO NPs reveal 100 nm sized diameter has a monoclinic structure and rod-like morphology. The study also concludes with 97.1% yield of T. chebula methyl ester (TCME) at molar ratio of (CH_3OH to oil) 9:1 and catalyst loading of 3 wt.% for 60 min at 60 ° C. The reusability experiment proves good catalytic stability of CuO NPs when used four times in a cycle and negligible yield whereas the kinetic study analysis predicts pseudo-first order reaction. Herein other significant properties when compared with ASTM standards predicts an excellent product quality.

Yatish et al. 2020b also reported the use of calcium titanate ($CaTiO_3$ NPs), a synthesized nanoparticles by combustion synthesis (SCS) for biodiesel production using *Ochrocarpus longifolius* leaves extract. The synthesized nano-catalyst was used in biodiesel synthesis from dairy waste scum oil (DWSO). $CaTiO_3$ NPs (15 nm size) using RSM (Response Surface Methodology) an optimization tool predicts the highest yield of 97.7% at operating condition of molar ratio of 9:1 (methanol to DWSO), 1.80 wt% nano-catalyst loading for 45 min at 65 °C and 650 rpm stirring speed. The reusability test depicts good catalytic stability of $CaTiO_3$ NPs up to five cycles with a small amount of loss in yield. Herein also pseudo-first order reaction fit well according to the kinetic study for biodiesel synthesis.

In another investigation carried out by Correia et al. 2017 CaO was derived from natural quail eggshell and act as an effective catalyst to give biodiesel using sunflower oil via transesterification. Calcination was done at 900 ° C for 3 to modify the structural change to enhance the nano-catalyst activity. The experiment carried out at a molar ratio of 1:10.5 (sunflower oil/methanol), 2 wt. % nano-catalyst loading for 2 h at 60 °C and at 1000 rpm produces excellent production of 99.00 ± 0.02 wt. %.

The MnO_2 nanoparticles also show a potential application in the field of biofuel generation. Stegarescu et al. 2020, used MnO_2 catalyst to enhance transesterification's reaction rate for biodiesel production. The particle was synthesized using the dry leaf of *Origanum vulgare* (*Rosmarinus officinalis,* oregano) and *Artemisia dracunculus* (tarragon) which act as reducing agent to reduce Mn^{7+} in $KMnO_4$ to Mn^{4+} state to form MnO_2 NPs as a final product. The transesterification reaction rate was enhanced via microwave-assisted using grapes residues and seeds in the ambience of MnO_2 nano-catalyst to obtain biofuel. The results conclude that

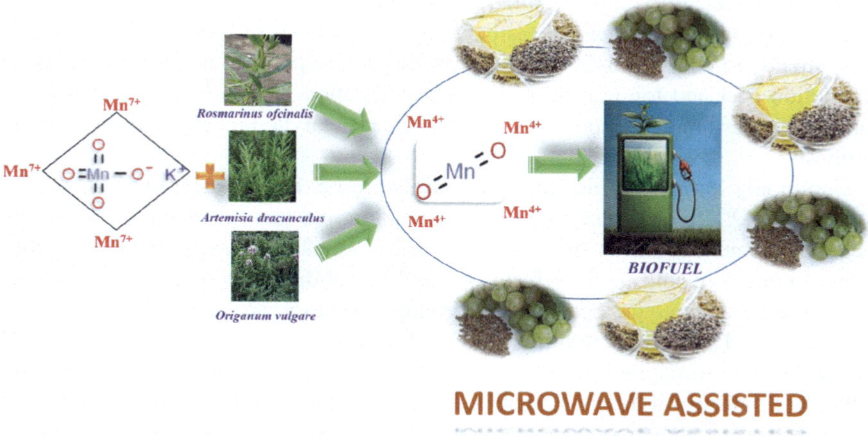

Fig. 8.5 Schematic representation of MnO_2 particle formation and biofuel production

the surface area of biochemically synthesized MnO_2 nanoparticles was four times higher when compared with the chemically synthesized particles. Among the three leaves, MnO_2 nanoparticles obtained from *Rosmarinus Officinalis,* oregano presents the best catalytic activity in biodiesel production. The results concluded a 3.5 times improvement in that reaction rate in presence of MnO_2-oregano nano-catalyst. Figure 8.5 represents the particle synthesis for biofuel production.

Salam et al. 2020 had done a comparative study using synthesized calcite (CaCO/ $CaCO_3$) nano-catalysts formed by two different known routes namely, sol–gel and thermal method from eggshell. Biodiesel yield of 93 and 98% were evaluated using a catalyst prepared from thermal and sol–gel method, respectively. Reusability test demonstrated a yield of 88% in presence of catalyst from thermal method while sol–gel derived catalyst gives a yield of 81% only. Furthermore, efficiency of the produced biodiesel was evaluated from the Cetane Index (CI) and the results suggested a range of 45–54, which gives an indication that CaCO/$CaCO_3$ nano-catalyst is potentially a good candidate to produce biodiesel from cooking oil efficiently. This report justifies the benefits of using a green catalyst (CaO (thermal method) derived from the eggshells) for generating alternative green diesel for its efficient performance in the application of diesel-powered engine.

8.8.1 *Magnetic Nano-Sized Materials*

To enhance the production efficiency along with a reduction in energy consumption in the production process, rapid recovery and catalyst reusability after the reaction is a significant parameter for sustainable process management. The utilization of green synthesized magnetic nanoparticles is an effective approach for segregation of heterogeneous catalysts and reusing. The integrated approach of using a magnetic catalyst and low-cost biomass feedstocks for the synthesis of biodiesel has received great attention (Wang et al. 2020a, b). This integrated approach leads to reduction in production cost and minimization of waste generation thus providing a promising approach toward biodiesel production using green catalytic processes for large-scale synthesis.

Commonly known ferromagnetic materials are Fe, Fe_3O_4, γ-Fe_2O_3, Co, and Ni (Wang et al. 2020a, b), after suitable modification and functionalization, retain the magnetic properties and at the same time allow quick and easy separation ability without compromising their performance and reusability. Magnetic CaO and ZnO show excellent catalytic performance giving biodiesel production yield >91%. According to Baskar and Soumiya (2016) transesterification reaction was carried out between ferromagnetic ZnO nanocomposites prepared via the coprecipitation method and castor oil as raw material. Under the optimal conditions, FAME yield was greater than 90% and at the same time the catalyst remains active under mild conditions due to the nanostructure of the catalyst. Similarly, Salimi and Hosseini (2019) using the coprecipitation method produce a cheap composite of ZnO/$BiFeO_3$ nano-catalyst and catalyses rapeseed oil via transesterification process to produce biodiesel.

Several studies have also reported carbon-based magnetic nanoparticle synthesized via green route in the biofuel application. Zhang et al. (2015) demonstrated the synthesis of Na_2SiO_3@Ni/C catalyst using a bamboo powder via precipitation and thermal decomposition followed by loading of Na_2SiO_3. The synthesized catalyst has super magnetic property (15.7 emu/g) due to nickel as this will accelerate separation rate and good alkalinity (3.18 mmol/g) from Na_2SiO_3 which will promote the transesterification reaction rate. Zhang et al. (2017a, b) also use *Jatropha curcas* shell to synthesize carbon solid base (Na_2SiO_3@Ni/JRC), a magnetic catalyst. Their study reported a biodiesel yield of 96.7%. Also, the recovery rate of catalyst was >85% and even after rinsing with ethanol and reusing it 10 times, the catalyst regains its activity.

Magnetic sulfonic acid, a carbon-based catalyst prepared via hydrothermal route is also receiving attention by the researchers because of its numerous advantages. Zhang et al. (2015) make use of glucose and ferric chloride to synthesize carbonized catalyst by hydrothermal precipitation method at a temperature of 180 °C followed by pyrolysis and then sulfonation at 400–800 °C and 150 °C, respectively. The obtained results show acid content in catalyst is 2.79 mmol/g and possess magnetism (14.4 emu/g) which ultimately produces a good yield of biodiesel production from Jatropha oil without losing its catalytic efficiency even after using it thrice. The work by Liu et al. (2015) used waste banana peel to synthesize the same catalyst for its wide application and research value. A comparative study concludes that the catalyst activity of sulfonated activated carbon was excellent than commercialized Amberlyst-15. Zhang et al. (2017a, b) prepared a composite of carbon/Fe_3O_4 of size less than 20 nm from the *Jatropha curcas* shell. Synthesized catalyst possesses acid content and magnetism of 2.69 mmol/g and 40.3 emu/g, respectively. Similarly, Ibrahim et al. (2019) prepared a bio-based catalyst by sulfonation with a char as a precursor composed of an empty fruit bundle. The developed magnetic catalyst shows the highest acidity (29.52 mmol/g). List of other green synthesized nano-catalyst for biodiesel application is illustrated in Table 8.3.

8.9 Future Prospects

In our contemporary society, low energy savings is a major problem and therefore it becomes crucially important to switch to renewable energy to reduce the consumption of conventional fossil fuels. Biomass-based fuel such as "biodiesel" production is considered as a sustainable and viable substitute but extensive investigation is required for its commercialization including long-term application. Advances have been implemented to enhance biodiesel yield through the use of different types of catalysts as they play an important role in the catalytic reaction of biodiesel because of their efficient catalytic activity. In the series of different advancement, employing nano-catalyst in the biodiesel production technology field is an advanced and effective approach. Numerous literatures have demonstrated that during the reaction, nano-catalyst application significantly improves biodiesel production. However, the yield and purity involved in the synthesis of nanomaterials via green route synthesis seem to be one of the major hurdles to its potential application in biodiesel production. Another drawback of this technology is the low recyclability and reusability of the nanomaterials, utilized at each step as this will have a negative impact on the cost economy. Nevertheless, it is also reported and anticipated that the use of nano-catalyst synthesized from biomass waste in biodiesel production overcomes the limitations persisting in biological routes for production. Further exploration needs to be done to understand the mechanisms and resolve issues of the synthesis step to make it more viable and economical.

Low-cost, easily recyclable nano-catalyst is an important direction for future research. At the same time research on scale-up is still existing in its preliminary stage and more research is needed on the use of continuous reactors.

8.10 Conclusion

The unstoppable population rise and industrial globalization worldwide has a direct impact on increases in demand for petroleum fuels. Research and development for sustainable renewable fuel become the center of attraction taking into account the limited resources of fossil fuels. In this context, the production of FAME from different renewable sources shows a promising pathway. Among different methods proposed for biodiesel production, transesterification is considered as a priority choice and makes use of catalysts that may be homogeneous or heterogeneous. Although the use of homogeneous catalysts is effective when considering biodiesel production rate, certain limitations persist such as wastewater generation due to

Table 8.3 Nano-catalyst: Green approach for synthesis and its application in production of biodiesel

Nano-catalyst	Source	Oil feedstock	Results	References
MgO	*Ziziphus mauritiana* leaves	*Ziziphus mauritiana* seeds oil	**Particle shape and size:** spongy, spherical, and hedgehog and 20 ± 10-nm **Biodiesel:** Ester content—98.2%, Kinematic viscosity—3.7 mm^2/s at 40 °C, flashpoint, cloud and pour points—98 °C, 0 °C, and− 5 °C, Cetane Number-54	Saman et al. (2021)
CaO	Cockle shells	Palm oil	**Particle Property:** Surface Area, pore volume, and diameter are 13.9113 m^2/g, 0.0318 cm^3/g, 33.17 nm, respectively **Biodiesel:** Yield—94.13% Utilization of the green nano CaO catalyst significantly reduced activation energy to 3786.69 J/mol and accelerating the reaction rate to 27.3% fatty acid methyl ether (FAME) yield/h	Chooi et al. (2021)
Nickel oxide	*Ficus elastic* extract	*Brachychiton populneus* seed oil (BPSO)	**Particle property:** 22–26 nm in diameter and spherical-cubic in shape **Biodiesel:** Yield 97.5% at an optimum value of oil to methanol 1:9 ratio, catalyst—2.5 wt%, and 85 °C as reaction temperature	Dawood et al. (2021)
CaO/Au (III)	Eggshell	Soyabean oil	**Particle property:** Rod-shaped with microchannel; Au particle size ranges between 2 and 4 nm **Biodiesel:** Yield 88.9% obtained at the optimum condition of oil to CH_3OH 1:12, catalyst weight—1 wt%, 70 °C as reaction temperature and retains its activity when recycled and reused 5 times	Liu et al. (2022)
ZnO	Razma seeds	Pongamiapinnata oil	**Particle shape and size:** Structure: Hexagonal wurtzite size: 30–35 nm **Biodiesel:** Yield 84.5%, Viscosity—5.1 mm^2/s, flashpoint points—168 °C	Yadav et al. (2018a, b)
CuO	*Centella Asiatica*	Coconut oil	**Particle crystallite size**—40 nm having small spherical flower shape **Biodiesel:** Maximum FAME conversion was 78.58%	Varghese et al. (2017)

(continued)

Table 8.3 (continued)

Nano-catalyst	Source	Oil feedstock	Results	References
CaO-based/Au	Eggshell, Mussel shell, Calcite, and Dolomite	Sunflower oil	**Biodiesel:** Yield 96.055%, 96.026%, 97.586, and 94.311% using catalyst synthesized from egg, mussel shell, calcite, dolomite, respectively. Achieved oil conversion 90–97% optimum value of oil to CH_3OH ratio 1:9, catalyst weight—3 wt%, 65 °C as reaction temperature	Bet-Moushoul et al. (2016)

unreacted chemicals along with high energy consumption adding cost to the production process. Thus, it becomes essential to develop efficient catalysts, so in this context nanotechnology, the most emerging branch of science gives new hope for having promising applications in the catalysis field. It was also reported that synthesized nano-catalyst have the capability to meet the condition as desired for applications, and can circumvent the issues related to conventional catalysts. Lastly, it can be concluded that nano-catalysts can be recovered easily and reused without compromising their efficiency using simple and cheap conventional filtration and centrifugation techniques.

In recent years, nanomaterials through recycling from waste resources or biomass-derived has drawn attention since the materials can be generated via easy and economical ways. In this chapter, we provided insights into the synthesis of nano-catalyst from various biomass waste but being the smallest size particle, difficulty in recycling is the major disadvantage associated with it. The introduction of magnetic nanoparticles can overcome the problem of recycling and promotes easy recovery by using a magnetic field thereby reducing the overall cost. Literature studies imply that the rate of biodiesel production enhances when using nano-catalyst. This property of the new nano-catalysts will be a milestone to produce biodiesel in an economical and eco-friendly manner. Nevertheless, the developments of homogeneous, heterogeneous, and nano-catalysts give a promising future but still more efforts are to be made to develop more effective and cheap catalysts with high catalytic efficiency.

References

Agarwal H, Kumar SV, Rajesh Kumar S (2017) A review on green synthesis of zinc oxide nanoparticles—An eco-friendly approach. Resour Eff Technol 3:406–413

Ahmed S, Saifullah, Ahmad M, Swami B, Ikram S (2016) Green synthesis of silver nanoparticles using Azadirachtaindica aqueous leaf extract. J Radiat Res Appl Sci 9:1–7

Akia M, Yazdani F, Motaee E, Han D, Arandiyan H (2014) A review on conversion of biomass to biofuel by nanocatalysts. Biofuel Res J 1:16–25

Amer W, Awwad AM (2021) Green synthesis of copper nanoparticles by *Citrus limon* fruits extract, characterization and antibacterial activity. Chem Int 7(1):1–8

Arif Z, Sethy NK, Mishra PK, Verma B (2020a) Development of eco-friendly, self-cleaning, antibacterial membrane for the elimination of Cr (VI) from tannery wastewater. Int J Environ Sci Technol 17:4265–4280

Arif Z, Sethy NK, Mishra PK, Verma B (2020b) Green approach for the synthesis of ultrafiltration photocatalytic membrane for tannery wastewater: modeling and optimization. Int J Environ Sci Technol 17:3397–3410

Arzamendi G, Campo I, Arguinarena E, Sanchez M, Montes M, Gandía LM (2007) Synthesis of biodiesel with heterogeneous NaOH/alumina catalysts: comparison with homogeneous NaOH. Chem Eng J 134:123–130

Baskar G, Soumiya S (2016) Production of biodiesel from castor oil using iron (II) doped zinc oxide nanocatalyst. Renew Energy 98:101–107

Bet-Moushoul E, Farhadi K, Mansourpanah Y, Nikbakht AM, Molaei R, Forough M (2016) Application of CaO-based/Au nanoparticles as heterogeneous nanocatalysts in biodiesel production. Fuel 164:119–127

Bilai S, Ali L, Khan AL, Shahzad R, Asaf S, Imran M, Kang SM, Kim SK, Lee IJ (2018) Endophytic fungus Paecilomycesformosus LHL10 produces sester-terpenoid YW3548 and cyclic peptide that inhibit urease and α-glucosidase enzyme activities. Arch Microbiol 200(10):1493–1502

Bray CD, Battye WH, Aneja VP (2019) The role of biomass burning agricultural emissions in the Indo-Gangetic Plains on the air quality in New Delhi, India. Atmos Environ 218:116983

Bridgwater AV (2012) Review of fast pyrolysis of biomass and product upgrading. Biomass Bioenergy 38:68–94

Changmai B, Vanlalveni C, Ingle AP, Bhagat R, Rokhum L (2020) Widely used catalysts in biodiesel production: a review. RSC Adv 10:41625–41679

Chen H, Wang W, Martin JC, Oliphant AJ, Doerr PA, Xu JF, DeBorn KM, Chen C, Sun L (2012) Extraction of lignocellulose and synthesis of porous silica nanoparticles from rice husks: a comprehensive utilization of rice husk biomass. ACS Sustain Chem Eng 1:254–259

Chooi CY, Sim JH, Tee SF, Lee ZH (2021) Waste-derived green nanocatalyst for biodiesel production: kinetic-mechanism deduction and optimization studies. Sustainability 13:5849

Correia LM, Cecilia JA, Rodríguez-Castellón E, Cavalcante CL, Vieira RS (2017) Relevance of the physicochemical properties of calcined quail eggshell (CaO) as a catalyst for biodiesel production. J Chem 2017:5679512. https://doi.org/10.1155/2017/567951

Dawood S, Koyande AK, Ahmad M, Mubashir M, Asif S, Klemeš JJ, Bokhari A, Saqib S, Lee M, Qyyum MA, Show PL (2021) Synthesis of biodiesel from non-edible (*Brachychiton populneus*) oil in the presence of nickel oxide nanocatalyst: parametric and optimisation studies. Chemosphere 278:130469

Du Y, Yin Z, Zhu J, Huang X, Wu X, Zeng Z, Yan Q, Zhang H (2012) A general method for the large-scale synthesis of uniform ultrathin metal sulphide nanocrystals. Nat Commun 3:1177

Ebrahimian J, Mohsennia M, Khayatkashani M (2020) Photocatalytic-degradation of organic dye and removal of heavy metal ions using synthesized SnO_2 nanoparticles by *Vitex agnus-castus* fruit via a green route. Mater Lett 263:127255

Ghorbani F, Sanati AM, Malek M (2015) Production of silica nanoparticles from rice husk as agricultural waste by environmental friendly technique. Environ Stud Persian Gulf 2(1):56–65

Ghotekar S (2019) A review on plant extract mediated biogenic synthesis of CdO nanoparticles and their recent applications. Asian J Green Chem 3:187–200

Goutam SP, Saxena G, Singh V, Yadav AK, Bharagava RN, Thapa KB (2018) Green synthesis of TiO_2 nanoparticles using leaf extract of *Jatropha curcas* L. for photocatalytic degradation of tannery wastewater. Chem Eng J 336:386–396

Gu S, Zhou J, Yu C, Luo Z, Wang Q, Shi Z (2015) A novel two-staged thermal synthesis method of generating nanosilica from rice husk via pre-pyrolysis combined with calcination. Ind Crop Prod 65:1–6

Hariharan V, Sivakumar G (2013) Studies on synthesized nanosilica obtained from bagasse ash. Int J ChemTech Res 5(3):1263–1266

Hassan AF, Abdelghny AM, Elhadidy H, Youssef AM (2014) Synthesis and characterization of high surface area nanosilica from rice husk ash by surfactant-free sol–gel method. J Sol-Gel Sci Technol 69:465–472

Hussein AK (2015) Applications of nanotechnology in renewable energies—a comprehensive overview and understanding. Renew Sustain Energy Rev 42:460–476

Ibrahim NA, Rashid U, Taufiq-Yap YH, Yaw TCS, Ismail I (2019) Synthesis of carbonaceous solid acid magnetic catalyst from empty fruit bunch for esterification of palm fatty acid distillate (PFAD). Energy Convers Manag 195:480–491

Janta R, Sekiguchi K, Yamaguchi R, Sopajaree K (2022) Ambient PM2.5, polycyclic aromatic hydrocarbons and biomass burning tracer in Mae Sot District, western Thailand. Atmos Pollut Res 11:27–39

Kamath SR, Proctor A (1998) Silica gel from rice husk ash: preparation and characterization. Cereal Chem 75:484–487

Karimi J, Mohsenzadeh S (2016) Effects of silicon oxide nanoparticles on growth and physiology of wheat seedlings. Russ J Plant Physiol 63:119–123

Krishnaswamy K, Vali H, Orsat V (2014) Value adding to grape waste: green synthesis of gold nanoparticles. J Food Eng 142:210–220

Ladole MR, Mevada JS, Pandit AB (2017) Ultrasonic hyperactivation of cellulase immobilized on magnetic nanoparticles. Bioresour Technol 239:117–126

Lakshmipathy R, Palakshi Reddy B, Sarada NC, Chidambaram K, Pasha SK (2015) Watermelon rind-mediated green synthesis of noble palladium nanoparticles: catalytic application. Appl Nanosci 5:223–228

Lin HN, Hu BB, Zhu MJ (2016a) Enhanced hydrogen production and sugar accumulation from spent mushroom compost by *Clostridium thermocellum* supplemented with PEG8000 and JFC-E. Int J Hydrog Energy 41(4):2383–2390

Lin R, Cheng J, Ding L, Song W, Liu M, Zhou J, Cen K (2016b) Enhanced dark hydrogen fermentation by addition of ferric oxide nanoparticles using *Enterpbacteraerogenes*. Bioresour Technol 207:213–219

Liu RL, Gao XY, An L, Ma J, Zhang JF, Zhang ZQ (2015) Fabrication of magnetic carbonaceous solid acids from banana peel for the esterification of oleic acid. RSC Adv 5(114):93858–93866

Liu T, He G, Lau AKH (2019) Statistical evidence on the impact of agricultural straw burning on urban air quality in China. Sci Total Environ 711:134633

Liu J, Liu M, Chen S, Wang B, Chen J, Yang D-P, Zhang S, Du W (2022) Conversion of Au(III)-polluted waste eggshell into functional CaO/Au nanocatalyst for biodiesel production. Green Energy Environ 7:352–359. https://doi.org/10.1016/j.gee.2020.07.019

Manjula-Rani K, Palanisamy PN, Sivakumar P (2014) Synthesis and characterization of amorphous nano-silica from biomass ash. Int J Adv Technol Eng Sci 2(10):71–76

Marciniuk LL, Hammerb P, Pastore HO, Schuchardt U, Cardoso D (2014) Sodium titanate as basic catalyst in transesterification reactions. Fuel 118:48–54

Mohammadinejad R, Karimi S, Iravani S, Varma RS (2016) Plant-derived nanostructures: types and applications. Green Chem 18(1):20–52

Mohanraj K, Kannan S, Barathan S, Sivakumar G (2012) Preparation and characterization of nano SiO2 from corn cob ash by precipitation method. Optoelectron Adv Mater 6:394–397

Mohd NK, Wee NNNA, Azmi AA (2017) Green synthesis of silica nanoparticles using sugarcane bagasse. AIP Conf Proc 1885:020123

Muramatsu H, Kim YA, Yang KS, Cruz-Silva R, Toda I, Yamada T, Terrones M, Endo M, Hayashi T, Saitoh H (2014) Rice husk-derived graphene with nano-sized domains and clean edges. Small 10(14):2766–2770

Naseer M, Aslam U, Khalid B, Chen B (2020) Green route to synthesize Zinc Oxide Nanoparticles using leaf extracts of Cassia fistula and Melia azadarach and their antibacterial potential. Sci Rep 10(1):1–10

Novoselov KS, Fal'ko VI, Colombo L, Gellert PR, Schwab MG, Kim K (2012) A roadmap for graphene. Nature 490:192–200

Okoronkwo EA, Imoisili PE, Olubayode SA, Olusunle SOO (2016) Development of silica nanoparticle from corn cob ash. Adv Nanopart 5:135–139

Omid A, Keyvan B, Ali M (2014) Synthesis of graphene from natural and industrial carbonaceous wastes. RSC Adv 4:20441–20448

Palanisamy S, Subramaniam BS, Thangamuthu S, Nallusamy S, Rengasamy P (2021) Review on agro-based nanotechnology through plant-derived green nanoparticles: synthesis, application and challenges. J Environ Sci Public Health 5(1):77–98

Pandit PR, Fulekar MH (2017) Egg shell waste as heterogeneous nanocatalyst for biodiesel production: optimized by response surface methodology. J Environ Manag 198:319–329

Patel KG, Misra NM, Vekariya RH, Shettigar RR (2017) One-pot multicomponent synthesis in aqueous medium of 1,4-dihydropyrano [2,3- c] pyrazole-5-carbonitrile and derivatives using a green and reusable nano-SiO_2 catalyst from agricultural waste. Res Chem Intermed 44:289–304

Patumsawad S (2011) Second generation biofuels: technical challenge and R and D opportunity in Thailand. J Sustain Energy Environ Special Issue: 47–50

Pawar V, Kumar M, Dubey PK, Singh MK, Sinha ASK, Singh P (2019) Influence of synthesis route on structural, optical and electrical properties of TiO_2. Appl Phys A 125:657

Phairuang W, Suwattiga P, Chetiyanukornkul T, Hongtieab S, Limpaseni W, Lkemori F, Hata M, Furuuchi M (2019) The influence of the open burning of agricultural biomass and forest fires in Thailand on the carbonaceous components in size-fractionated particles. Environ Pollut 247: 238–247

Qadri SB, Gorzkowski E, Rath BB, Feng J, Qadri SN, Kim H, Caldwell JD, Imam MA (2015) Nanoparticles and nanorods of silicon carbide from the residues of corn. J Appl Phys 117: 044306

Rajesh K, Rajesh KS, Dinesh PS (2016) Natural and waste hydrocarbon precursors for the synthesis of carbon-based nanomaterials: graphene and CNTs. Renew Sustain Energy Rev 8:976–1006

Salam AM, Hazim FA, Hajer A, Fazal M, Emad Y, Ali AA, Zainab H, Khamis Al R (2020) Assessment of nanocalcite synthesis routs on green biodiesel production: comparative study. Biointerf Res Appl Chem 10(5):6139–6147

Salimi Z, Hosseini SA (2019) Study and optimization of conditions of biodiesel production from edible oils using $ZnO/BiFeO_3$ nano magnetic catalyst. Fuel 239:1204–1212

Salinas D, Guerrero S, Cross A, Araya P, Wolf E (2016) Potassium titanate for the production of biodiesel. Fuel 166:237–244

Saman S, Balouch A, Talpur FN, Memon AA, Mousavi BM, Verpoort F (2021) Green synthesis of MgO nanocatalyst by using *Ziziphus mauritiana* leaves and seeds for biodiesel production. Appl Organomet Chem 5:e6199

Sana NO, Kurs C, Tümtas UY, Yas Ö, Ortac B, Tekinay T (2014) Novel one-step synthesis of silica nanoparticles from sugarbeet bagasse by laser ablation and their effects on the growth of freshwater algae culture. Particuology 17:29–23

Satyanarayana KG, Mariano AB, Vargas JVC (2011) A review on microalgae, a versatile source for sustainable energy and materials. Int J Energy Res 35(4):291–311

Schwierz F (2010) Graphene transistors. Nat Nanotechnol 5:487–496

Sethy NK, Arif Z, Mishra PK, Kumar P (2019) Synthesis of SiO_2 nanoparticle from bamboo leaf and incorporated in PDMS membrane to enhance its separation properties. J Polym Eng

Shanmugam S, Hari A, Pandey A, Mathimanie T, Felix LO, Pugazhendhi A (2020) Comprehensive review on the application of inorganic and organic nanoparticles for enhancing biohydrogen production. Fuel 270:117453

Sharma D, Kanchi S, Bisetty K (2019) Biogenic synthesis of nanoparticles: a review. Arab J Chem 12:3576–3600

Singh PS, Vidyasagar GM (2014) Biosynthesis, characterization, and antidermatophytic activity of silver nanoparticles using Raamphal plant (Annona reticulata) aqueous leaves extract. Ind J Mater Sci 2014:412452. https://doi.org/10.1155/2014/412452

Sinsinwar S, Sarkar MK, Suriya KR, Nithyanand P, Vadivel V (2018) Use of agricultural waste (coconut shell) for the synthesis of silver nanoparticles and evaluation of their antibacterial activity against selected human pathogens. Microb Pathog 124:30–37
Somanathan T, Prasad K, Ostrikov K, Saravanan A, Krishna VM (2015) Graphene oxide synthesis from agro waste. Nanomaterials 5(2):826–834
Srikar SK, Giri DD, Pal DB, Mishra PK, Upadhyay SN (2016) Green synthesis of silver nanoparticles: a review. Green Sustain Chem 6:34–56
Srinivasan M, Venkatesan M, Arumugam V, Natesan G, Saravanan N, Murugesan S, Ramachandran S, Ayyasamy R, Pugazhendhi A (2019) Green synthesis and characterization of titanium dioxide nanoparticles (TiO_2 NPs) using *Sesbania grandiflora* and evaluation of toxicity in zebrafish embryos. Process Biochem 80:197–202
Srivastava V, Gusain D, Sharma YC (2015) Critical review on the toxicity of some widely used engineered nanoparticles. Ind Eng Chem Res 54:6209–6233
Stegarescu A, Lung I, Leoştean C, Kacso I, Opriş O, Lazăr MD, Copolovici L, Guţoiu S, Stan M, Popa A, Pană O, Porav AS, Soran ML (2020) Green synthesis, characterization and test of MnO_2 nanoparticles as catalyst in biofuel production from grape residue and seeds oil. Waste Biomass Valor 11:5003–5013
Sun D, Hussain HI, Yi Z, Rookes JE, Kong L, Cahill DM (2016) Mesoporous silica nanoparticles enhance seedling growth and photosynthesis in wheat and lupin. Chemosphere 152:81–91
Suriyaprabha R, Karunakaran G, Kavitha K, Yuvakkumar R, Rajendran V, Kannan N (2014) Application of silica nanoparticles in maize to enhance fungal resistance. IET Nanobiotechnol 8(3):133–137
Taherdanak M, Zilouei H, Karimi K (2015) Investigating the effect of iron and nickel nanoparticles on dark hydrogen production from starch using central composite design. Int J Hydrog Energy 40(38):12956–12963
Thandapani K, Kathiravan M, Namasivayam E, Padiksan IA, Natesan G, Tiwari M, Giovanni B, Perumal V (2018) Enhanced larvicidal, antibacterial, and photocatalytic efficacy of TiO_2nanohybrids green synthesized using the aqueous leaf extract of *Parthenium hysterophorus*. Environ Sci Pollut Res 25:10328–10339
Upadhyay AK, Mandotra SK, Kumar N, Singh NK, Singh L, Rai UN (2016) Augmentation of arsenic enhances lipid yield and defense responses in alga Nannochloropsis sp. Bioresour Technol 221:430–437
Varghese R, Joy PH, Johnson I (2017) Green synthesis and characterizations of flower shaped CuO nanoparticles for biodiesel application. Sensors Transducers 210(3):38–41
Verma ML, Barrow CJ, Puri M (2013) Nanobiotechnology as a novel paradigm for enzyme immobilization and stabilization with potential applications in biodiesel production. Appl Microbiol Biotechnol 97:23–39
Wang A, Sudarsanam P, Xu Y, Zhang H, Li H, Yang H (2020a) Functionalized magnetic nanosized materials for efficient biodiesel synthesis via acid-base/enzyme catalysis. Green Chem 22: 2977–3012. https://doi.org/10.1039/D0GC00924E
Wang Z, Que B, Gan J, Chen Q, Zheng L, Marraiki N, Elgorban AM, Zhang Y (2020b) Zinc oxide nanoparticles synthesized from *Fraxinus rhynchophylla* extract by green route method attenuates the chemical and inflammatory pain models in mice. J Photochem Photobiol B Biol 202: 111668
Wei Z, Zeng G, Huang F, Kosa M, Huang D, Ragauskas AJ (2015) Bioconversion of oxygen pre-treated Kraft lignin to microbial lipid with oleaginous Rhodococcusopacus DSM 1069. Green Chem 17(5):2784–2789
Wen X, Chen X, Tian N, Gong J, Liu J, Rümmeli MH, Chu PK, Mijiwska E, Tang T (2014) Nanosized carbon black combined with Ni_2O_3 as "Universal" catalysts for synergistically catalyzing carbonization of polyolefin wastes to synthesize carbon nanotubes and application for supercapacitors. Environ Sci Technol 48:4048–4055

Xiao Y, Long C, Zheng M, Dong H, Lei B, Zhang H, Liu Y (2014) High-capacity porous carbons prepared by KOH activation of activated carbon for super capacitors. Chin Chem Lett 25(6): 865–868

Yadav LSR, Raghavendra M, Kumar KHS, Dhananjay N, Nagaraju G (2018a) Biosynthesised ZnO: Dy^{3+} nanoparticles: biodiesel properties and reusable catalyst for N-formylation of aromatic amines with formic acid. Eur Phys J Plus 133:153

Yadav LSR, Raghavendra M, Udaybhanu, Manjunath K, Nagaraju G (2018b) Photocatalytic, biodiesel, electrochemical sensing properties and formylation reactions of ZnO nanoparticles synthesized via eco-friendly green synthesis method. J Mater Sci Mater Electron 29:8747–8759

Yadav LSR, Pratibha S, Manjunath K, Shivanna M, Ramakrishnappa T, Dhananjaya N, Nagaraju G (2019) Green synthesis of Ag-ZnO nanoparticles: structural analysis, hydrogen generation, formylation and biodiesel applications. J Sci Adv Mater Dev 4:425–431

Yahya NY, Ngadi N, Jusoh M, Halim NAA (2016) Characterization and parametric study of mesoporous calcium titanate catalyst for transesterification of waste cooking oil into biodiesel. Energy Convers Manag 129:275–283

Yaro SA, Olajide OS, Asuke F (2017) Synthesis of groundnut shell nanoparticles: characterization and particle size determination. Int J Adv Manuf Technol 91:1111–1116

Yatish KV, Prakash RM, Ningaraju C, Sakar M, GeethaBalakrishna R, Lalithamba HS (2020a) *Terminalia chebula* as a novel green source for the synthesis of copper oxide nanoparticles and as feedstock for biodiesel production and its application on diesel engine. Energy 215:119165

Yatish KV, Lalithamba HS, Suresh R, Latha HKE (2020b) *Ochrocarpus lingifolius* assisted green synthesis of $CaTiO_3$ nanoparticle for biodiesel production and its kinetic study. Renew Energy 147:310–321

Zhang F, Fang Z, Wang YT (2015) Biodiesel production directly from oils with high acid value by magnetic $Na_2SiO_3@Fe_3O_4/C$ catalyst and ultrasound. Fuel 150:370–377

Zhang F, Tian XF, Fang Z, Shah M, Wang YT, Jiang W, Yao M (2017a) Catalytic production of Jatropha biodiesel and hydrogen with magnetic carbonaceous acid and base synthesized from Jatropha hulls. Energy Convers Manag 142:107–116

Zhang F, Tian X, Shah M, Yang W (2017b) Synthesis of magnetic carbonaceous acids derived from hydrolysates of Jatropha hulls for catalytic biodiesel production. RSC Adv 7(19):11403–11413

The manufacturer's authorised representative in the EU is Springer Nature Customer Service Centre GmbH, Europaplatz 3, 69115 Heidelberg, Germany. If you have any concerns regarding our products, please contact ProductSafety@springernature.com

Printed and bound by CPI Group (UK) Ltd, Croydon, CR0 4YY
01/10/2026
02226185-0002